SOLIDWORKS Simulation 2022:

A Tutorial Approach
(3rd Edition)

CADCIM Technologies
525 St. Andrews Drive
Schererville, IN 46375, USA
(www.cadcim.com)

Contributing Author
Sham Tickoo
Professor
Department of Mechanical Engineering Technology
Purdue University Northwest
Hammond, Indiana, USA

CADCIM Technologies

SOLIDWORKS Simulation 2022: A Tutorial Approach
Sham Tickoo

CADCIM Technologies
525 St Andrews Drive
Schererville, Indiana 46375, USA
www.cadcim.com

ISBN 978-1-64057-176-1

www.cadcim.com

DEDICATION

To teachers, who make it possible to disseminate knowledge
to enlighten the young and curious minds
of our future generations

To students, who are dedicated to learning new technologies
and making the world a better place to live in

THANKS

To the faculty and students of the MET Department of
Purdue University Northwest for their cooperation

To employees of CADCIM Technologies for their valuable help

Online Training Program Offered by CADCIM Technologies

CADCIM Technologies provides effective and affordable virtual online training on various software packages including Computer Aided Design, Manufacturing and Engineering (CAD/CAM/CAE), computer programming languages, animation, architecture, and GIS. The training is delivered 'live' via Internet at any time, any place, and at any pace to individuals as well as the students of colleges, universities, and CAD/CAM/CAE training centers. The main features of this program are:

Training for Students and Companies in a Classroom Setting

Highly experienced instructors and qualified engineers at CADCIM Technologies conduct the classes under the guidance of Prof. Sham Tickoo of Purdue University Northwest, USA. This team has authored several textbooks that are rated "one of the best" in their categories and are used in various colleges, universities, and training centers in North America, Europe, and in other parts of the world.

Training for Individuals

CADCIM Technologies with its cost effective and time saving initiative strives to deliver the training in the comfort of your home or work place, thereby relieving you from the hassles of traveling to training centers.

Training Offered on Software Packages

CADCIM provides basic and advanced training on the following software packages:

***CAD/CAM/CAE**: CATIA, Creo Parametric, Creo Direct, SOLIDWORKS, Autodesk Inventor, Solid Edge, NX, AutoCAD, AutoCAD LT, AutoCAD Plant 3D, Customizing AutoCAD, EdgeCAM, and ANSYS*

***Architecture and GIS**: Autodesk Revit (Architecture, Structure, MEP), AutoCAD Civil 3D, AutoCAD Map 3D, Primavera, and Bentley STAAD Pro*

***Animation and Styling**: Autodesk 3ds Max, Autodesk Maya, Autodesk Alias, The Foundry NukeX, and MAXON CINEMA 4D*

***Computer Programming**: C++, VB.NET, Oracle, AJAX, and Java*

For more information, please visit the following link: ***https://www.cadcim.com***

Note

If you are a faculty member, you can register by clicking on the following link to access the teaching resources: ***https://www.cadcim.com/Registration.aspx***. The student resources are available at ***https://www.cadcim.com***. We also provide **Live Virtual Online Training** on various software packages. For more information, write us at ***sales@cadcim.com***.

Table of Contents

This page is intentionally left blank

Preface

SOLIDWORKS Simulation 2022

SOLIDWORKS Simulation, developed by Dassault Systems' SolidWorks Corporation, USA, is one of the leading FEA software packages that efficiently evaluates performance, improves quality, and boosts product innovation. Using SOLIDWORKS Simulation, you can set up virtual real-world environments to test your product designs before finalizing them for manufacturing process. It enables you to test against a broad range of parameters during the design process such as durability, static and dynamic response, assembly motion, heat transfer, fluid dynamics, and also helps solve complex simulations such as vibration impact and make the validation process easier. Based mainly on the user feedback, this FEA tool is remarkably user-friendly and allows you to be productive from day one.

SOLIDWORKS Simulation enables you to simulate and analyze design performance with fast, easy-to-use SOLIDWORKS Simulation CAD-embedded analysis solutions. You can quickly and easily employ advanced simulation techniques to optimize design performance and thus cuts down on costly prototypes, eliminating rework and delays, and saving time and product development costs

SOLIDWORKS Simulation provides simulation tools to test your designs and make the decisions to improve quality and functionality.

SOLIDWORKS Simulation 2022: A Tutorial Approach textbook has been written to help the users learn the basics of FEA. In this textbook, the author has used the tutorial point of view and the learn-by-doing theme to explain the tools and concepts of FEA using SOLDWORKS Simulation. Real-world mechanical engineering industry examples and tutorials have been used to ensure that the users can relate the knowledge gained through this book with the actual mechanical industry designs. The main features of the book are as follows:

- **Tutorial Approach**
  The author has adopted the tutorial point-of-view and the learn-by-doing approach throughout the book. This approach guides the users to learn the tools and concepts faster.

- **Real-world Mechanical Engineering Projects as Tutorials**
  The author has used the real-world mechanical engineering projects as tutorials in this textbook so that the readers can correlate the tutorials with the real-time models in the mechanical engineering industry.

- **Coverage of Major Simulation Studies**
  All major studies of SOLIDWORKS Simulation are covered in this book. These include the **Static, Frequency, Thermal, Fatigue, Nonlinear** and **Buckling studies**.

- **Tips and Notes**
  Additional information related to various topics is provided to the users in the form of tips and notes.

- **Learning Objectives**
  The first page of every chapter summarizes the topics that are covered in the chapter.

- **Self-Evaluation Test, Review Questions, and Exercises**
  Each chapter ends with Self-Evaluation Test that enables the users to assess their knowledge of the chapter. The answers to Self-Evaluation Test are given at the end of the chapter. Also, the Review Questions and Exercises are given at the end of the chapters and they can be used by the instructors as test questions and exercises.

Symbols Used in the Textbook

Note

The author has provided additional information to the users about the topic being discussed in the form of Notes.

Tip

Special information on various techniques is provided in the form of Tips that will increase the efficiency of the users.

Formatting Conventions Used in the Textbook

Please refer to the following list for the formatting conventions used in this textbook.

Names of tools, buttons, options, toolbars, and are written in boldface.

Example: The **Extrude Boss/Base** tool, the **Mid-Plane** option, the **OK** button, the **Features** toolbar, and so on.

Names of CommandManager, PropertyManager rollouts, dialog box, drop-down lists, spinners, selection boxes, areas, edit boxes, check boxes, and radio buttons are written in boldface.

Example: The **Simulation CommandManager,** the **Study PropertyManager,** the **Open** dialog box, the **End Condition** drop-down list, the **Depth** spinner, the **Direction of Extrusion** selection box, the **Draft outward** check box, and so on.

Values entered in edit boxes are written in boldface.

Example: Enter **5** in the **Radius** edit box.

Names and paths of the files are written in italics

Example: *C:\Documents\SOLIDWORKS\c08\c08_tut01*

Naming Conventions Used in the Textbook

Tool

If you click on an item in a toolbar and a command is invoked to create/edit an object or perform some action, then that item is termed as tool.

For example:
To Create: **Line** tool, **Smart Dimension** tool, **Extruded Boss/Base** tool
Action: **Zoom to Fit** tool, **Pan** tool, **Copy** tool

If you click on an item in a toolbar and a dialog box is invoked wherein you can set the properties to create/edit an object, then that item is also termed as tool, refer to Figure 1.

Flyout

A flyout is the one in which a set of tools are grouped together. You can identify a flyout with a down arrow on it. A flyout is given a name based on the types of tools grouped in it. For example, **Line** flyout, **View Settings** flyout, **Fillet** flyout, and so on; refer to Figure 1.

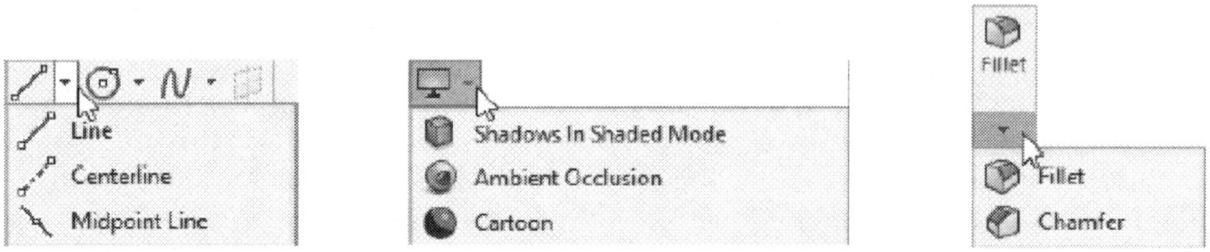

Figure 1 *The **Line**, **View Settings**, and **Fillet** flyouts*

PropertyManager

The naming conventions for the components in a PropertyManager are mentioned in Figure 2, Figure 3 and Figure 4.

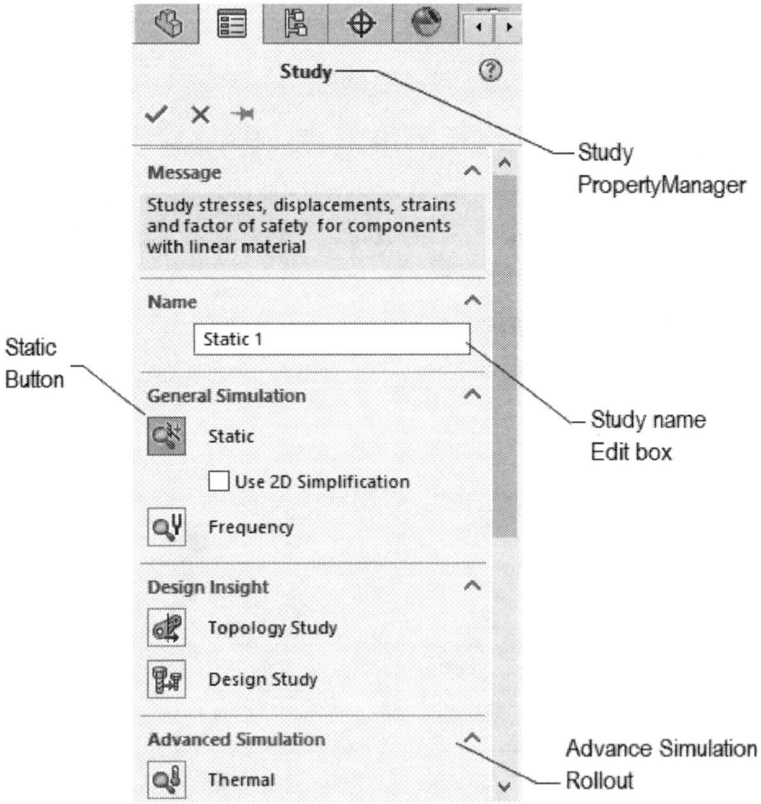

Figure 2 *The PropertyManager, edit box, rollout, and button*

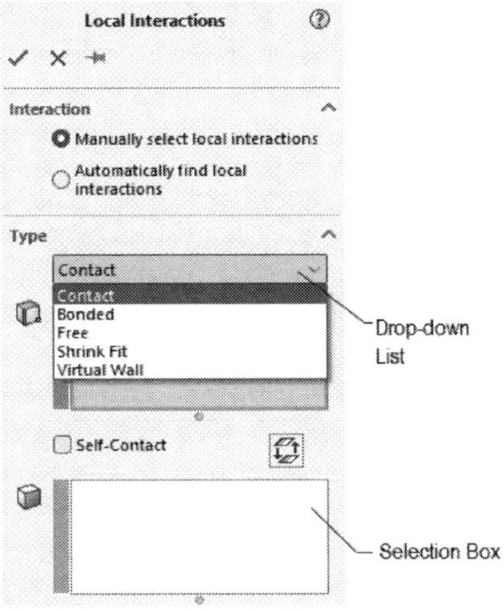

Figure 3 *The selection box and the drop-down list*

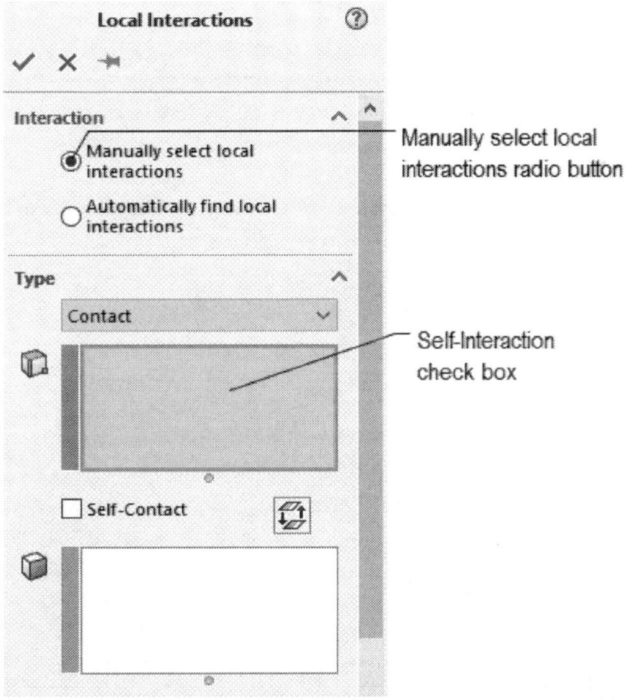

Figure 4 *The check box and radio button*

Button

The items in a dialog box that has a 3D shape like a button is termed as **Button**. For example, **OK** button, **Cancel** button, and so on.

Free Companion Website

It has been our constant endeavor to provide you the best textbooks and services at affordable price. In this endeavor, we have come out with a Free Companion website that will facilitate the process of teaching and learning of SOLIDWORKS 2022. If you purchase this textbook, you will get access to the files on the Companion website.

The following resources are available for the faculty and students in this website:

Faculty Resources

- **Technical Support**
  You can get online technical support by contacting ***techsupport@cadcim.com***.

- **Instructor Guide**
  Solutions to all review questions and exercises in the textbook are provided in the instructor guide to help the faculty members test the skills of the students. Note that it is compulsory to perform exercises as they comprise some of the parts of the final model.

- **Part Files**
  The part files used in illustrations, examples, and exercises are available for free download. Note that it is compulsory to perform exercises as they comprise some of the parts of the final model.

You can get additional information at ***https://allaboutcadcam.blogspot.com***

If you face any problem in accessing these files, please contact the publisher at ***sales@cadcim.com*** or the author at ***Stickoo@pnw.edu*** or ***tickoo525@gmail.com***.

Student Resources

- **Technical Support**
  You can get online technical support by contacting ***techsupport@cadcim.com***.

- **Part Files**
  The part files used in illustrations and examples are available for free download.

Video Courses

CADCIM offers video courses in CAD, CAE Simulation, BIM, Civil/GIS, and Animation domains on various e-Learning/Video platforms. To enroll for the video courses, please visit the CADCIM website using the link **https://www.cadcim.com/video-courses.**

Stay Connected

You can now stay connected with us through Facebook and Twitter to get the latest information about our textbooks, videos, and teaching/learning resources. To stay informed of such updates, follow us on Facebook *(www.facebook.com/cadcim)* and Twitter (***@cadcimtech***). You can also subscribe to our YouTube channel *(www.youtube.com/cadcimtech)* to get the information about our latest video tutorials.

This page is intentionally left blank

Chapter 1

Introduction to FEA and SOLIDWORKS Simulation

Learning Objectives

After completing this chapter, you will be able to:

- *Understand basic concepts and general working of FEA*
- *Understand the advantages and limitations of FEA*
- *Understand important terms and definitions used in FEA*
- *Understand SOLIDWORKS simulation*
- *Understand the type of analysis performed using SOLIDWORKS Simulation*
- *Basic steps in SOLIDWORKS Simulation*

INTRODUCTION TO FEA

The Finite Element Analysis (FEA) is a computing technique that is used to obtain approximate solution to boundary value problems. It is a numerical procedure to find solution of engineering problems through various analyses such as structural, thermal, fluid flow, electrical, and so on. The most common use of FEA is in structural analysis. These problems are basically mathematical models for physical situation. These mathematical models are differential equations with a set of boundary values and initial conditions. The method used to derive these equations is called Finite Element Method (FEM).

To understand the concept of FEA, consider a circle whose perimeter is to be calculated. To measure the perimeter of the circle without using the conventional formula, divide it into equal segments, as shown in Figure 1-1. Next, join the start and end points of these segments with a straight line. Now, you can measure the length of the straight line very easily, and thus, the perimeter of the circle by adding the length of these straight lines.

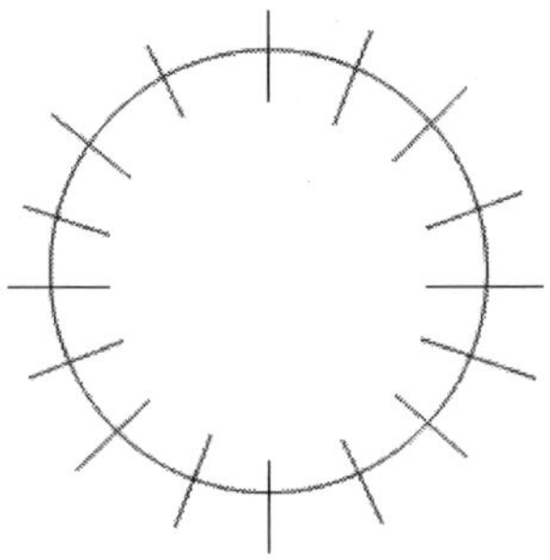

Figure 1-1 *The circle divided into small equal segments*

If the number of segments into which the circle is divided is less, you will not get accurate results. For accuracy, divide the circle into more number of segments. However, with more segments, the time required will be more. The same concept can be applied to FEA also, and therefore, there is always a compromise between accuracy and speed while using this method. This compromise between accuracy and speed makes it an approximate method.

The FEA was first developed to be used in the aerospace and nuclear industries where the safety of structures is critical. Today, even the simplest of products rely on FEA for design evaluation.

The FEA simulates the loading conditions of a design and determines the design response in those conditions. It can be used in new product design as well as in existing product refinement. A model is divided into a finite number of regions/divisions called elements. These elements can be of predefined shapes, such as triangular, quadrilateral, hexahedron, tetrahedron, and so on. The predefined shape of an element helps define the equations that describe how the element will respond to certain loads. The sum of the responses of all the elements in a model gives the total response of the complete model.

GENERAL WORKING OF FEA

A better knowledge of FEA helps in creating more accurate finite element models. Also, it helps in understanding the backend working of SOLIDWORKS Simulation. Here, a simple model is discussed to give you a brief overview of the working of FEA.

Figure 1-2 shows a spring assembly that represents a simple two-spring element model. In this model, two springs are connected in series and one of the springs is fixed at the left most end point, refer to Figure 1-2. In this figure, the stiffness of the springs has been represented by the spring constants K1 and K2. The movement of end points of each spring is restricted in X direction only. The change in position from the undeformed state of each end point can be defined by X1 and X2 variables. The two forces acting on the end points of the springs are represented by F1 and F2.

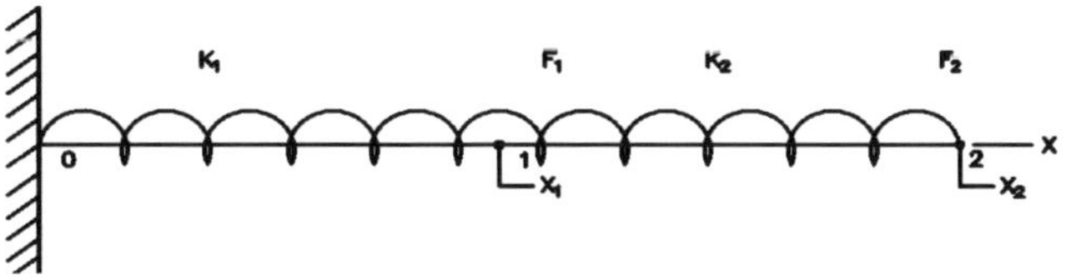

***Figure 1-2** Representation of a two-spring assembly*

To develop a model that can predict the state of this spring assembly, you can use the linear spring equation given below:

$$F = KX$$

where,
F = force applied,
X = displacement, and
K = spring constant

If you use the spring parameters defined above and assume a state of equilibrium, the following equations can be written for each endpoint:

$$F_1 - X_1K_1 + (X_2 - X_1)K_2 = 0$$
$$F_2 - (X_2 - X_1)K_2 = 0$$

Therefore,

$$F_1 = (K_1 + K_2)X1 + (-K_2)X_2$$
$$F_2 = (-K_2)X_1 + K_2X_2$$

If the set of equation is written in matrix form, it will be represented as follows:

$$\begin{bmatrix} F_1 \\ F_2 \end{bmatrix} = \begin{bmatrix} K_1 + K_2 & -K_2 & X_1 \\ -K_2 & K_2 & X_2 \end{bmatrix}$$

In the above mathematical model, if the spring constants (K_1 and K_2) are known and the deformed shapes (X_1 and X_2) are defined, then the resulting forces (F_1 and F_2) can be determined. alternatively, if the spring constants (K_1 and K_2) are known and the forces (F_1 and F_2) are defined, then the resulting deformed shapes (X_1 and X_2) can be determined. The terminologies that are used in the previous example are discussed next.

Stiffness Matrix

In the previous equation, the following part represents the stiffness matrix (K):

$$\begin{bmatrix} K_1 + K_2 & -K_2 \\ -K_2 & K_2 \end{bmatrix}$$

This matrix is relatively simple because it comprises only one pair of springs, but it turns complex when the number of springs increases.

Degrees of Freedom

Degrees of freedom is defined as the least number of independent coordinates required to define the configuration of a system in space. In the previous example, you were only concerned about displacement and forces. By making one endpoint fixed, you will restrict all degrees of freedom for that particular node. This means there will be no translational or rotational degrees of freedom for that node. But, still there are two nodes with some degrees of freedom. As these two nodes are allowed to translate along the X axis only, they have 1 degree of freedom each considering that no rotational degree of freedom exists in them. The number of degrees of freedom on free nodes in a model determines the number of equations required to solve a mathematical model.

Boundary Conditions

The boundary conditions are used to eliminate the unknowns in a system. A set of equations that is solvable is meaningless without the input. In the previous example, the boundary condition $X_0 = 0$, and the input forces are F_1 and F_2. Either ways, the displacements could have been specified in place of forces as boundary conditions and the mathematical model could have been solved for the forces. In other words, the boundary conditions help you reduce or eliminate the unknowns in the system.

Note

The solutions generated by using FEA are always approximate.

Elements and Element Shapes

Elements and element shapes are the building blocks of Finite Element Method (FEM). These are discussed next.

Elements

Element is an entity into which the system under study is divided. An element shape is specified by nodes. The shape (area, length, and volume) of an element depends on the nodes. A node and an element (triangular shaped) are shown in Figure 1-3.

Element Shapes

There are many types of element shapes that are further divided into various classes depending on their usage. The following are some basic element shapes:

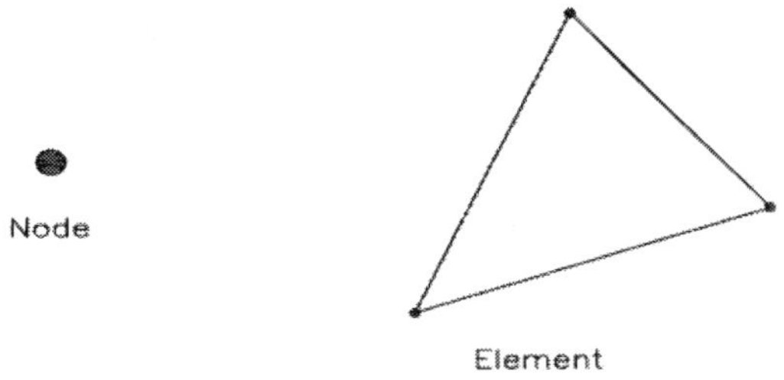

Figure 1-3 A node and an element

Line Element

A line element has the shape of a line or a curve. Therefore, a minimum of two nodes are required to define it. There can be higher order elements that have additional nodes (at the middle of the edge of an element). An element that does not have a node in between its edges is called a linear element. The elements that have nodes in between the edges are called quadratic or second order elements. Figure 1-4 shows some line elements.

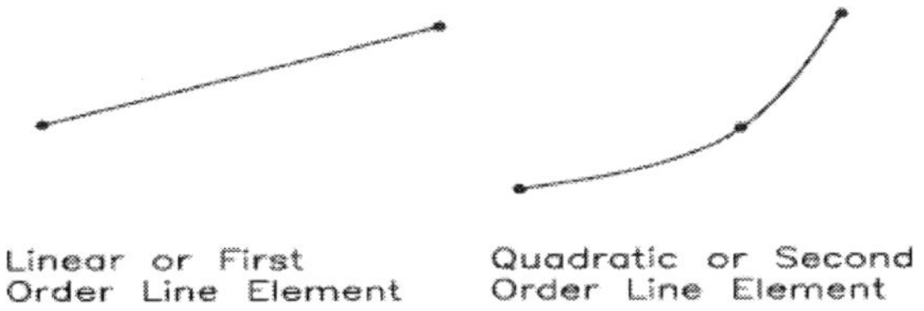

Figure 1-4 Line elements

Area Element

An area element has the shape of a triangle or a quadrilateral; therefore, it requires a minimum of three or four nodes to define it. Some area elements are shown in Figure 1-5.

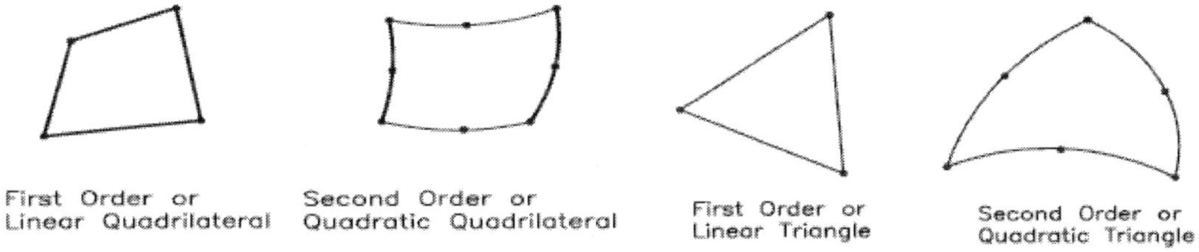

Figure 1-5 The area elements

Volume Element

A volume element has the shape of a hexahedron, wedge, tetrahedron, or a pyramid. Some of the volume elements are shown in Figure 1-6.

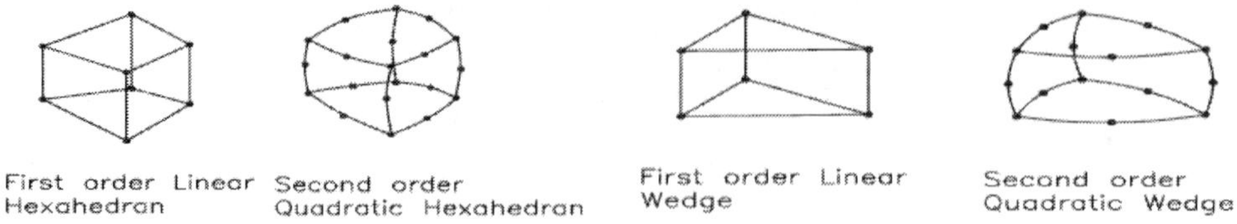

***Figure 1-6** The volume elements*

General Procedure to Conduct Finite Element Analysis

To conduct the finite element analysis, you need to follow certain steps that are given next.

1. Set the type of analysis to be used.
2. Create the model.
3. Define the element type.
4. Divide the given geometry into nodes and elements (mesh the model).
5. Apply material properties and boundary conditions.
6. Derive element matrices and equations.
7. Assemble element equations.
8. Solve the unknown parameters at nodes.
9. Interpret the results.

The general process of FEA by using software is divided into three main phases: preprocessing, solution, and postprocessing, refer to Figure 1-7.

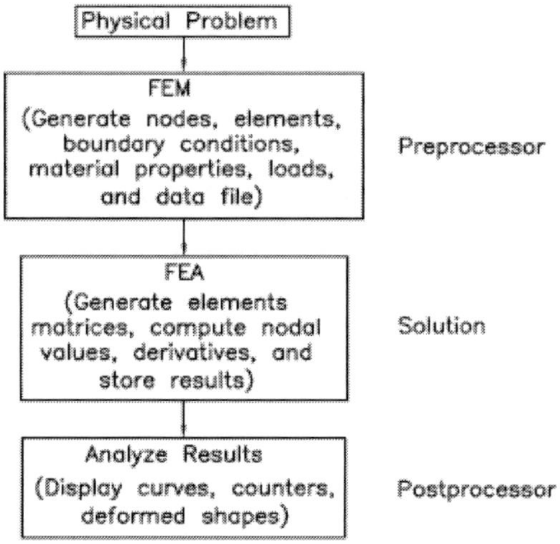

***Figure 1-7** FEA Process*

Preprocessor

In the preprocessor phase, the input data is processed to produce output which is further used as input in the subsequent phase (solution). The input data required for the preprocessor phase is given next:

1. Type of analysis (structural or thermal, static or dynamic, and linear or nonlinear).
2. Element type.
3. Real constants for elements (Cross-sectional area, Moment of Inertia, Shell thickness, and so on).
4. Material properties (Young's Modulus, Poisson's ratio, Spring Constant, Thermal Conductivity, Coefficient of Thermal Expansion, and so on).
5. Geometric model (either created in the FEA software or imported from other CAD packages).
6. FEA model (discretizing the geometric model into small elements).
7. Loading and boundary conditions (defining loads, pressures, moments, temperature, conductivity, convection, constraints such as fixed, pinned, or frictionless/symmetrical), and so on.

The input data are preprocessed for the output data and the preprocessor generates the data files automatically with the help of users. These data files are used in the subsequent phase (solution), refer to Figure 1-7.

Solution

The solution phase is completely automatic. The FEA software generates element matrices, computes nodal values and derivatives, and stores the result data in files. These files are further used in the subsequent phase (postprocessor) to review and analyze the results through the graphic display and tabular listings, refer to Figure 1-7.

Postprocessor

The output from the solution phase (result data files) is in numerical form and consists of nodal values of the field variable and its derivatives. For example, in structural analysis, the output of the postprocessor is nodal displacement and stress in elements. The postprocessor processes the result data and displays them in graphical form to check or analyze the result. The graphical output gives the detailed information about the required result data. The postprocessor phase is automatic and generates graphical output in the specified form, refer to Figure 1-7.

FEA SOFTWARE

A variety of commercial FEA software packages are available in the market. Every CAE software provides various modules for various analysis requirements. Depending on your requirement, you can select a required module for your analysis. Some firms use one or more CAE software and others develop customized version of commercial software to meet their requirements.

Advantages and Limitations of FEA Software

Following are some of the advantages and limitations of FEA software:

Advantages

1. It reduces the amount of prototype testing, thereby saving the cost and time.
2. It gives graphical representation of the result of analysis.
3. The finite element modeling and analysis are performed in the preprocessor and solution phases, which if done manually would consume a lot of time and in some cases, might be impossible to perform.
4. Variables such as stress and temperature can be measured at any desired point of the model.
5. It helps optimize a design.
6. It is used to simulate the designs that are not suitable for prototype testing.
7. It helps you create more reliable, high quality, and competitive designs.

Limitations

1. It does not provide exact solution.
2. FEA packages are costly.
3. An inexperienced user can deliver incorrect answers upon which expensive decisions will depend.
4. Results give solutions but not remedies.
5. Features such as bolts, welded joints, and so on cannot be accommodated in a model. This may lead to approximation and errors in the result.
6. For more accurate results, more hard disk space, RAM, and time are required.

KEY ASSUMPTIONS IN FEA

There are four types of key assumptions that must be considered while performing the finite element analysis. These assumptions are not comprehensive but cover a wide variety of situations applicable to the problem. Moreover, by no means do all the following assumptions apply to all situations. Therefore, you need to consider only those assumptions that are applicable for analysis problem.

Assumptions Related to Geometry

1. Displacement values will be small so that a linear solution is valid.
2. Stress behavior outside the area of interest is not important. Therefore, geometric simplifications in those areas do not affect the outcome.
3. Only internal fillets in the area of interest will be included in the solution.
4. Local behavior at the corners, joints, and intersection of geometries is of primary interest, therefore, no special modeling of these areas is required.
5. Decorative external features will be assumed insignificant for the stiffness and performance of the part and these external features will be omitted from the model.
6. Variation in the mass due to suppressed features is negligible.

Assumptions Related to Material Properties

1. Material properties will remain in the linear region and the nonlinear behavior of the material property cannot be accepted.
2. Material properties are not affected by the load rate.
3. The component is free from surface imperfections that can produce stress concentration.
4. All simulations will assume room temperature unless otherwise specified.
5. The effect of relative humidity or water absorption on the material used will be neglected.
6. No compensation will be made to account for the effect of chemicals, corrosives, wears, or other factors that may have an impact on the long term structural integrity.

Assumptions Related to Boundary Conditions

1. Displacements will be small so that the magnitude, orientation, and distribution of the load remains constant throughout the process of deformation.
2. Frictional loss in the system is considered to be negligible.
3. All interfacing components will be assumed rigid.
4. The portion of the structure being studied is assumed as separate from the rest of the system, so that any reaction or input from the adjacent features is neglected.

Assumptions Related to Fasteners

1. Residual stresses due to fabrication, pre loading on bolts, welding, or other manufacturing or assembly processes will be neglected.
2. All welds between components will be considered as ideal and continuous.
3. The failure of fasteners will not be considered.
4. The load on the threaded portion of the part is supposed to be evenly distributed among the engaged threads.
5. The stiffness of bearings, both in radial and in axial directions, will be considered as infinite or rigid.

IMPORTANT TERMS AND DEFINITIONS

There are some important terms and definitions used in FEA software that are discussed next.

Strength

When a material is subjected to an external load, the system undergoes deformation. The material, in turn, offers resistance against this deformation. This resistance is offered by the material by virtue of its strength.

Load

The external force acting on a body is called load.

Stress

The force of resistance offered by a body per unit area against the deformation is called stress. The stress is induced in the body while the load is being applied on the body. The stress is calculated as load per unit area.

$$p = F/A$$

Where,

p = Stress in N/mm^2
F = Applied Force in Newton
A = Cross-Sectional Area in mm^2

The material can undergo various types of stresses which are discussed next.

Tensile Stress

If the resistance offered by a body is against the increase in size, the body is said to be under tensile stress.

Compressive Stress

If the resistance offered by a body is against the decrease in size, the body is said to be under compressive stress. Compressive stress is just the reverse of tensile stress.

Shear Stress

The shear stress exists when two layers of materials tend to slide across each other in any typical plane of shear on the application of force parallel to that plane. In other words, shear stress is generated in the body when force is applied parallel to the cross-section of the body.

Shear Stress = Shear resistance (R) / Shear area (A)

Strain

When a body is subjected to a load (force), its length undergoes change. The ratio of change in length to the original length of the member is called strain. If the body returns to its original shape on removing the load, the strain is called elastic strain. If the body remains distorted after removing the load, the strain is called plastic strain. The strain can be of three types, tensile, compressive, and shear strain.

Strain (e) = Change in Length (dl) / Original Length (l)

Strain is further categorized as Longitudinal strain and Lateral strain. The ratio of deformation in the body to the original length in the direction of the applied load is known as longitudinal strain or linear strain.

The strain at right angle to the direction of applied load is known as lateral strain or transverse strain.

Elastic Limit

The maximum stress that can be applied to a material without causing permanent deformation is known as the elastic limit of the material. If the stress is within the elastic limit, the material returns to its original shape and size on removing the external load.

Hooke's Law

This law states that the stress is directly proportional to the strain within the elastic limit.

Stress / Strain = Constant (within the elastic limit)

This constant is known as Young's modulus or Modulus of Elasticity, and denoted by E.

Hence, **E = p/e**

Shear Modulus or Modulus of Rigidity

In case of shear loading, the ratio of shear stress to the corresponding shear strain is constant. This ratio is called shear modulus, and it is denoted by G.

Ultimate Strength

The maximum stress that a material withstands when subjected to an applied load is called its ultimate strength.

Yield Strength

The maximum stress that can be developed in a material without causing plastic deformation is called its yield strength.

Factor of Safety

The ratio of the absolute strength to the working stress (design stress) is known as factor of safety. It is necessary that the value of factor of safety should be equal or greater than unity for the design to be safe.

Factor of Safety (FOS) = Absolute strength/Working stress

Poisson's Ratio

The ratio of the lateral strain to the longitudinal strain is constant within the elastic limit. This ratio is called the Poisson's ratio and is denoted by μ. The value of 'μ' lies between 0.0 to 0.5.

Poisson's ratio (μ) = Lateral Strain / Longitudinal Strain

Bulk Modulus

If a body is subjected to equal stresses along three mutually perpendicular directions, the ratio of the direct stresses to the corresponding volumetric strain is found constant for a given material, when the deformation is within a certain limit. This ratio is called the Bulk Modulus and is denoted by K.

Stress Concentration

The value of stress changes abruptly in the regions where the cross-section or profile of a structural member changes abruptly. The phenomenon of this abrupt change in stress is known as stress concentration and the region of the structural member affected by stress concentration is known as the region of stress concentration. The region of stress concentration needs to be meshed densely to get accurate results.

Bending

When a force is applied perpendicular to the longitudinal axis of a body, the body starts deforming. This phenomenon is known as bending. In case of bending, strains vary linearly from the c line of beam to the circumference. In case of pure bending, the value of strain is zero at the center line.

Bending Stress

When a non-axial force is applied on a structural member, some compressive and tensile stresses are developed in the member. These stresses are known as bending stresses.

Creep

At elevated temperature and constant load, many materials continue to deform but at a slow rate. This behavior of materials is called creep. At constant stress and temperature, the rate of creep is approximately constant for a long period of time. After a certain amount of deformation, the rate of creep increases, thereby causing fracture in the material. The rate of creep is highly dependent on both stress and temperature.

Degrees of Freedom (DOF)

The degrees of freedom is defined as the freedom allowed to a given object to move and rotate in any direction in space.

There are six DOFs for any object in 3-dimensional (3D) space: 3 translational DOFs (one each in the X,Y and Z directions) and 3 rotational DOFs (one rotation about each of the X,Y, and Z axes).

INTRODUCTION TO SOLIDWORKS SIMULATION

SOLIDWORKS Simulation is a Computer Aided Engineering(CAE) tool which works on Finite Element Analysis (FEA) technology. SOLIDWORKS Simulation was originally developed by Structural Research and Analysis Corporation (SRAC), USA and its first version came in 1995. SRAC joined with SOLIDWORKS and created COSMOSWorks. SRAC was acquired by Dassault Systemes in 2001. In 2008, COSMOSWorks was renamed to SOLIDWORKS Simulation.

SOLIDWORKS Simulation is fully integrated with SOLIDWORKS and supports multi-CAD environment products by providing direct file exchange with other CAD applications such as Creo Parametric, CATIA V5, NX (Unigraphic), Solid Edge, Autodesk Inventor, and so on. It makes iterative design change without redefining material, loads, constraints, or other simulation data when working with the native CAD format. You can also import geometry of universal file formats such as ACIS, IGES, STEP, and STL for solid models and CDL, DXF, and IGES for wireframe models.

Now, SOLIDWORKS Simulation is available in different packages and solvers which are discussed next.

SOLIDWORKS Simulation Standard

SOLIDWORKS Simulation Standard can perform static and stress analysis on assembly and parts only. It is limited to basic contacts and connections.

SOLIDWORKS Simulation Professional

SOLIDWORKS Simulation Professional covers the analysis portion discussed in SOLIDWORKS Simulation Standard. And apart from this, advanced contacts and connections can also be applied during the analysis. The frequency, buckling, thermal, drop test, fatigue, and pressure vessel design analysis can also be performed with professional suite.

SOLIDWORKS Simulation Premium

SOLIDWORKS Simulation Premium covers all analysis tools available in SOLIDWORKS Simulation Standard and Professional suite as well as the tools used for non linear and dynamic simulation.

Note

1. Note that all these three packages of SOLIDWORKS Simulation do not run as stand-alone software. They are integrated with SOLIDWORKS to run on the PC machine.

2. SOLIDWORKS package comes with a simulation tool known as SimulationXpress. But, SimulationXpress is limited to part analysis and simple load and connections can be applied.

Analysis Solvers

In SOLIDWORKS Simulation, two direct solvers and one iterative solver are available for the solution of the set of equations. Infinite element analysis, a problem is represented by a set of algebraic equations that must be solved to find out solutions. There are two classes of solution methods; direct and iterative. Direct method solves the equations using exact numerical techniques. Iterative method solves the equations using approximate techniques where in each iteration, a solution is assumed and the associated errors are evaluated. The iterations continue until the errors become acceptable.

Automatic
This solver option is the default option for static, frequency, buckling, and thermal studies.

Direct Sparse
In the case of multi-area contact problems where the area of contact is found through several contact iterations, the Direct Sparse solver is preferred.

FFEPlus
In general, FFEPlus is faster in solving problems with degrees of freedom (DOF) over 100,000. It becomes more efficient as the problem gets larger.

Large Problem Direct Sparse
The Direct Sparse solver requires about 10 times more RAM and is faster than the FFEPlus solver. It leverages multicore processing capability and improves solution speed for static and nonlinear studies. Note that it is available for static and nonlinear studies.

Types of Analyses Performed Using SOLIDWORKS Simulation
You can perform different types of analysis using SOLIDWORKS Simulation. Few of them are discussed next.

Static Analysis
In static analysis, the load or field conditions do not vary with respect to time, and therefore, it is assumed that the load or field conditions are applied gradually. When loads are applied to a body, the body deforms and the effect of loads is transmitted throughout the body. The external loads induce internal forces and reactions to make the body into a state of equilibrium. Linear Static Analysis is used to calculate the displacements, strains, stresses, and reaction forces under the effect of applied loads in equilibrium state. It helps to avoid failure due to high stress.

Dynamic Analysis
This type of study assumes that the materials are linear and the loads are either time dependent, frequency dependent or defined by limiting spectra. Also, the mass and inertia effects are included and damping is available. The options in SOLIDWORKS Simulation are Drop Test (also known as Direct Time History Analysis), Modal Time History Analysis (Mode Superposition Analysis), Harmonic Analysis (Harmonic Response Analysis), and Random Vibration Analysis (Response Spectra Analysis). The last three analysis types require a Frequency Analysis to be completed to supply the eigenvalues (natural frequencies) and eigenvectors (mode shapes) needed as inputs.

Thermal Analysis
Thermal study calculates temperatures, temperature gradients, and heat flow based on heat generation, conduction, convection, and radiation conditions. Thermal analysis can help you avoid undesirable thermal conditions like overheating and melting.

Frequency Analysis

When a structure is put under a dynamic load with a frequency that coincides with one of its natural frequencies, the structure undergoes large displacements and stresses. This phenomenon is known as resonance. For undamped systems, resonance theoretically causes infinite motion. Damping, however, puts a limit on the response of the structures due to resonant loads and calculates natural frequencies and associated mode shapes.

Buckling Analysis

Buckling refers to sudden large displacements due to axial loads. Slender structures subject to axial loads can fail due to buckling at load levels lower than those required to cause material failure. Buckling can occur in different modes under the effect of different load levels. Buckling studies can help to avoid failure due to buckling. Usually the lowest buckling load is the subject of interest and is lower than those required to cause material failure.

Fatigue

Fatigue study evaluates the consumed life of an object based on a very large number of fatigue events (cycles). Repeated loading weakens materials over a period of time even when the induced stresses are low. The number of cycles that can cause failure depends on the material and the stress fluctuations. The data is provided by the material S-N curve which depicts the number of cycles that cause failure due to different stress levels. Fatigue study determines the consumed life of an object, in relation to fatigue events, based on fatigue calculations of stress intensity, von-Mises stress or Maximum principal stress.

Pressure Vessel Design

Pressure vessel design study combines the results of multiple static studies with the desired load factors. This study combines the results algebraically using a linear combination or the square root of the sum of the squares. Engineering plastics have been used for pressure vessel applications for a long time. Lighter bodies, ballcock valves, and spray paint containers are few examples of successful developments in this area.

Drop Test

Drop test study evaluates the impact of dropping body on a rigid floor. You can specify the dropping distance or the velocity at the time of analysis in addition to gravity. In this test no other loads or restraints are allowed.

Nonlinear Study

Nonlinear study enables to quickly and efficiently analyze stresses and deformations under general conditions to ensure high quality, performance, and safety. Nonlinear stress analysis calculates the stresses and deformations under the most general loading and material conditions such as Dynamic (time dependent) loads, Large component deformation, and Nonlinear materials.

Topology Study

In SOLIDWORKS Simulation, topology study allow you to explore design iterations of a component that satisfy a given optimization goal and geometric constraints. A Topology study performs nonparametric topology optimization of parts. It starts with maximum design space which represents the maximum allowed size for a component with all applied loads, fixtures, and manufacturing constraints. The topology optimization seeks a new material layout within the boundaries of the maximum allowed geometry by redistributing the material. The optimized component satisfies all the required functional and manufacturing requirements.

Starting SOLIDWORKS Simulation

To start SOLIDWORKS Simulation, activate SOLIDWORKS Simulation by clicking on the downward arrow beside the **Options** button in the menu bar; a flyout will be displayed, as shown in Figure 1-8. Next, choose the **Add-Ins** options from the flyout; the **Add-Ins** dialog box will be displayed, as shown in Figure 1-9. You can also invoke this dialog box by choosing the Add-Ins option from the **Tools** menu. In this dialog box, if you select the check box located on the right of the add-ins name, then SOLIDWORKS Simulation will be activated whenever SOLIDWORKS is started. If you prefer to activate the SOLIDWORKS Simulation program only when it is needed for an analysis, select the check box available on the left of **SOLIDWORKS Simulation** in the **Add-Ins** dialog box and choose the **OK** button.

***Figure 1-8** Flyout displayed on clicking the Options button arrow*

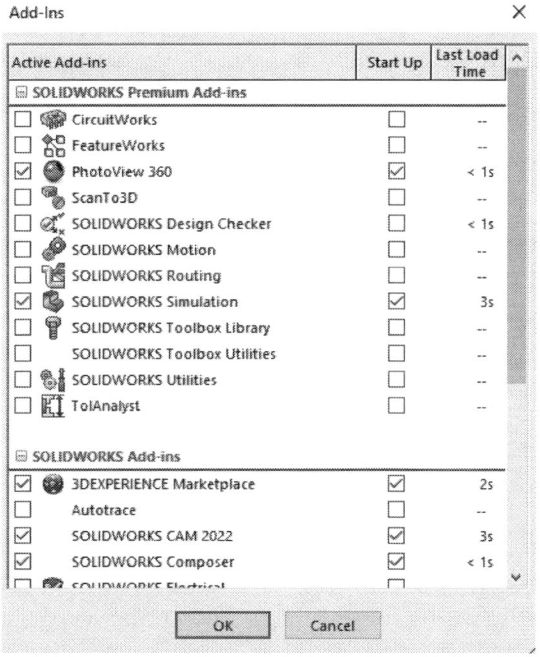

***Figure 1-9** The **Add-ins** dialog box*

Next, start a new part document in SOLIDWORKS by using the **New SOLIDWORKS Document** dialog box; the **Simulation** menu is added in the SOLIDWORKS menus and the **Simulation** tab is added to the CommandManager, as shown in Figure 1-10.

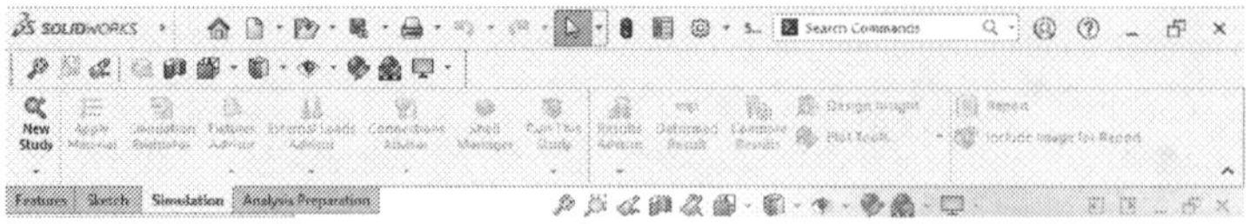

Figure 1-10 Simulation CommandManager

To start a new study, click on the downward arrow below the **New Study** button; a drop-down menu will be displayed. Now, choose the **New Study** option from this drop-down menu; the **Study PropertyManager** will be displayed, as shown in Figure 1-11. Next, select the type of study which you want to perform and then select the **OK** button from the **Study PropertyManager**; the Simulation Tree will be displayed, as shown in Figure 1-12.

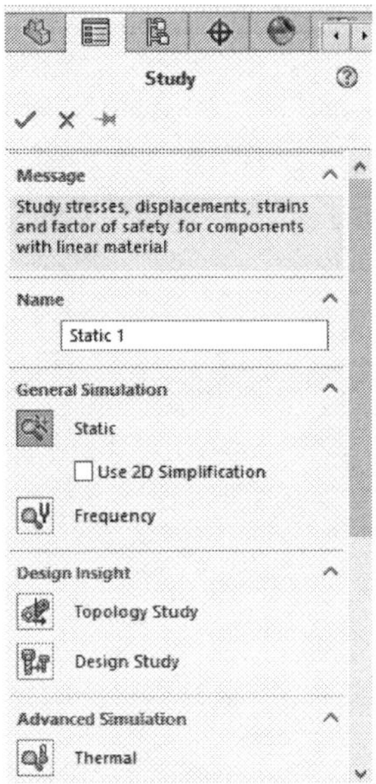

Figure 1-11** The **Study PropertyMananger

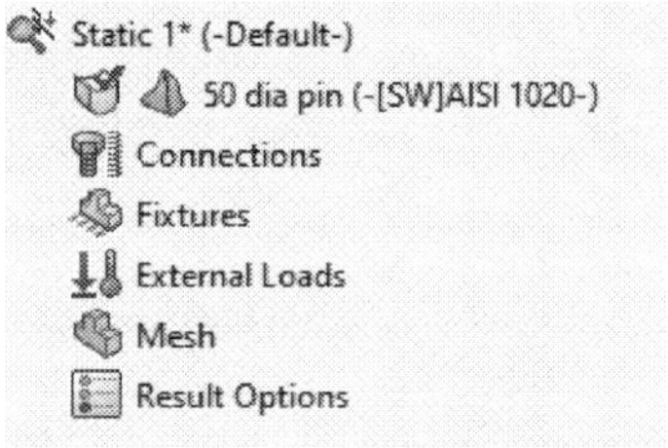

***Figure 1-12** Simulation Tree*

BASIC STEPS REQUIRED TO PERFORM SIMULATION IN SOLIDWORKS SIMULATION

You need to perform the following basic steps to complete the finite element analysis in SOLIDWORKS Simulation. The steps are discussed next.

Prepare the Model

Finite element analysis (FEA) requires a well described model for study, in order to perform correct analysis which in return, should offer plausible results. In FEA, different assumptions and simplifications are made such as 3D geometrical model and proposed type of study. Simplifications and repairing of design are required so that it take less computing time during the analysis. By combining the right assumptions with a good robust model, a successful technical analysis can be achieved.

Assign Material Properties

When you are modelling a component in SOLIDWORKS, you can assign material to the model. You can also assign material to the model imported in simulation environment. To apply the material in simulation environment, right-click on the name of the model in the Simulation Tree; a shortcut menu will be displayed, as shown in Figure 1-13. Choose the material from the **Apply Favorite Material** list or choose the **Apply/Edit Material** option; the **Material** dialog box will be displayed, as shown in Figure 1-14. Select the required material (Alloy Steel); the properties of the selected material will be displayed, as shown in Figure 1-14. You will notice that the material properties will be displayed in red, blue, and black colors. These colors describe the mandatory and optional properties for the selected study. The description in red color (Elastic Modulus, Poisson's Ratio, Mass Density, Yield Strength) indicates the property is mandatory for the active study type and the material of the model. A blue description (Tensile Strength, Compressive Strength, Thermal Expansion Coefficient) indicates an optional property. A black description (Shear Modulus, Thermal Conductivity, Specific Heat, Material Damping Ratio) indicates properties are not applicable to the current study.

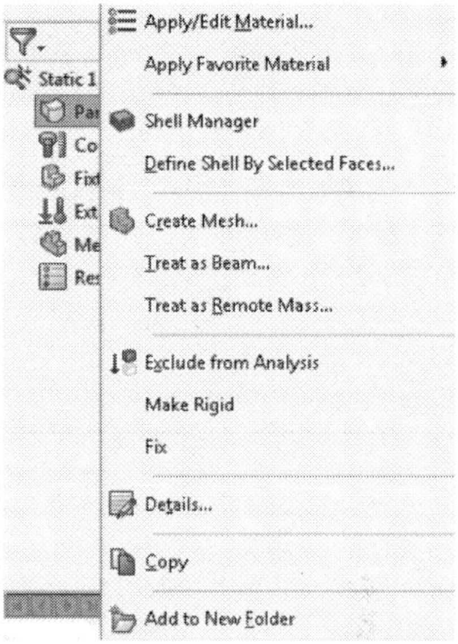

Figure 1-13 *The shortcut menu displayed*

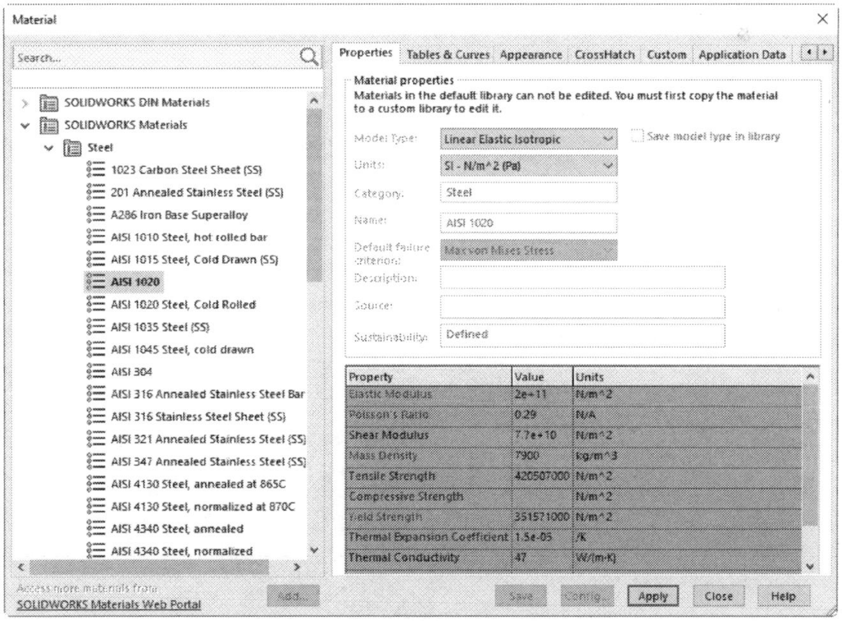

Figure 1-14 *The* ***Material*** *dialog box*

Note

1. *If you are working with an assembly or with composite material, you need to expand the part node and then apply material by right-clicking on the part.*

2. *Material applied in SOLIDWORKS Simulation will not affect the material applied on CAD model in SOLIDWORKS.*

Apply Loads and Restraints

Loads and restraints are applied to define actual environmental conditions. The applied loads and restraints are represented by an icon in Simulation Tree. SOLIDWORKS Simulation provides context-sensitive options for defining restraints. For example, if reference axis is selected or the selected faces are cylindrical, the program expects you to define radial, circumferential, and axial restraints.

If you make geometry changes such that the entity is no longer defined, then after applying a restraint or load to the entity, the restraint becomes dangled. SOLIDWORKS Simulation gives you a message that lists dangled restraints or load for each study you defined.

Meshing

FEA provides a reliable numerical technique for analyzing engineering designs. The process starts with the creation of a geometric model. Then, the program subdivides the model into small pieces of simple shapes called elements, connected at common points called nodes. The process of subdividing a model into small pieces is called meshing.

Meshing is a crucial step in design analysis. SOLIDWORKS Simulation allow you to create a mesh of solid elements (3D tetrahedral), shells (2D triangular), or beam (1D linear). The solid mesh is appropriate for bulky or complex 3D models. Shell elements are suitable for thin parts (such as sheet metals). The accuracy of the results depends on the quality of the mesh. In general, accurate and finer the mesh, better the accuracy. The resultant mesh depends on the following factors:

- Type of mesh (solid, shell using midsurfaces, or shell using surfaces).
- Active mesh preferences.
- Mesh control.
- Contact conditions for static and thermal assembly problems.
- Global element size and mesh tolerance.

Run Studies

After assigning materials, defining loads and restraints, and meshing the model, you can run the study.

SOLIDWORKS Simulation has four options for solvers: Auto, FFEPlus, Direct Sparse, and Large Problem Direct Sparse. By default, SimulationXpress and SOLIDWORKS Simulation use FEEPlus solver. Solver is selected when you define the properties of a study. In some cases, the program switches to another solver automatically if the selected solver does not support all options used in a study. All solvers provide the same results provided the same mesh is used. However, the performance and speed vary depending on the type and size of the problem.

View Results

After the completion of study, you can automatically generate standard plots for each type of analyses. The standard plots for an analysis type are the most commonly used results. For example, after running a static study, SOLIDWORKS Simulation creates result folders containing default plots for stress, displacement, and strain. You can view a plot by double-clicking on its node in the Simulation Tree.

You can also view the other plots by right-clicking on the **Results** node and then selecting the required option from the shortcut menu displayed. You can also generate reports that include all available plots automatically. To do so, choose the **Report** tool in **Simulation Command Manager;** the **Report Options** dialog box will be displayed. Fill the required input and choose the **Publish** button; the report for the study will be generated in Word Document.

Note

It is not compulsory that the above sequence should be maintained to define the parameters. You can define these parameters in any order but make sure that you have applied all necessary parameters before you run the study.

Self-Evaluation Test

Answer the following questions and then compare them to those given at the end of this chapter:

1. Which of the following options represents the shape of a face of first order tetrahedral finite element?

   (a) Rectangle (b) Triangle
   (c) Pentagon (d) None of these

2. What is the maximum number of nodes that can be available in a hexahedral element?

   (a) 6 (b) 4
   (c) 8 (d) None of these

3. Stress induced in a model due to an external load, gets changed due to changing material of the model. (T/F)

4. Stiffness constant of a model does not depend upon the material of the model. (T/F)

5. Stress is defined as the resistance offered by a body against deformation. (T/F)

6. A design is assumed to be safe if the factor of safety value for that model lies between 0 to 1. (T/F)

7. A model can have infinite degrees of freedom. (T/F)

8. Material applied to a model in part environment of SOLIDWORKS can be used for SOLIDWORKS Simulation. (T/F)

Review Questions

Answer the following questions:

1. How many faces does a tetrahedral element have?

   (a) 6 (b) 4
   (c) 3 (d) 8

2. What is the minimum number of nodes required to define a line element?

   (a) 2 (b) 4
   (c) 3 (d) None of these

3. What is the maximum number of degrees of freedom that a tetrahedral element can have?

   (a) 6 (b) 4
   (c) 3 (d) 8

4. Factor of safety of a model depends upon the material of the model. (T/F)

5. Factor of safety for mild steel is defined as the ratio of ultimate strength to working stress. (T/F)

6. An area element can have a maximum of six nodes. (T/F)

7. Yield point can not be defined for brittle materials. (T/F)

8. In Finite Element Analysis, you can assign materials, apply fixtures and loads, and run the study to visualize the results. (T/F)

Answers to Self-Evaluation Test

1. a, **2**. d, **3**. F, **4**. F, **5**. F, **6**. F, **7**. F, **8**. T

Chapter 2

Defining Material Properties

Learning Objectives

After completing this chapter, you will be able to:

- *Apply different materials to models*
- *Understand mandatory and optional properties for the selected study*
- *Create custom materials*
- *Apply and save custom materials*

INTRODUCTION

Generally, the industrial components are made-up of metals. This is the reason most of the FEA calculations involve metallic components. The analysis of metallic components can be carried out by either linear or nonlinear stress analysis. The analysis approach to be used depends upon the design and applications of the load.

If the geometry returns to its original shape and remains in the linear elastic range once the load is removed, then linear stress analysis may be applied to the geometry. In this case, rotations and displacements are relative supports to the geometry. For such analysis, Factor of Safety (FoS) is a common design goal to validate the design.

If you want to evaluate the effects of post-yield load cycling on the geometry, a nonlinear stress analysis should be carried out. In this case, the impact of strain hardening on the residual stresses and permanent deformation are important analysis interest. The analysis of nonmetallic components, such as plastic or rubber components, should be carried out using nonlinear stress analysis methods, due to their complex load deformation relationship.

Most Finite Element Analysis (FEA) programs use energy methods to find a numerical solution. In these methods, first the displacement is calculated for the load applied. The displacement can then be used to calculate strain, and then the strain can be found out by using the Young's modulus (E= Stress/Strain). Displacement is the primary result for FEA study. Note that the stress is not a function of material properties unless the stresses are located close to the constraints or the model is indeterminate. Otherwise, the stress should be the same or maybe slightly different due to stiffness (k numerical error).

To define the material of an FEA model is one of the most important steps in the analysis process. Based on the analysis type chosen, you must define all the necessary material properties. For example, the modulus of elasticity is required for static, frequency, and buckling studies, while thermal conductivity is needed for thermal studies. SOLIDWORKS Simulation is provided with almost all the standard materials that are available in SOLIDWORKS Material library. You can also create and edit custom materials or libraries in SOLIDWORKS Simulation and then apply them any time before running the analysis. Note that the material applied to the CAD model using Simulation does not update the material assigned to the CAD model before importing.

APPLYING A MATERIAL

In SOLIDWORKS, you can apply material to a part, parts of an assembly, or multi body part. When you apply a material to a part, all physical properties of the applied material will be assigned to the part. To apply a material to a model (part or assembly), right-click on the **Part(s)** node and then choose the **Apply Material to All** option from the shortcut menu displayed, as shown in Figure 2-1; the **Material** dialog box will be displayed, refer to Figure 2-2. Select the desired material from the SOLIDWORKS Materials library. For example, here Alloy Steel has been selected in the library and the property of the selected material is displayed in the

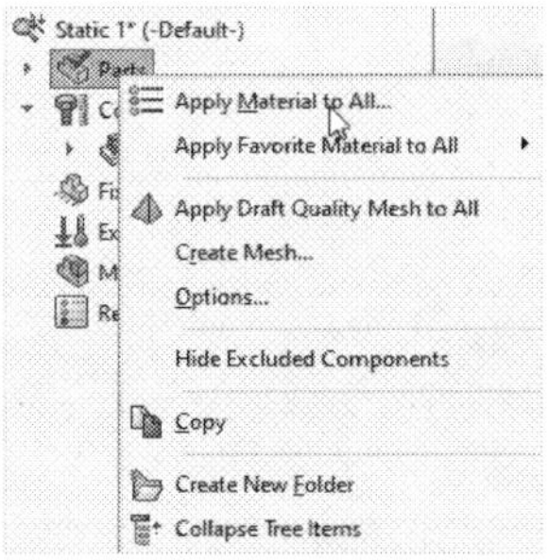

***Figure 2-1** Selecting the **Apply Material to All** options*

Material dialog box, as shown in Figure 2-2. You will notice that the material properties will be displayed in red, blue, and black colors. These colors describe the mandatory and optional properties for the selected study. Red color (Elastic Modulus, Poisson's Ratio, Mass Density, Yield Strength) indicates that the property is mandatory based on the active study type and the material model.Blue color (Tensile Strength, Compressive Strength, Thermal Expansion Coefficient) indicates an optional property. Black color (Shear Modulus, Thermal Conductivity, Specific Heat, Material (Damping Ratio) indicates the properties not applicable to the current study. Next, choose the **Apply** button and then choose **Close** from the **Material** dialog box; the selected material will be applied to all components of the model.

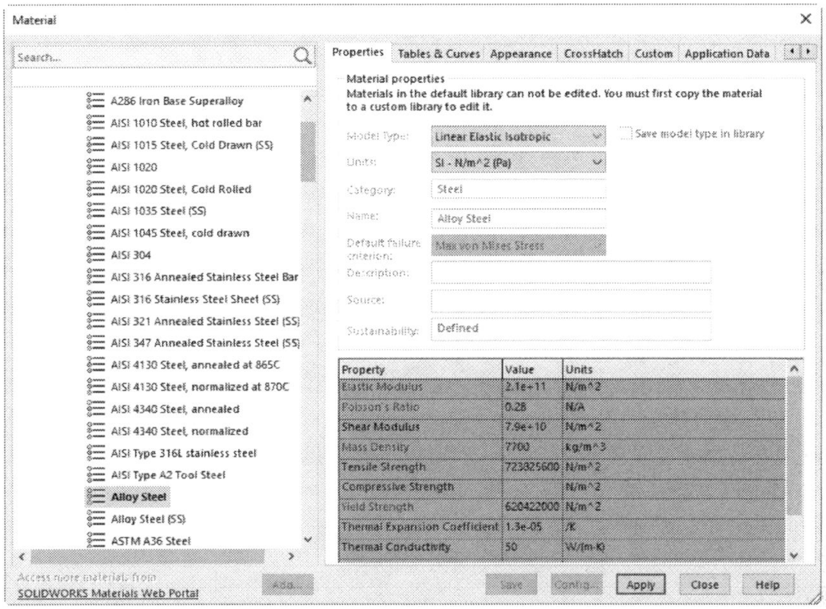

***Figure 2-2** The **Material** dialog box*

Note

1. *When you are applying material to a part, you will get the **Apply/Edit Material** option in the shortcut menu in place of the **Apply Material to All** option.*

2. *Material applied in simulation can not be removed but can be changed by applying another material.*

3. *The material properties in the SOLIDWORKS Materials library are read-only. These properties can not be edited.*

The left side of the dialog box contains a tree of available material types and materials. The right side area contains different tabs which display information about the selected material. These tabs are discussed next.

Properties Tab

The options in the **Properties** tab are used to display all properties of the material in SOLIDWORKS Library. When you create a custom material using this property tab then you can assign properties to the custom material. Various options under this tab are discussed next.

Model Type

This drop-down list is used to define the material type for the study. The available list of material types depends upon the active study type.

Units

This drop-down list is used to specify a unit for material properties.

Include creep effect

This check box will be available for nonlinear studies and is used to activate the creep effect for the selected material. This option will not be available for linear elastic orthotropic and viscoelastic material.

Number of constants

This drop-down list will be available only for hyperelastic materials in nonlinear studies. The options of this drop-down list are used to define the hyperelastic constant. You can enter the material constants manually, or let the program calculate them from test data.

Use Curve Data to Compute Material Constants

This check box will be available only for hyperelastic and viscoelastic material models in nonlinear studies. Select this check box to let the program calculate the material constants from available test data. You can enter up to three data curves for simple tension, planar tension or pure shear, and biaxial tension.

Reference Geometry

This selection box will be available only for the linear elastic orthotropic material model and is used to define reference geometry for the directions of the orthotropic material properties. The reference geometry can be a plane, an axis, or a coordinate system.

Create Stress-Strain Curve

This button will be available only for nonlinear elastic, plasticity - von Mises, and plasticity - Tresca material models. When you choose this button, the **Table and Curves** tab of the **Material** dialog box gets activated where you can define stress-strain curve.

Category

This edit box displays the name of the category into which the material is organized and will be activated in case of custom materials. So, you can change the category name by entering new name in this edit box.

Name

This edit box displays the name of the selected material. This edit box works similar to the **Category** edit box.

Default Failure Criterion
This drop-down list is used to specify the failure criterion to be used for computing the factor of safety when the **Automatic** option is selected from the **Criterion** drop-down list in the **Factor of Safety PropertyManager.**

Description
This edit box is used to add a comment about the material.

Source
This edit box is used to specify a source of reference for custom defined materials.

Sustainability
This display box indicates whether the material has defined sustainability. For SOLIDWORKS materials it displays, **Defined** or **Undefined**. But in case of custom materials, the **Undefined** option is displayed. Also the **Select** button is displayed on the right of the display box. The **Select** button allows the users to specify a link to the material in the **SOLIDWORKS Material** library.

Properties Table
This table displays the properties of the selected material. In case of a custom material, you can change these properties.

Tables & Curves
This tab is used to define stress-strain curves and stretch ratio-stress curves for nonlinear material models. You can also define temperature-dependent material properties for linear material models. The options under this tab are discussed next.

Type
The **Type** drop-down list in the **Type** area is used to list material properties that are allowed to be temperature-dependent.

Table data
The options in this area are used to set the unit for temperature curve and the curve data.

Preview
This area displays the graph as you enter curve data points.

File
This button is used to import the curve data from a curve data points file (*.dat).

View
This button is used to display a graph for material property variation with respect to temperature.

Save
This button is used to save the custom material to the selected material library file.

Appearance

This tab is used to assign a new texture or color to the selected material. This tab does not change the physical properties of the material. Note that the appearance of standard SOLIDWORKS materials can not be changed.

Cross Hatch

Using the options of this tab, you can define cross-hatch of section view in the drawing document.

Custom

The area under this tab will be available for editing when the selected material is in a custom library. This tab is used to add non-standard properties to a material.

Application Data

The area under this tab will be available for editing when the selected material is in a custom library. This tab is used to record notes about the selected material, such as document processing, manufacturing, or test information.

Favorites

This tab is used to manage the list of favorite materials displayed when you right-click on the part node available in Simulation Tree. You can assign a material to a part by selecting a material from the favorite materials list. The different options under this tab are discussed next.

Add

This button is used to add a new material to the favorite material list.

Remove

This button is used to remove a material from the favorite material list.

Up

This button is used to move the selected material one position up in the favorite material list.

Down

This button is used to move the selected material one position down in the favorite material list.

CREATING A CUSTOM MATERIAL

As discussed earlier, SOLIDWORKS comes with a material library which contains almost all standard materials. However, if the material you need is not available in the SOLIDWORKS material library then you can add additional materials to your custom library in SOLIDWORKS. To create a custom material, right-click on the **Part** node in the Simulation Tree and choose the **Apply/Edit Material** from the shortcut menu, as shown in Figure 2-3; the **Material** dialog box will be displayed. Right-click on the **Custom Materials** library at the bottom of the list of material libraries; a shortcut menu will be displayed. Choose the **New Category** option from the shortcut menu, as shown in Figure 2-4; a new material category will be added to the **Custom Materials** library as **New Category.** Right-click on the **New Category**; a shortcut menu will be displayed. Choose the **New Material** from the shortcut menu, as shown in Figure 2-5.

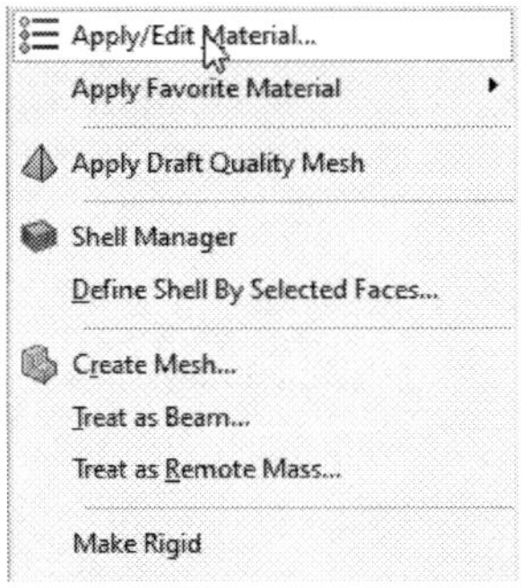

Figure 2-3 *Choosing the **Apply/Edit Material** option*

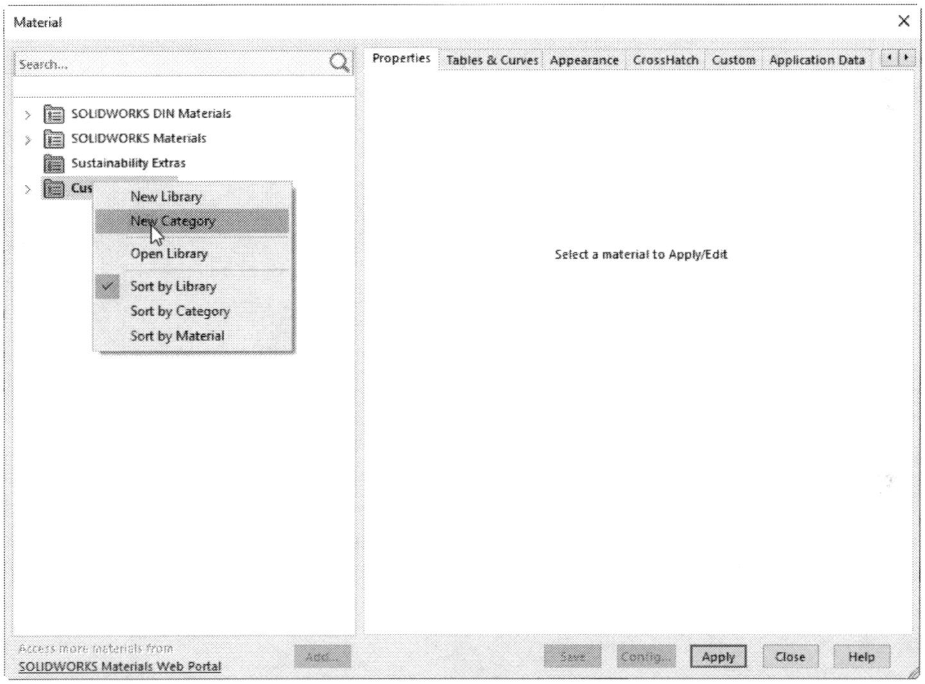

Figure 2-4 *Choosing the **New Category** option from the shortcut menu*

After choosing the **New Material** option from the shortcut menu, a custom material will be added with the name **Default** and the properties of this newly added material will be displayed in the **Properties** tab of the **Material** dialog box, as shown in Figure 2-6. These properties are editable and you need to change them as per your material properties.

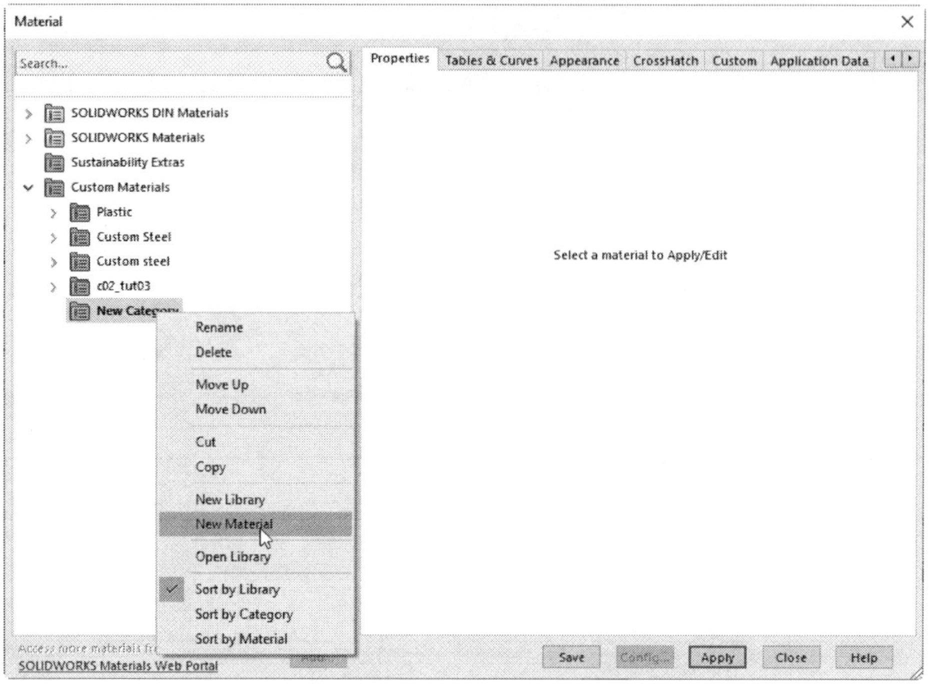

*Figure 2-5 The **New Material** option chosen in the shortcut menu*

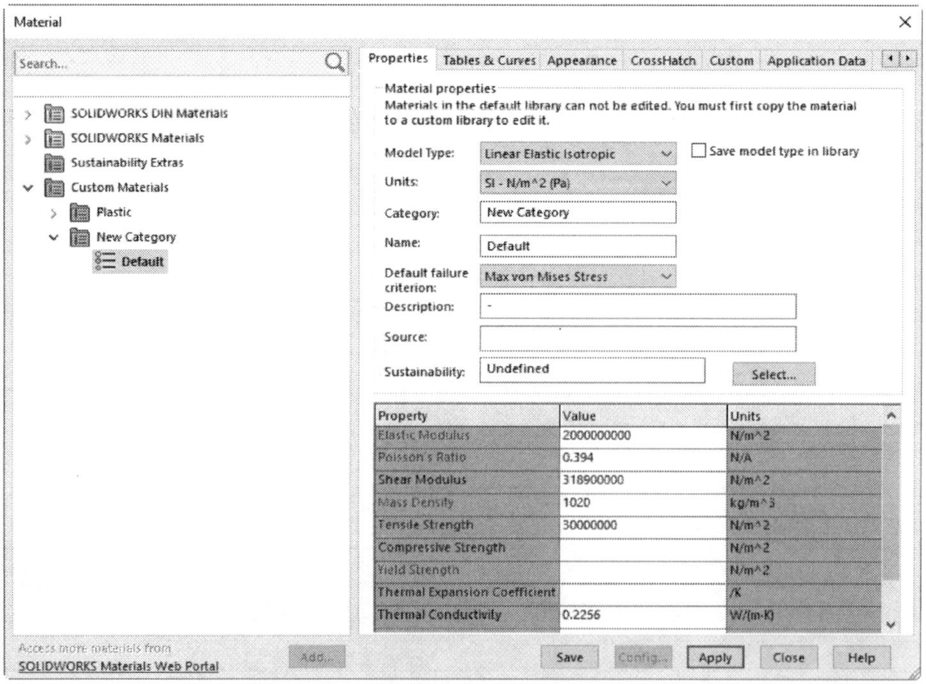

*Figure 2-6 The properties of the **Default** material in the **Material** dialog box*

You can also assign custom material appearance to the custom material by selecting it from one of the preset appearances available, as shown in Figure 2-7. Next, choose the **Apply** button from the **Material** dialog box; the custom material properties will be applied to the part and the custom material will be saved in SOLIDWORKS Material library.

Note

When you are adding custom material, default names get assigned to the material category and to the material. But, you need to rename them according to your custom material.

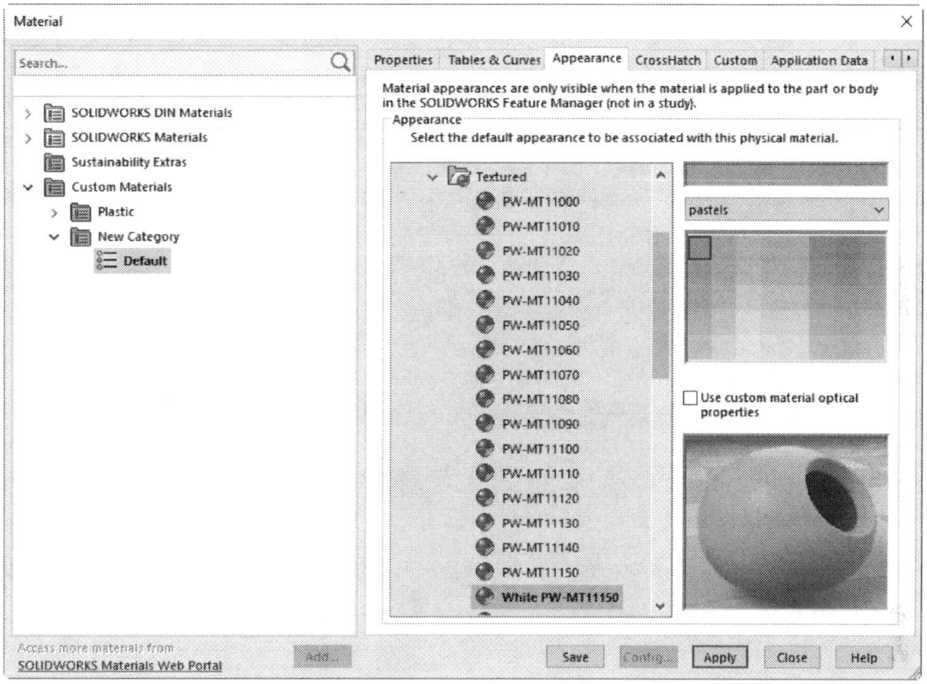

***Figure 2-7** Different preset appearances available in the **Appearance** tab*

TUTORIALS

To perform the tutorials, you need to download the input files from *www.cadcim.com*. The complete path for downloading the files is as follows:

Textbooks > CAE Simulation > SOLIDWORKS Simulation > SOLIDWORKS Simulation 2022: A Tutorial Approach > Input Files > c02_simulation_2022_input.zip

Extract the downloaded zipped file and save it.

Tutorial 1

In this tutorial, you will open *c02_tut01* file and assign Plain Carbon Steel from the SOLIDWORKS Material library to the model. Figure 2-8 shows the model for the tutorial. Next, start the linear static study and assign material from the SOLIDWORKS Material library to the model.

(Expected time: 30 min)

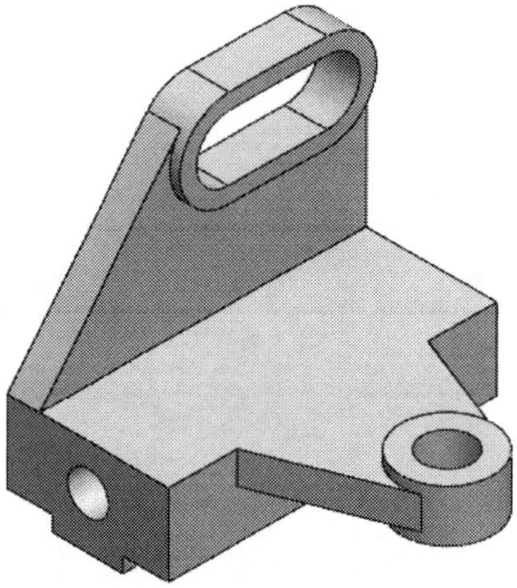

Figure 2-8 *Model for Tutorial 1*

The following steps are required to complete this tutorial:

a. Activate SOLIDWORKS Simulation environment.
b. Open the model.
c. Start static study.
d. Create a custom material as Steel.
e. Apply the material to the model.
f. Save the model.

Adding the SOLIDWORKS Simulation to SOLIDWORKS and Starting a New Part File

Now, you need to add the SOLIDWORKS Simulation to SOLIDWORKS to perform the analysis.

1. Start SOLIDWORKS and then choose **Tools > Add-Ins** from the SOLIDWORKS menus; the **Add-Ins** window is displayed. Next, select both check boxes on the left and right side of SOLIDWORKS Simulation and then choose the **OK** button.

2. Start a new part document by using the **New SOLIDWORKS Document** dialog box. Note that the Simulation menu is added to the **SOLIDWORKS** menus and the **Simulation** tab is added to the CommandManager.

Opening the Model

Start the SOLIDWORKS Simulation and open the model *(c02_tut01)*.

1. Start SOLIDWORKS and choose the **Open** button from the menu bar; the **Open** dialog box is displayed.

2. Browse to the folder where the model has been downloaded and then select the model.

3. Choose the **Open** button in the **Open** dialog box; the model is opened in SOLIDWORKS part mode.

Starting Linear Static Study

In this section, you will start linear static study.

1. Choose the **Simulation** tab in the **CommandManager** tab.

2. Choose the **New Study** tool from the **New Study** flyout in the **Simulation CommandManager**; the **Study PropertyManager** is displayed.

3. Choose the **Static** button from the **General Simulation** rollout and assign it the name as **Linear Static** in the **Study name** edit box of the **Name** rollout.

4. Choose the **OK** button from the **Study PropertyManager**; a static study is started and Simulation Tree is displayed

Applying Material

In this section, you will apply material to the model from the SOLIDWORKS Materials library.

1. Right-click on the part node **(c02_tut01)** in the Simulation Manager tree and choose the **Apply/Edit Material** option from the shortcut menu; the **Material** dialog box is displayed.

2. Expand the **SOLIDWORKS Materials** node if not expended by default; different nodes with the name of material are displayed under the **SOLIDWORKS Materials** node.

3. Expand the **Steel** sub-node; a list of steel materials is displayed under the **Steel** sub-node.

4. Select the **Plain Carbon Steel** from the list of steel materials.

5. Choose the **Apply** button in the **Material** dialog box; the material is applied to the model.

6. Close the **Material** dialog box by choosing the **Close** button in the **Material** dialog box.

7. Choose the **Save** button; the model is saved.

Tutorial 2

In this tutorial, you will open the *c02_tut02* file that you have downloaded. Next, you will start the linear static study on the model and assign Steel material to it. The properties of the Steel material are given next. Figure 2-9 shows the model for the tutorial.

(Expected time: 30 min)

Figure 2-9 *Model for Tutorial 2*

Properties of Steel

Mass density: 8100 kg/m^3
Elastic modulus: 1.9e+011 N/m^2
Poisson's ratio: 0.27
Yield strength: 2.1e+008 N/m^2
Tensile strength: 3.6e+008 N/m^2

The following steps are required to complete this tutorial:

a. Activate SOLIDWORKS Simulation environment.
b. Open the model.
c. Start static study.
d. Create a custom material as Steel.
e. Apply the material to the model.
f. Save the model

Adding the SOLIDWORKS Simulation to SOLIDWORKS and Starting a New Part File

Now, you need to add the SOLIDWORKS Simulation to SOLIDWORKS to perform the analysis.

1. Start SOLIDWORKS and then choose **Tools > Add-Ins** from the SOLIDWORKS menus; the **Add-Ins** window is displayed. Next, select both check boxes on the left and right side o SOLIDWORKS Simulation and then choose the **OK** button.

2. Start a new part document by using the **New SOLIDWORKS Document** dialog box. Note that the Simulation menu is added to the **SOLIDWORKS** menus and the **Simulation** tab is added to the CommandManager.

Starting SOLIDWORKS Simulation and Opening the Model

Open the model *(c02_tut02)* in SOLIDWORKS Simulation.

1. Start SOLIDWORKS, choose the **Open** button from the menu bar; the **Open** dialog box is displayed.

2. In this dialog box, browse to the folder where the downloaded model is saved and then select the model.

3. Choose the **Open** button in the **Open** dialog box; the model is opened in the part mode of SOLIDWORKS.

Starting Linear Static Study

In this section, you will start the linear static study before creating custom material.

1. Choose the **Simulation** tab in the CommandManager tab.

2. Choose the **New Study** tool from the **New Study** flyout; the **Study PropertyManager** is displayed.

3. Choose **Static** from the **General Simulation** rollout and assign **Linear Static** as the name of the study in the **Study name** edit box of the **Name** rollout.

4. Choose the **OK** button from the **Study PropertyManager**; the linear static study is started.

Creating a Custom Material

In this section, you will create custom steel material and apply that to the model.

1. Right-click on the part node *(c02_tut02)* in the Simulation Manager tree and choose the **Apply/Edit Material** option from the shortcut menu; the **Material** dialog box is displayed.

2. Right-click on the **Custom Materials** at the bottom of the list of material libraries; a shortcut menu is displayed.

3. Select the **New Category** option from the shortcut list; a folder with the name **New Category** is added to the **Custom Materials** library.

4. Right-click on the **New Category** folder and rename it as **Custom Steel**.

5. Right-click on the **Custom Steel** folder and choose the **New Material** option from the shortcut menu; a node is added under the Custom Steel folder with the name as **Default**.

6. Right-click on the **Default** node and rename it as **Steel**.

7. Choose the **Properties** tab in the **Material** dialog box if not chosen by default, and select the **SI - N/m2 (Pa)** option from the **Units** drop-down list.

8. Delete all the properties data present in the **Value** column of the **Material** dialog box.

9. Enter **1.9e+011** in the **Value** column of the **Elastic Modulus** row.

10. Enter **0.27** in the **Value** column of the **Poisson's Ratio** row.

11. Enter **8100** in the **Value** column of the **Mass Density** row.

12. Enter **3.6e+008** in the **Value** column of the **Tensile Strength** row.

13. Enter **2.1e+008** in the **Value** column of the **Yield Strength** row.

14. Choose the **Apply** button in the **Material** dialog box and close the dialog box; the custom steel material is applied to the model and also saved in the SOLIDWORKS Material libraries for further use.

15. Choose the **Save** button; the material is saved.

Tutorial 3

In this tutorial, you will open the *c02_tut03* file in SOLIDWORKS. Next, you will start linear static study and apply different materials to different components of the Anti Vibration Mount assembly. The Bolt and Nut are made of Grey Cast Iron, the Yoke Plate and Body are made up of Mild Steel and the Bush is made up of Natural Rubber. The properties of Mild Steel and Gray Cast Iron are given next. Figure 2-10 shows the model (Anti Vibration Mount assembly) for the tutorial and Figure 2-11 shows various components of the Anti Vibration Mount assembly.

(Expected time: 30 min)

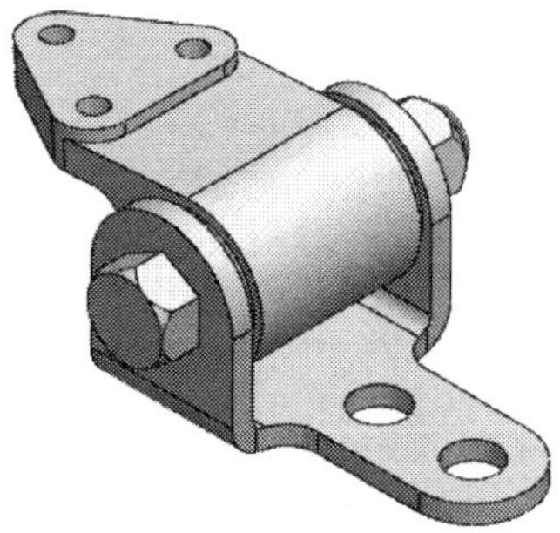

Figure 2-10 *Anti Vibration Mount assembly*

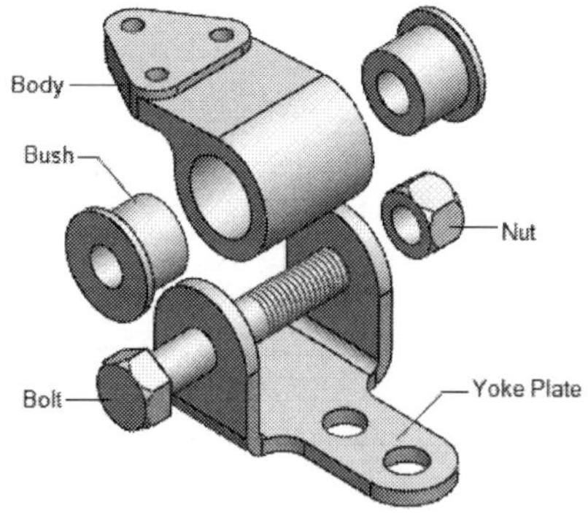

Figure 2-11 *Various components of the Anti Vibration Mount assembly*

Properties of Mild Steel:

Mass density:	7850 kg/m^3
Elastic modulus:	1.9e+011 N/m^2
Poisson's ratio:	0.27
Yield strength:	2.3e+008 N/m^2
Tensile strength:	3.99e+008 N/m^2

Properties of Gray Cast Iron:

Mass density:	7300 kg/m^3
Elastic modulus:	1.9e+011 N/m^2
Poisson's ratio:	0.26
Yield strength:	2.41e+008 N/m^2
Tensile strength:	4.48e+008 N/m^2

The following steps are required to complete this tutorial:

a. Start SOLIDWORKS Simulation.
b. Start a static study.
c. Create a custom material as Gray Cast Iron and apply it on Bolt and Nut.
d. Create another custom material as Mild Steel and apply it on Yoke Plate and Body.
e. Apply Natural Rubber material on bush.
f. Save the model

Adding the SOLIDWORKS Simulation to SOLIDWORKS and Starting a New Part File

Now, you need to add the SOLIDWORKS Simulation to SOLIDWORKS to perform the analysis.

1. Start SOLIDWORKS and then choose **Tools > Add-Ins** from the SOLIDWORKS menus; the **Add-Ins** window is displayed. Next, select both check boxes on the left and right side of SOLIDWORKS Simulation and then choose the **OK** button.

2. Start a new part document by using the **New SOLIDWORKS Document** dialog box. Note that the Simulation menu is added to the **SOLIDWORKS** menus and the **Simulation** tab is added to the CommandManager.

Starting SOLIDWORKS Simulation and Opening the Model

Start the SOLIDWORKS Simulation and open the model *(c02_tut03)*.

1. Start SOLIDWORKS and choose the **Open** button from the menu bar; the **Open** dialog box is displayed.

2. Browse to the folder in the **Open** dialog box where the model has been downloaded and then select the model.

3. Choose the **Open** button from the **Open** dialog box; the model is opened in SOLIDWORKS assembly mode.

Starting Linear Static Study

Now, you need to start a linear static study before creating custom material.

1. Choose the **Simulation** tab in the **CommandManager** tab.

2. Choose the **New Study** tool from the **New Study** flyout; the **Study PropertyManager** is displayed.

3. Choose the **Static** button from the **General Simulation** rollout and enter **Linear Study** in the **Study name** edit box of the **Name** rollout.

4. Choose the **OK** button from the **Study PropertyManager**; a linear static study is added.

Creating Custom Materials

In this section, you will create custom Gray Cast Iron and Mild Steel and then apply them to their respective parts.

1. Expand the **Parts** node by clicking on the symbol in the Simulation Manager tree; all components of the assembly are displayed under the **Parts** node.

2. Select **Bolt** and **Nut** in Simulation Tree by holding the Ctrl key and then right-clicking; a shortcut menu is displayed.

3. Choose the **Apply/Edit Material** option from the shortcut menu; the **Material** dialog box is displayed.

4. Right-click on the **Custom Materials** library at the bottom of the list of material libraries; a shortcut menu is displayed.

5. Select the **New Category** option from the shortcut menu; a folder with the name **New Category** is added to the **Custom Materials** library.

6. Right-click on the **New Category** folder and rename it as **Custom Steel.**

7. Right-click on the **Custom Steel** folder and choose the **New Material** option from the shortcut menu; a node is added under the **Custom Steel** folder with the name as **Default.**

8. Right-click on the **Default** node and rename it as **Gray Cast Iron.**

9. Choose the **Properties** tab in the **Material** dialog box if not chosen by default, and select the **SI (N/m2)** (Pa) option from the **Units** drop-down list.

10. Delete all the properties data present in the **Value** column of the **Material** dialog box.

11. Enter **1.9e+011** in the **Value** column of the **Elastic Modulus** row.

12. Enter **0.26** in the **Value** column of the **Poisson's Ratio** row.

13. Enter **7300** in the **Value** column of the **Mass Density** row.

14. Enter **4.48e+008** in the **Value** column of the **Tensile Strength** row.

15. Enter **2.41e+008** in the **Value** column of the **Yield Strength** row.

16. Choose the **Apply** button in the **Material** dialog box and close the dialog box; custom steel material is applied to the Nut and Bolt and also saved in the SOLIDWRKS libraries for further use.

17. Close the **Material** dialog box by choosing the **Close** button in it.

    Next, you have to create Mild Steel custom material and apply it to Yoke Plate and Body.

18. Select **YokePlate** and **Body** in Simulation Manager tree by holding the Ctrl key and right-click; a shortcut menu is displayed.

19. Choose the **Apply/Edit Material** option from the shortcut menu; the **Material** dialog box is displayed.

20. Expand the **Custom Materials** library node in the **Material** dialog box.

21. Right-click on the **Custom Steel** folder and choose the **New Material** option from the shortcut menu; a node is added under the Custom Steel folder with the name as **Default**.

22. Right-click on the **Default** node and rename it as **Mild Steel**.

23. Choose the **Properties** tab in the **Material** dialog box if it is not chosen by default.

24. Delete all the properties data present in the **Value** column of the **Material** dialog box.

25. Enter **1.9e+011** in the **Value** column of the **Elastic Modulus** row.

26. Enter **0.27** in the **Value** column of the **Poisson's Ratio** row.

27. Enter **7850** in the **Value** column of the **Mass Density** row.

28. Enter **3.99e+008** in the **Value** column of the **Tensile Strength** row.

29. Enter **2.3e+008** in the **Value** column of the **Yield Strength** row.

30. Choose the **Apply** button in the **Material** dialog box and close the dialog box; custom steel material is applied to the Yoke Plate and Body and also saved in the SOLIDWORKS libraries for further use.

31. Close the **Material** dialog box by choosing the **Close** button in the **Material** dialog box.

Applying Material from SOLIDWORKS Materials Library

In this section, you will apply Natural Rubber to Bush from the SOLIDWORKS Materials library.

1. Select both the **Bushes** in Simulation Tree by holding the Ctrl key and right-click; a shortcut menu is displayed.

2. Choose the **Apply/Edit Material** option from the shortcut menu; the **Material** dialog box is displayed.

3. Expand the **SOLIDWORKS Materials library** node if not expanded already; a list of different material categories is displayed under this node.

4. Expand the node of the **Rubber** from the list of the list of the material category; different materials are displayed under this node.

5. Select the **Natural Rubber** from the list of these materials and choose the **Apply** button from the **Material** dialog box; the selected material is applied.

6. Close the **Material** dialog box by choosing the **Close** button in the **Material** dialog box.

7. Choose the **Save** button; the model is saved.

Self-Evaluation Test

Answer the following questions and then compare them to those given at the end of this chapter:

1. You can access standard materials in the __________ dialog box.

2. In SOLIDWORKS Simulation, you can not apply custom material to a model. (T/F)

3. The material properties available in the SOLIDWORKS **Materials library** can not be modified. (T/F)

4. Material applied to a model in modelling (SOLIDWORKS) environment can not be used for analysis. (T/F)

5. In SOLIDWORKS Simulation, you can not perform studies by assigning different materials on the same model. (T/F)

Review Questions

Answer the following questions:

1. Properties of a material for creating a custom material is assigned by using the __________ tab of the **Material** dialog box.

2. Material appearance for custom material is assigned by using the _________ tab in the **Material** dialog box.

3. When you are applying material to a part then you will get the _________ option in the shortcut menu in place of the **Apply Material to All** option.

4. In SOLIDWORKS Simulation, individual material can not be applied to the components of an assembly. (T/F)

5. Material applied in SOLIDWORKS Simulation will not affect the material applied to a CAD model in SOLIDWORKS. (T/F)

EXERCISES

To perform the exercises of this chapter, you need to download the input files from *www.cadcim.com.*

The complete path for downloading the file is:

Textbooks > CAE Simulation > SOLIDWORKS Simulation > SOLIDWORKS Simulation 2022: A Tutorial Approach > Input Files > c02_simulation_2022_input.zip

Extract the downloaded zipped file at the location *C:\SOLIDWORKS Simulation\c02.*

Exercise 1

In this exercise, you will open the downloaded *c02_exr01* file in SOLIDWORKS. Start a linear Static study and assign AISI 1045 Steel, cold drawn material from the **SOLIDWORKS Material library** to the model. Figure 2-12 shows the model for the exercise.

(Expected time: 30 min)

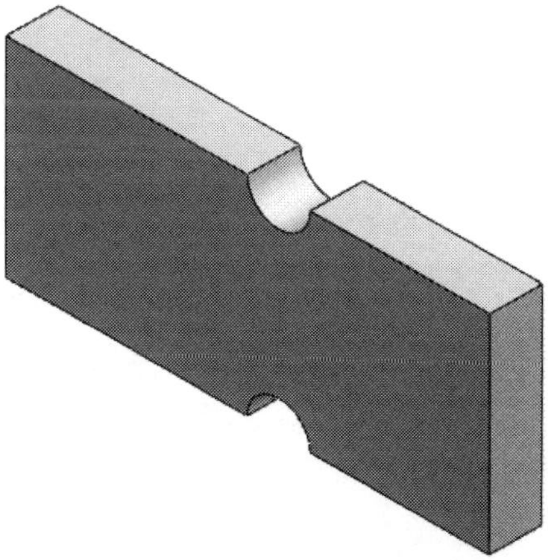

Figure 2-12 *Model for Exercise 1*

Exercise 2

In this exercise, you will open the downloaded *c02_exr02* file in SOLIDWORKS and apply different materials to different components of the assembly. Figure 2-13 shows the Bench vice assembly. The Vice Body, Base Plate, Jaw Screw, Screw Bar and Bar Gloves are made up of Grey Cast Iron whereas the Vice Jaw and the Vice Jaw Base are made up of Mild Steel. The properties of Mild Steel and Gray Cast Iron are given next. Figure 2-14 shows various components of the Bench vice assembly.

(Expected time: 30 min)

Properties of Mild Steel

Mass density:	7850 kg/m^3
Elastic modulus:	1.9e+011 N/m^2
Poisson's ratio:	0.27
Yield strength:	2.3e+008 N/m^2
Tensile strength:	3.99e+008 N/m^2

Properties of Gray Cast Iron

Mass density:	7300 kg/m^3
Elastic modulus:	1.9e+011 N/m^2
Poisson's ratio:	0.26
Yield strength:	2.41e+008 N/m^2
Tensile strength:	4.48e+008 N/m^2

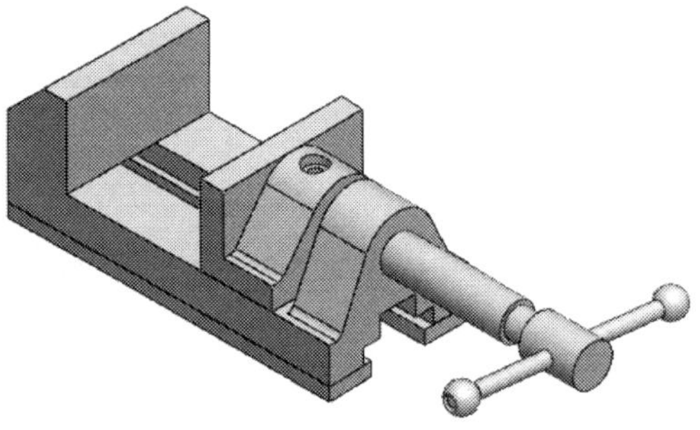

Figure 2-13 *Bench vice assembly*

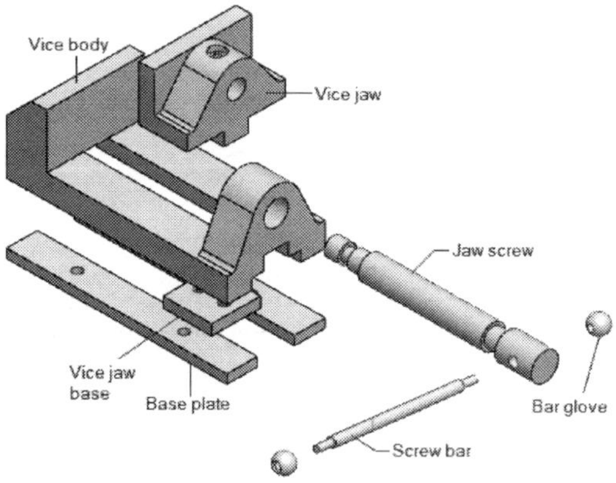

Figure 2-14 *Various components of the Bench vice assembly*

Answers to Self-Evaluation Test

1. Material, **2**. F, **3**. T, **4**. F, **5**. F

Chapter 3

Meshing

Learning Objectives

After completing this chapter, you will be able to:

- *Understand the concepts of generating a mesh*
- *Generate mesh for complex models*
- *Refine the mesh*
- *Apply local mesh control to a mesh*
- *Check the mesh quality*
- *Optimize the mesh quality of a model*

INTRODUCTION

Meshing is the process of subdividing a model, also called discretization, into a number of small pieces having simple shape, same size, and connected at common points. A component discretized into a number of small elements of defined size is referred to as mesh. These simple shapes are called elements and the common points where these elements are connected are known as nodes. Figure 3-1 shows a solid model and Figure 3-2 shows the mesh created on the solid model.

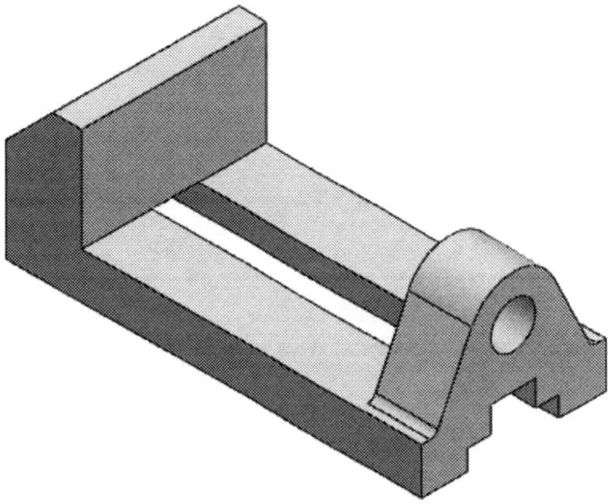

Figure 3-1 *A solid model*

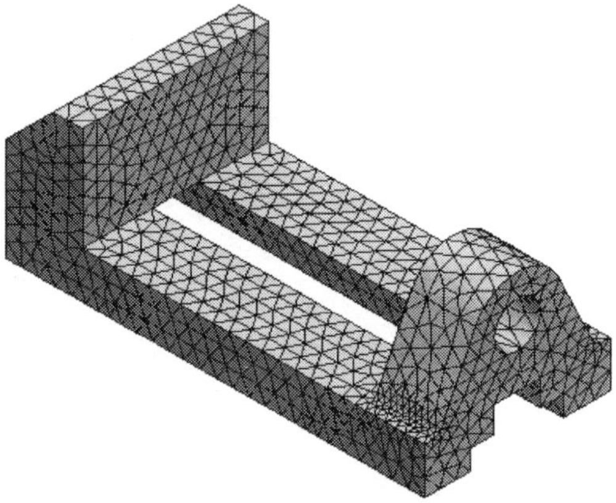

Figure 3-2 *Mesh created on the solid model*

Meshing is an important step in design analysis. The automatic mesher in SOLIDWORKS Simulation generates a mesh based on global element size, tolerance, local mesh control specifications, and contact specifications. Mesh control lets you specify different sizes of elements for components, faces, edges, and vertices. In FEA, element size plays an important role. Result of an analysis and the time consumed in it change with the changes done in the element size. Smaller elements may give more accurate results but increase the computation time.

SOLIDWORKS Simulation lets you create a mesh of solid elements (3D tetrahedral), shells (2D triangular), or beam (1D linear). A mesh consists of elements of one type unless a mixed mesh type is specified. Solid mesh is appropriate for bulky or complex 3D models. Shell elements are suitable for thin parts like sheet metal components. Beam and truss elements are suitable for modeling structural members. The accuracy of the solution depends on the quality of the mesh. Different methods of meshing are discussed next.

Solid Meshing

Solid meshing is generally used for bulky models. Each node in a solid element has three translational degrees of freedom in three orthogonal directions. SOLIDWORKS Simulation uses the X, Y, and Z directions of the global cartesian coordinate system in formulating FEA problems.

In solid meshing, the mesh elements have tetrahedral shape. The tetrahedral elements are classified into two types: linear solid elements and parabolic solid elements. The software generates one of the types of element based on the active mesh option specified for the analysis. Types of solid mesh are discussed next.

Draft Quality Solid Mesh

In a draft quality solid mesh, linear tetrahedral shaped elements are generated. Linear elements are also known as first order or lower order elements. A linear tetrahedral element is defined by four corner nodes connected by six straight edges. The shape of linear tetrahedral element is shown in Figure 3-3.

High Quality Solid Mesh

In a high quality solid mesh, parabolic tetrahedral shaped elements are generated. Parabolic elements are also known as second order or higher order elements. A parabolic tetrahedral element is defined by four corner nodes, six mid-side nodes, and six edges. The shape of parabolic tetrahedral element is shown in Figure 3-4.

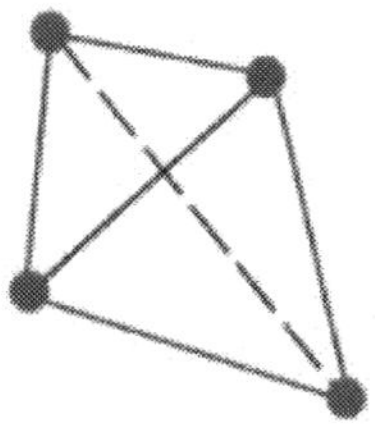

Figure 3-3 *Linear tetrahedral solid element*

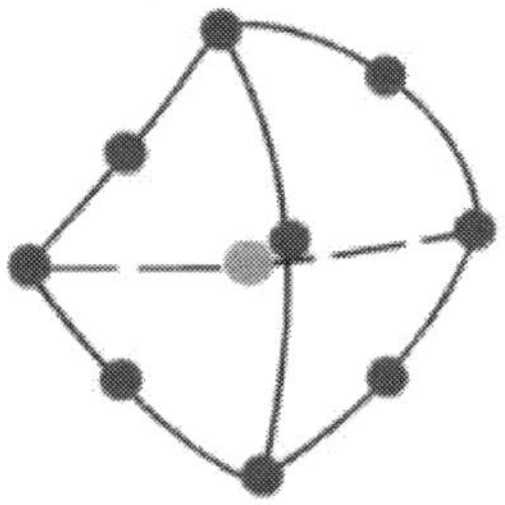

Figure 3-4 *Parabolic tetrahedral solid element*

In general, for the same mesh density, parabolic elements provide better results than linear elements. This is because parabolic elements represent curved boundaries better and hence produce better mathematical approximation. They, however, require more computation time than linear elements.

Note

It is highly recommended to use the high quality option for final results and for models with curved geometry. Draft quality meshing can be used for quick evaluation.

Shell Meshing

Shell elements are suitable for thin parts (like sheet metals, surface bodies). In shell meshing, the program determines the shell thickness of the sheet metal component and automatically creates mesh on the mid surface. Shell elements are 2D elements capable of resisting membrane and bending loads. SOLIDWORKS Simulation automatically generates triangular shell elements. These triangular elements may be linear or parabolic depending upon the active meshing options for the study. Types of shell meshing are discussed next.

Draft Quality Shell Mesh

In a draft quality shell mesh, linear triangular elements are generated, refer to Figure 3-5. A linear triangular shell element is defined by three corner nodes connected by three straight edges.

High Quality Shell Mesh

In a high quality shell mesh, parabolic triangular elements are generated, refer to Figure 3-6. A parabolic triangular element is defined by three corner nodes, three mid-side nodes, and three parabolic edges. The shape of parabolic triangular element is shown in Figure 3-6.

Note

A fine draft quality mesh and a high quality mesh provide similar results when the number of elements is same. The results differ if the model includes a curved geometry.

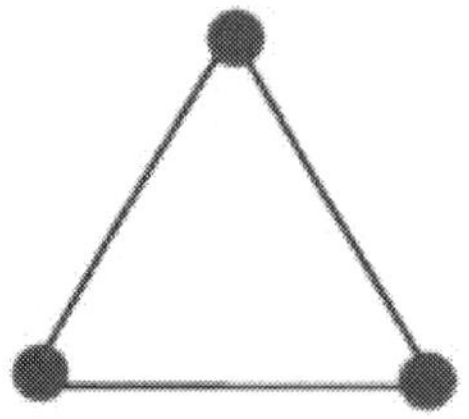

Figure 3-5 *Linear triangular shell element*

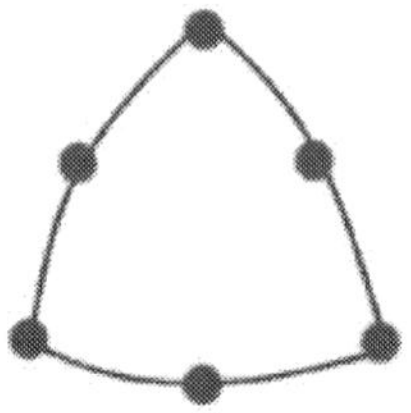

Figure 3-6 *Parabolic triangular shell element*

Beam Meshing

Beam meshing is used for meshing structural members and weldment structures. You can treat an extrusion (meshes with solid elements by default) as a beam by right-clicking on the part node in the Simulation Tree and choosing the **Treat as Beam** option from the shortcut menu displayed.

Mixed Meshing

SOLIDWORKS Simulation automatically uses a mixed mesh when different geometries are present in the model.

CREATING MESH

For creating a mesh, right-click on the **Mesh** node in the Simulation Tree; a shortcut menu will be displayed. Choosethe **Create Mesh** option from the shortcut menu; the **Mesh PropertyManager** will be displayed, as shown in Figure 3-7. Choose the **OK** button in the **Mesh PeropertyManager**;the mesh will be created depending upon the geometry ofthe model. The options in the **Mesh PropertyManager** are used to control the mesh parameters. These options are discussed next.

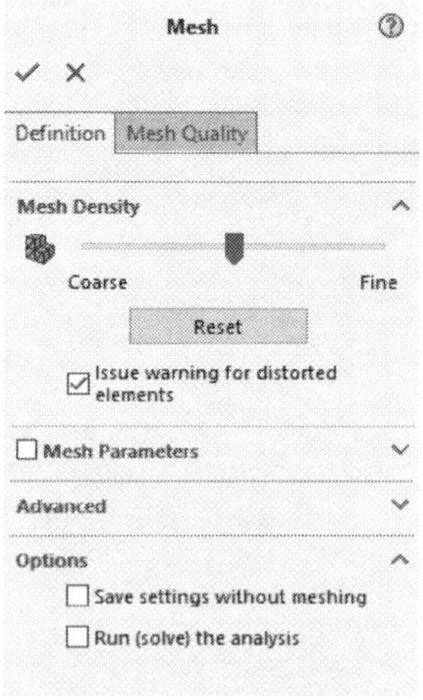

Figure 3-7 *The* ***Mesh PropertyManager***

Mesh Density Rollout

The options in the **Mesh Density** rollout are used to define the element size while meshing.

Mesh Factor

You can set the mesh factor by using the slider in the **Mesh Density rollout.** When you move it toward the right, the element size decreases and becomes half of the default size when the slider is at the right end point. When you move the slider toward left, the element size increases and doubles when the slider is on the left end point.

Reset
This button resets the **Mesh Factor** slider to default position.

Mesh Parameters Rollout
The options in this rollout are used to set the global mesh size while meshing.

Standard mesh
This radio button is selected by default. As a result, the Voronoi-Delaunay triangulation meshing scheme is used for subsequent meshing operations. In this method, the given geometry splits into triangular elements or tetrahedrons. It is faster than other processes and should be preferred.

Curvature-based mesh
This radio button is used to create curvature-based meshing scheme for subsequent meshing operations. On selecting this radio button, more elements are automatically created in higher-curvature areas (without need for mesh control). In the curvature-based meshing, the element size is determined mathematically by minimum number of elements that fit in a hypothetical circle while taking into account the user specified minimum and maximum element sizes.

Blended curvature-based mesh
The blended curvature-based mesh is an improved algorithm that provides a much higher quality surface mesh. The mesher is able to create high quality elements in certain regions of a model and deliver a more efficient mesh. By this, edges in areas of no interest have fewer elements, while more elements are added in areas which require more detail.

Unit
This drop-down list is used to set the units for mesh sizes and tolerance.

Global Size
This spinner will be available only for the standard mesh and is used to set global average element size. Based on the model volume and surface area, software automatically assigns a value for the element size.

Tolerance
Tolerance is defined as the minimum distance variation between two adjacent nodes. If the distance between the nodes is less than the minimum tolerance value specified, then both the nodes will get merged unless other conditions are specified. This spinner will be available only for the standard mesh and is used to set the tolerance value. By default, the value set in this spinner is 5% of the global element size. But if the mesh control is applied, it can be 30% of the smallest mesh element size defined. You can also change this value using the spinner.

Automatic transition
This check box will be available only for the standard mesh. By default, this check box remains cleared. When this check box is selected, mesh control gets applied to small features, holes, fillets, and other fine details of the model.

Maximum element size
This spinner will be available when the **Curvature based mesh** radio button is selected and is used to define the element size for boundaries with lowest curvature.

Minimum element size
This spinner will be available when the **Curvature based mesh** radio button is selected. This option is used to define the element size for boundaries with highest curvature.

Min number of elements in a circle
This spinner will be available when the **Curvature based mesh** radio button is selected. This spinner is used to specify the minimum number of elements that fit in a hypothetical circle while taking into account the user specified minimum and maximum element sizes.

Element size growth ratio
This spinner will be available only when the **Curvature based mesh** radio button is selected. It is used to specify global element size growth ratio starting from regions of high curvatures in all directions.

Advanced rollout
The options available is this rollout are used to set the parameters for the quality of mesh.

Jacobian points
This drop-down list will be available only when high quality mesh is used. This drop-down list is used to define the Jacobian points for Jacobian checks. Jacobian check is a measure of deformity in a given element. A perfectly straight sided element has an ideal Jacobian value of 1.0, and the Jacobian ratio increases as the curvature of the edges increase.

Draft Quality Mesh
The draft quality mesh is used for quick evaluation of results. This check box is selected by default and is used to specify four corner nodes for each solid element and three corner nodes for each shell element.

Automatic Trials for Solid
This check box will be available only for the standard mesh. When you select this check box, the model automatically gets remeshed using a smaller global element size. The number of trails performed for remeshing will be specified in the **Number of trials** spinner. The **Number of trials** spinner will be available when you select the **Automatic Trials for Solid** check box. The ratio by which the global element size and tolerance are reduced for each trial is 0.8.

Render Shell Thickness in 3D
This option is used to display the mesh of shell bodies using the 3D representation of the shell elements.

THE DECISION MAKING TO FIND OPTIMUM RESULTS

To achieve optimum results in an analysis, you need to discretize the model into required number of elements. To do so, decide the number of times upto which a mesh has to be refined to reduce the element size. The decision is very difficult and is based on the experience of the design engineer. For a real complex model where accuracy is the primary concern and not the time, an FEA engineer would refine the mesh as long as the analysis shows improved results. However, time is also an important factor to find a perfect mesh for a model. Figure 3-8 shows a model with its boundary and loading conditions and Figure 3-9 compares the data gathered from various analyses run on the same model. In this figure, X coordinate represents the element count and the Y coordinate represents the total deformation evaluated.

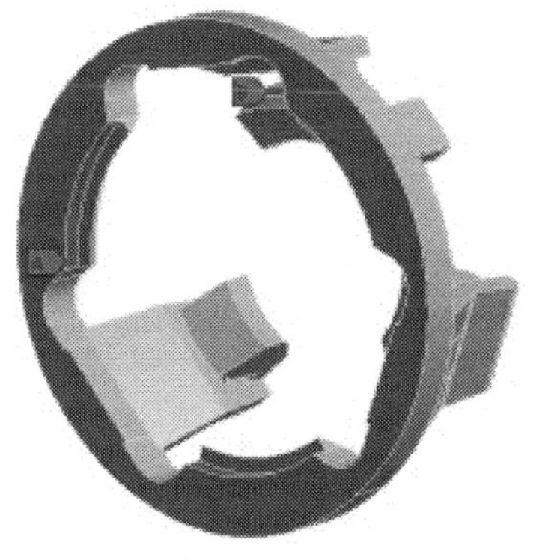

Figure 3-8 *Model with boundary and loading conditions*

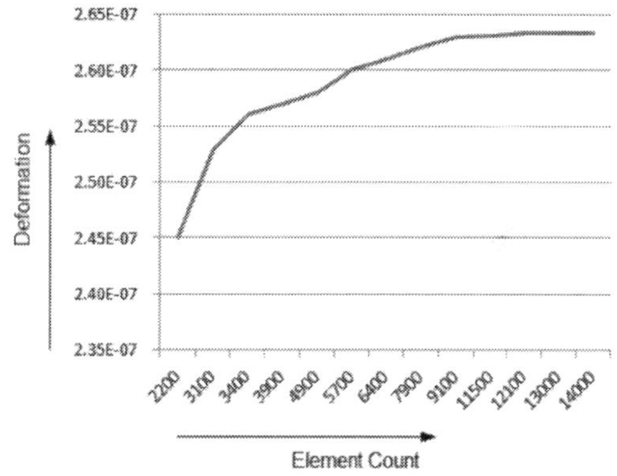

Figure 3-9 *Graph showing the relationship between element count and deformation*

In this graph, when the element count is 2200, the deformation is 2.45 E-07. As the number of elements increases, the deformation also increases. Note that the maximum value of deformation occurs where the element count is 11,500. Hence, it is obvious that meshing the component beyond this point is not required. Meshing the model further results increase of the runtime of the analysis.

TUTORIALS

To perform the tutorials, you need to download the input files from *www.cadcim.com*. The complete path for downloading the files is as follows:

Textbooks > CAE Simulation > SOLIDWORKS Simulation > SOLIDWORKS Simulation 2022: A Tutorial Approach > Input Files > c03_simulation_2022_input.zip

Extract the downloaded *c03_simulation_2022_input* zipped file and save it.

Tutorial 1

In this tutorial, you will import the model *(c03_tut01)* shown in Figure 3-10 to SOLIDWORKS Simulation. Next, you will mesh the model and then refine the mesh for further analysis. The parameters to be used for meshing the model are as follows:

Mesh type: Solid
Element Size: 1.5 mm

After generating thmwith specified mesh settings, you need to refine it further for better mesh density and quality.

(Expected time: 30 min)

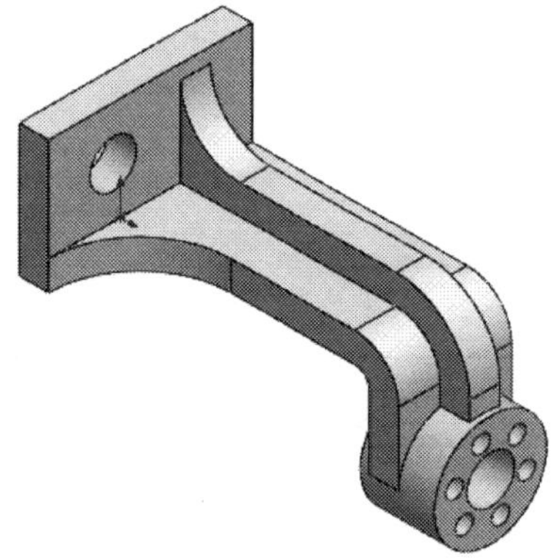

Figure 3-10 *A solid model*

The following steps are required to complete this tutorial:

a. Import the model.
b. Start a new Static study.
c. Create mesh.
d. Insert local mesh controls.
e. List the statics of the mesh.

Importing the Model

In this section, you will start SOLIDWORKS Simulation and import the input file from *www.cadcim.com* and open the file.

1. Start SOLIDWORKS, choose the **Open** button from the menu bar; the **Open** dialog box is displayed.

2. Browse to the folder in which the model has been downloaded and then select the model *(c03_tut01)*.

3. Choose the **Open** button in the **Open** dialog box; the model opens in SOLIDWORKS.

Starting Linear Static Study

In this section, you need to start linear static analysis in the SOLIDWORKS Simulation.

1. Choose the **New Study** tool from the **New Study** flyout in the **Simulation Command Manager**; the Study PropertyManager is displayed.

2. Choose the **Static** button from the **General Simulation** rollout and enter the name of the study as **Linear Static** in the **Study name** edit box in the **Name** rollout.

3. Choose the **OK** button from the **Study PropertyManager** to start the linear static analysis.

Generating Mesh for the Model

Now, you need to perform the meshing operations to generate the mesh for analysis.

1. Choose the **Create Mesh** tool from the **Run This Study** flyout in the **Simulation Command Manager**; the **Mesh PropertyManager** is displayed.

   Now, you need to set the mesh density to control global element size for the mesh by using the slider bar.

2. Move the slider to the right end of the slider bar.

3. Choose the **OK** button from the PropertyManager; the **Mesh Progress** window is displayed and the mesh with specified setting is generated, refer to Figure 3-11.

   By using the **Reset** button available in the **Mesh Density** rollout, you can reset the slider bar to its default position.

Figure 3-11 *Meshed solid model*

Inserting Local Mesh Controls

In this section, you will set local mesh controls for generating the mesh. An effective mesh can be generated by using the options available in the **Mesh Control PropertyManager.**

1. Right- click on the **Mesh** node in the Simulation Tree; a shortcut menu is displayed, refer to Figure 3-12.

2. Choose the **Apply Mesh Control** option from the shortcut menu; the **Mesh Control PropertyManager** is displayed.

3. Select all the holes surfaces in the **Selected Entities** rollout; selected entities are highlighted in the selection box.

4. Enter 1.5 in the **Element Size** edit box in the **Mesh Parameters** rollout and select the **mm** option in the Unit drop-down. Make sure that the use per part size check box is cleared to set the element size for the mesh. Accept other default settings in the dialog box to create the mesh.
.

5. Choose the **Create Mesh** button from the **Selected Entities** rollout; the **Simulation** message box is displayed informing you about the mesh controls of the selected entities.

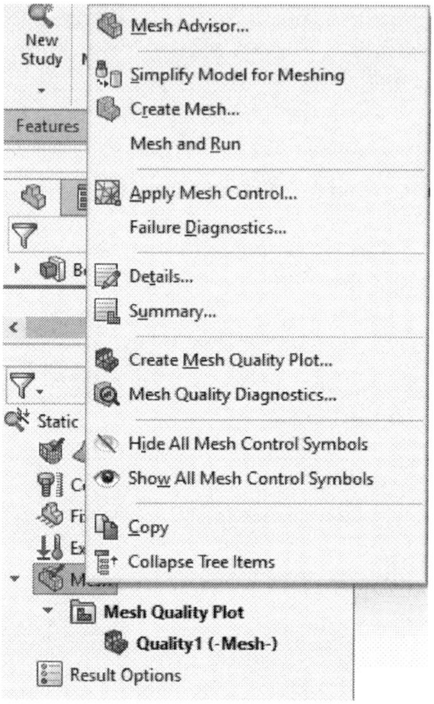

***Figure 3-12** The shortcut menu displayed on right-clicking the **Mesh** node*

6. Choose the **Yes** button from the message box; the **Mesh Progress** window is displayed and mesh is generated. Figure 3-13 shows the model after applying the mesh controls on the holes of the model.

***Figure 3-13** Model after applying the mesh control on the holes*

Listing the Statics of the Mesh

After generating the mesh and applying the mesh control, list the number of elements and nodes on the mesh.

1. Right-click on the **Mesh** node in the simulation Tree; a shortcut menu is displayed.

2. Choose the **Details** option from the shortcut menu; the **Mesh Details** window is displayed.

   All the details of the elements, nodes, and mesh type and so on for the model get displayed under this window and you can see the total number of elements and nodes created for the mesh.

3. Next, close the **Mesh Details** window.

Save the Tutorial and Exit SOLIDWORKS Simulation

After generating the mesh, you need to save the tutorial and exit SOLIDWORKS Simulation.

1. Save the tutorial at the desired location by using the **Save** button.

Tutorial 2

In this tutorial, you will create a mixed mesh on the model shown in Figure 3-14. To perform the tutorial, open the *c03_tut_02.SLDASM* file in SOLIDWORKS Simulation. Next, you will generate the mesh and then optimize the model for further use. The model after importing into SOLIDWORKS Simulation is shown in Figure 3-14. **(Expected time: 30 min)**

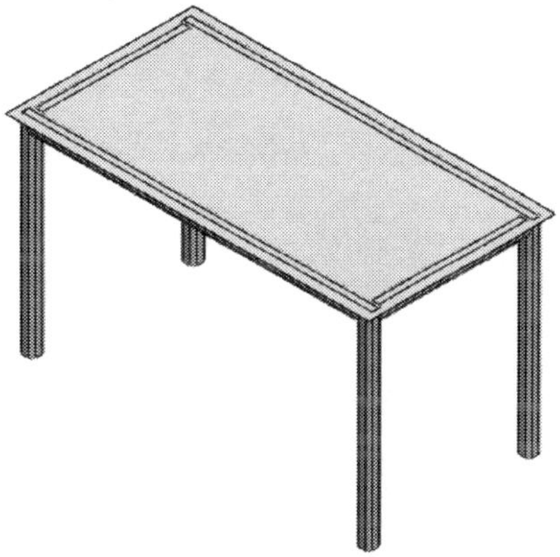

Figure 3-14 *A solid model*

The following steps are required to complete this tutorial:

a. Import the model.
b. Start a new Static study.
c. Create mesh.
d. Insert local mesh controls.
e. List the statics of the mesh.
f. Check the mesh quality.

Importing the Model

You will start SOLIDWORKS Simulation and then import the model for analysis.

1. Start SOLIDWORKS and choose the **Open** button from the menu bar; the **Open** dialog box is displayed.

2. Browse the folder in which the model has been downloaded and then select the Table model.

3. Choose the **Open** button in the **Open** dialog box; the model gets opened in SOLIDWORKS.

Starting Linear Static Study

In this section, you need to start linear static analysis in the SOLIDWORKS Simulation.

1. Choose the **New Study** tool from the **New Study** flyout in the **Simulation Command Manager**; the **Study PropertyManager** is displayed.

2. Choose the **Static** button from the **General Simulation** rollout and enter the name of the study as **Mixed Meshing** in the **Study name** edit box in the **Name** rollout.

3. Choose the **OK** button from the **Study PropertyManager** to start the linear static study.

Generating Mixed Mesh for the Model

Now, you need to perform the meshing operations by choosing the **Create Mesh** button in the **Simulation Command Manager.**

1. Select the **Mesh** node in the Simulation Tree and then right-click; a shortcut menu is displayed.

2. Choose the **Create Mesh** option from this menu; the **Mesh PropertyManager** is displayed, as shown in Figure 3-15. Now, create the mesh with default settings in the PropertyManager.

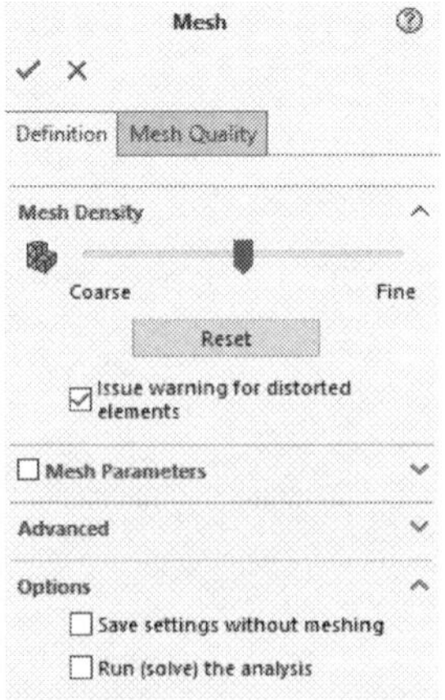

*Figure 3-15 The **Mesh PropertyManager***

3. Expand the **Parts** node under the **Mixed Meshing** node in the Simulation Tree. Next, expand the subnode of the **Parts** node in the Simulation Tree.

   Figure 3-16 shows the Table Top and Table legs have shell and beam bodies, respectively. The resultant mixed mesh model is shown in Figure 3-17.

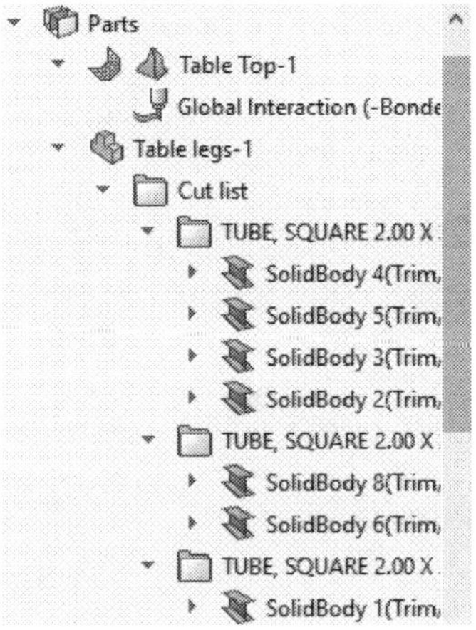

***Figure 3-16** Expanded **Parts** node with shell and beam bodies*

***Figure 3-17** The resultant mixed mesh model*

Checking the Mesh Quality

The quality of mesh plays a key role in the accuracy of the analysis results. SOLIDWORKS Simulation uses aspect ratio and jacobian points checks to measure the quality of elements in a mesh.

Best solid mesh is achieved by a mesh with uniform perfect tetrahedral elements whose edges are equal in length. But generating uniform mesh of perfect tetrahedral elements is not possible. Due to small edges, curved geometry, thin features, and sharp corners in a geometry, elements can have some of their edges much longer than others.

When the edges of an element are of different length, the accuracy of the results is affected. The aspect ratio of a tetrahedral element is defined as the ratio between the longest edge and the shortest normal dropped from a vertex to the opposite face. The aspect ratio of a perfect tetrahedral element is 1.0. The aspect ratio check assumes the edges connecting the four corner nodes.

Parabolic elements can map curved geometry much more accurately than linear elements of the same size. The Jacobian of an extremely distorted element becomes negative and stops the analysis. The jacobian ratio of a parabolic tetrahedral element with all mid-side nodes located exactly at the middle of the straight edges, is 1.0. The Jacobian ratio increases as the curvature of the edges increase. It is recommended to set the jacobian check on the nodes when using the p-method to solve static problems. For high order shells, the jacobian check uses 6 points located at the nodes. To check the mesh quality, you need to perform the following steps.

1. Select the **Mesh** node in the Simulation tree and then right-click; a shortcut menu is displayed.

2. Choose the **Create Mesh Quality Plot** option from the shortcut menu; the **Mesh Quality Plot PropertyManager** is displayed.

3. Select the **Aspect ratio** radio button from the **Display** rollout and then choose the **OK** button from the PropertyManager to display a aspect ratio plot, as shown in Figure 3-18.

Based on the aspect ratio, you can check your mesh quality. The aspect ratio of an ideal tetrahedral element is 1.0. This is a ratio of the longest edge to the shortest normal dropped from a vertex to the opposite face.

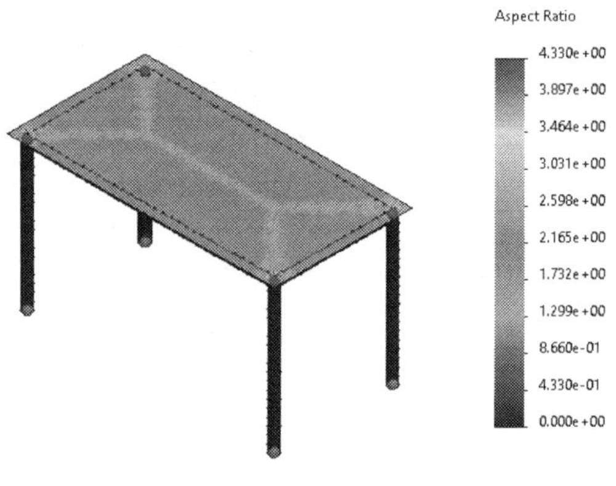

Figure 3-18 *Aspect ratio plot*

It is evident from Figure 3-18 that the maximum aspect ratio for this mesh is 4.330 and is within acceptable limits. The second method used to determine mesh quality is the Jacobian ratio. This method is only available for second order (high quality) mesh elements.

Checking the Mesh Statics

In this section, you will check the mesh statics such as nodes and elements.

1. Right-click on the **Mesh** node in the Simulation tree; a shortcut menu is displayed.

2. Select the **Details** option from the shortcut menu; the **Mesh details** window is displayed. Using this window, you can check the statics of the mesh like mesh type, quality of mesh, total number of node, and total number of elements and so on.

Saving the Model

After generating the mesh, you need to save the tutorial and exit the SOLIDWORKS Simulation.

1. Save the model at the desired location by using the **Save** button.

Tutorial 3

In this tutorial, you will create a mix mesh on the rim model shown in Figure 3-19. To perform the tutorial, open the *A-Wheel Rim.SLDASM* file in SOLIDWORKS Simulation. Next, you will generate the mesh and then optimize the model for further use and better mesh quality. The model after importing into SOLIDWORKS Simulation is shown in Figure 3-19.

(Expected time: 30 min)

Figure 3-19 *A solid model*

The following steps are required to complete this tutorial:

a. Import the model.
b. Start a new Static study.
c. Create mesh.
d. Insert local mesh controls.
e. List the statics of the mesh.
f. Check the mesh quality.

Importing the Model

You will start SOLIDWORKS Simulation and then import the model for analysis.

1. Start SOLIDWORKS and choose the **Open** button from the menu bar; the **Open** dialog box is displayed.

2. Browse the folder in which the model has been downloaded and then select the Rim *(A-Wheel Rim.SLDASM)* model.

3. Choose the **Open** button in the **Open** dialog box; the model gets opened in SOLIDWORKS.

Starting Linear Static Study

In this section, you will start linear static study to discretize the model.

1. Choose the **New Study** tool from the **New Study** flyout in the **Simulation Command Manager**; **the Study PropertyManager** is displayed.

2. Choose the **Static** button from the **General Simulation** rollout and specify the name of the study as **Rim Mesh** in the **Study name** edit box of the **Name** rollout.

3. Choose the **OK** button from the **Study PropertyManager** to start a linear static study.

Generating Mixed Mesh for the Model

Now, you need to generate mixed mesh for performing the study.

1. Expand the **Parts** node under the **Rim Mesh** node in the Simulation Tree. Next, select **Rim Support -2-4,** refer to Figure 3-20, and right-click; a shortcut menu is displayed.

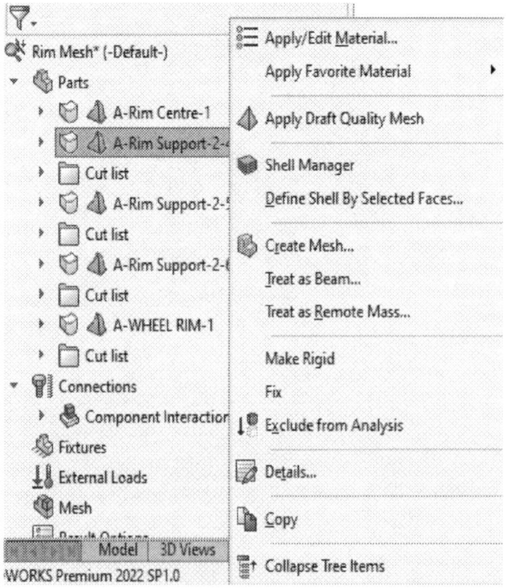

Figure 3-20 The shortcut menu

2. Choose the **Shell Manager** option from the shortcut menu; the **Shell Manager PropertyManager** is displayed.

3. Select all the rim supports such as Rim Support 2-4, Rim Support 2-5, and Rim Support 2-6 under the **Shell Entity** rollout in the **Properties** area of **Shell Manager PropertyManager.** Make sure the **Faces** button is chosen in the **Properties** rollout to convert the selected faces into shell.

4. Select the **Thick** and **Middle surface** options in the **Shell Type** and **Shell offset** drop-down lists, respectively.

5. Enter **5** in the **Thickness** edit box and then choose the **OK** button from the PropertyManager; all the rim supports get converted into the shell body.

6. Select the **Mesh** node in the Simulation Tree and then right-click; a shortcut menu is displayed.

7. Choose the **Create Mesh** option from the shortcut menu; the **Mesh PropertyManager** gets displayed.

8. Set the mesh refinement to coarse most by using the **Mesh Density** slider in the PropertyManager.

9. Select the **Mesh Parameters** check box to expand the rollout and then select the **curvature-based mesh** radio button.

   In this model, you have many curvature surfaces so it is recommended to select the **Curvature-based mesh** radio button to get a quality mesh.

10. Enter the **5** in the **Maximum element size** edit box and **1** in the **Minimum element size** edit box. Make sure the **mm** option is selected in the **Unit** drop-down list.

11. Enter **8** in the **Min number of element in a circle** edit box and **1.5** in the **element size growth ratio** edit box and then choose the **OK** button from the PropertyManager; the **Mesh Progress window** is displayed showing the progress of the mesh.

Apply Local Mesh Control to the Mesh

In SOLIDWORKS Simulation, you can control the local mesh size by assigning different element sizes to different regions in the model. A smaller element size in a region improves the accuracy of the results for that region. You can specify local mesh control at vertices, edges, faces, and components. Note that the local mesh control settings for maximum and minimum element size should be within the range you specify for global mesh settings. The procedure to select the element size is determined mathematically by minimum number of elements that fit in a hypothetical circle. When minimum number of elements in a circle is set to 8, the circle will contain 8 elements and each quarter of the circle will contain 2 elements. Standard mesh activates the Voronoi-Delaunay meshing scheme for subsequent meshing operations and curvature-based mesh activates curvature-based meshing scheme for subsequent meshing operations. The software creates more elements in higher-curvature areas automatically in this scheme. Blended curvature-based mesh activates the Blended curvature-based mesh. Use the Blended curvature-based mesh for the models that fail to mesh with standard or curvature-based mesh. There are cases where the Blended curvature based mesh can overcome mesh failure, generating a solid mesh with higher quality elements or lower aspect ratios than standard or curvature-based mesh.

1. Select the **Mesh** node in the Simulation Tree and then right-click; a shortcut menu is displayed.

2. Choose the **Apply Mesh Control** option from this shortcut menu; the **Mesh Control PropertyManager** is displayed.

3. Select all the Rim supports under the **Selected Entities** rollout. Next, enter **10.5** in the **Element Size** edit box in the **Mesh Parameters** rollout and then choose the **OK** button.

4. Select the **Mesh** node again in the Simulation Tree; a shortcut menu is displayed.

5. Choose the **Apply Mesh Contro**l option from this shortcut menu; the **Mesh Control PropertyManager** is displayed.

6. Select the **Rim center part** under the **Selected Entities** rollout from Design Tree. Next, enter **10** in the **Element Size** edit box in the **Mesh Parameters** rollout and then choose the **OK** button; the Control-1 and Control-2 subnodes get added under the **Mesh Controls** node in the Simulation Tree.

7. Select the **Mesh** node in the Simulation Tree and then right-click; a shortcut menu is displayed.

8. Choose the **Create Mesh** option from the shortcut menu; the **Mesh PropertyManager** is displayed. Choose the **OK** button from the PropertManager to complete the meshing operations.

Checking the Mesh Quality

In this section, you will check the mesh quality. The quality of mesh plays a key role in the accuracy of the analysis results. SOLIDWORKS Simulation uses aspect ratio and jacobian points checks to measure the quality of elements in a mesh.

1. Select the **Mesh** node in the Simulation Tree and then right-click; a shortcut menu is displayed.

2. Choose the **Create Mesh Quality Plot** option from the shortcut menu; the **Mesh Quality plot PropertyManager** is displayed.

3. Select the **Aspect ratio** radio button from the **Display** rollout and then choose the **OK** button from the PropertyManager to display the aspect ratio plot, as shown in Figure 3-21.

   On the basis of aspect ratio, you can check the mesh quality. The aspect ratio of an ideal tetrahedral element is 1.0. This is the ratio of the longest edge to the shortest normal dropped from a vertex to the opposite face.

4. Similarly, construct a jacobian plot using the **Create Mesh Plot** option, refer to Figure 3-22.

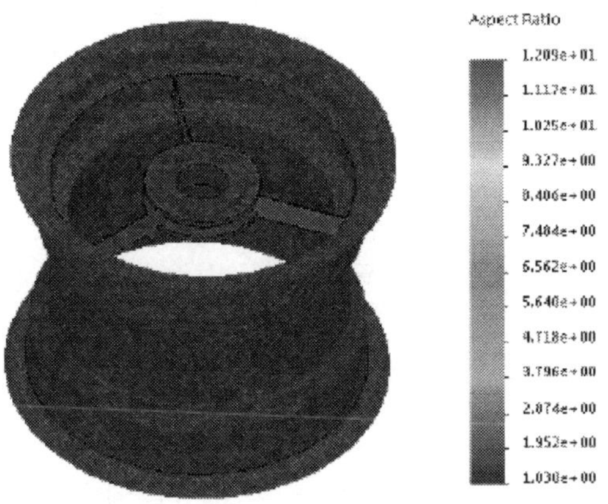

Figure 3-21 *Aspect ratio plot*

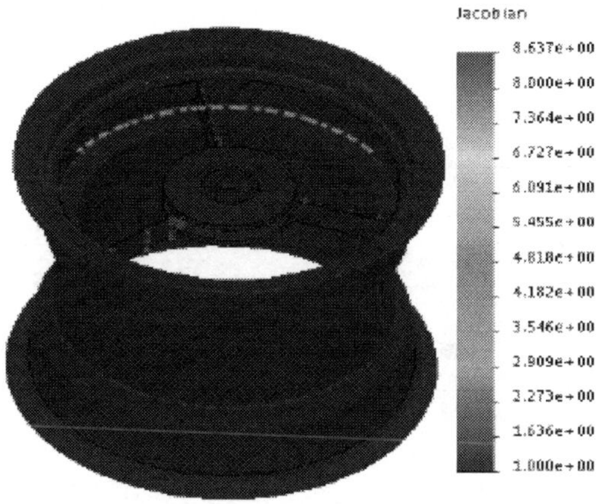

Figure 3-22 *Jacobian plot*

It is evident from Figure 3-21 and Figure 3-22 that the maximum aspect ratio and maximum jacobian values for this mesh are within acceptable limits. Jacobian plot is used to determine mesh quality of elements using jacobian points. This method is only available for second order (high quality) mesh elements.

Checking the Mesh Statics

In this section, you need to check the mesh statics like nodes and elements.

1. Right-click on the **Mesh** node in the Simulation Tree; a shortcut menu is displayed.
2. Choose the **Details** option from the shortcut menu; the **Mesh Details** window is displayed.

   Using this window, you can check the statics of the mesh like mesh type, quality of mesh, total number of node, and total number of elements.

Saving the Model

After generating the mesh, you need to save the tutorial and exit SOLIDWORKS Simulation.

1. Save the model at the desired location by using the **Save** button.

Self-Evaluation Test

Answer the following questions and then compare them to those given at the end of this chapter:

1. You can invoke the **Mesh Details** window by choosing the__________ option from the shortcut menu displayed on right-clicking on the **Mesh** node in Simulation Tree.

2. You can invoke the **Mesh Quality PropertyManager** by choosing the __________ option from the shortcut menu displayed on right-clicking in the Simulation Tree.

3. When you choose the __________ button from the **Run This Study** drop-down in the **Simulation Command Manager**, the **Mesh PropertyManager** is displayed.

4. The orthogonal quality of a perfect tetrahedral element is 1.0. (T/F)

5. In SOLIDWORKS Simulation, you can specify shell thickness in the **Shell Thickness** edit box of the **Shell Definition PropertyManager.** (T/F)

Review Questions

Answer the following questions:

1. If you choose the __________ option from the shortcut menu, then the selected SolidBody parts will be converted into a beam.

2. The aspect ratio of a perfect tetrahedral element is __________ .

3. You need to discretize a model into required number of elements to achieve optimum results in an analysis. (T/F)

4. Shell elements are 2D elements capable of resisting membrane and bending loads. (T/F)

5. A linear triangular shell element is defined by three corner nodes connected by three straight edges. (T/F)

EXERCISES

To perform the exercises of this chapter, you need to download the input files from www.cadcim.com. The complete path for downloading the file is:

Textbooks > CAE Simulation > SOLIDWORKS Simulation > SOLIDWORKS Simulation 2022: A Tutorial Approach > Input Files > c03_simulation_2022_input.zip

Extract the downloaded zipped file at the location *C:\SOLIDWORKS Simulation\c03*.

Exercise 1

In this exercise, you will generate an optimized mesh on the model *(c03_exr01)* shown in Figure 3-23. The dimensions of the model are given in Figure 3-24. Next, assign the AISI 1045 Steel, cold drawn material from the SOLIDWORKS material library to the model. After assigning the material, check the mesh quality, and elements and nodes in the mesh.

(Expected time: 30 min)

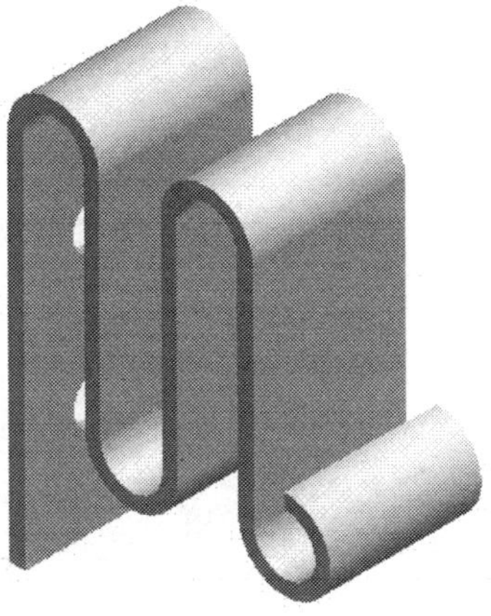

Figure 3-23 *The model for Exercise 1*

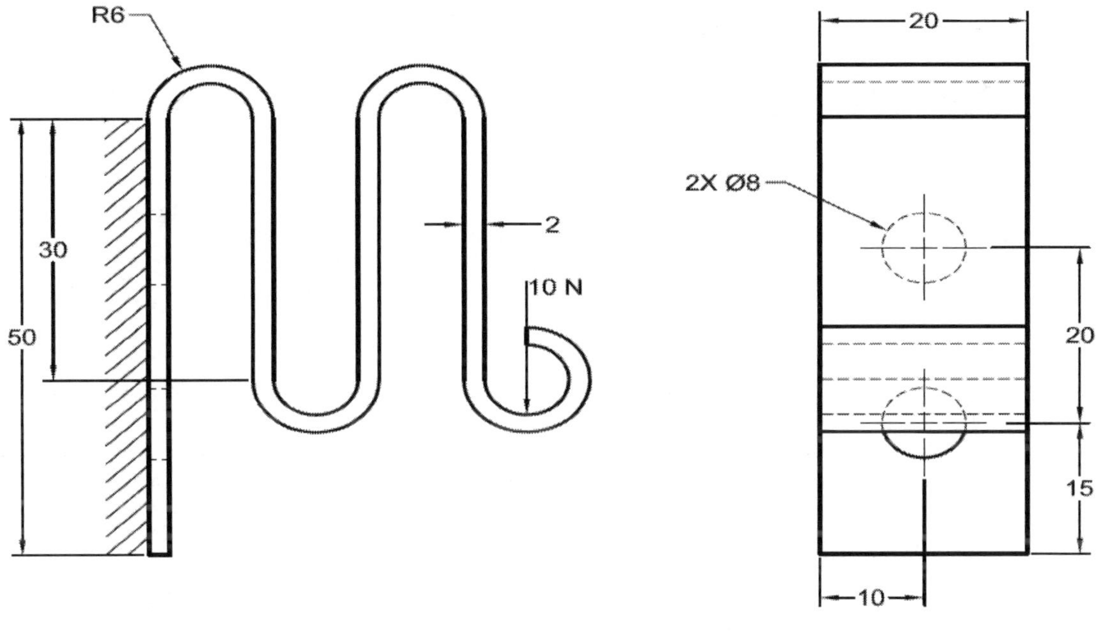

Figure 3-24 *The dimensions of the model*

Exercise 2

In this exercise, you will open the input file *(c03_exr02)* downloaded earlier for Exercise 2 and create a mesh on it. The mesh type is solid and use an appropriate method to generate a local mesh control in the model, as shown in Figure 3-25. After applying the mesh controls, check the mesh quality, and elements and nodes in the mesh. The model is made up of Alloy Steel material. **(Expected time: 45 min)**

Figure 3-25 The Model for Exercise 2

Exercise 3

In this exercise, you will create a surface mesh on the model shown in Figure 3-26. The dimensions and views of the model are shown in Figure 3-27 and 3-28. Figure 3-29 shows the boundary conditions of the model. The mesh type here is surface. Use an appropriate method to generate a local mesh control in the model. After applying the mesh controls, check the mesh quality, elements and nodes in the mesh. The model is made up of alloy steel material.

(Expected time: 30 min)

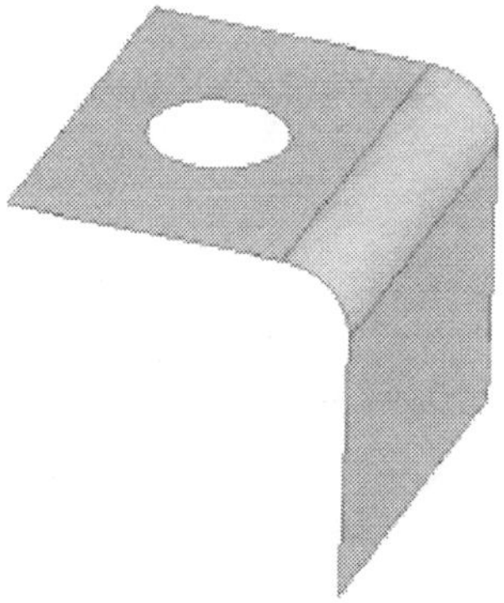

Figure 3-26 Surface model for Exercise 3

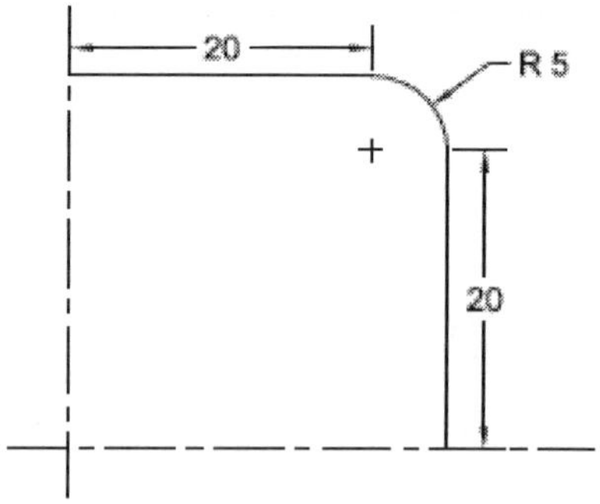

Figure 3-27 Front view of the model

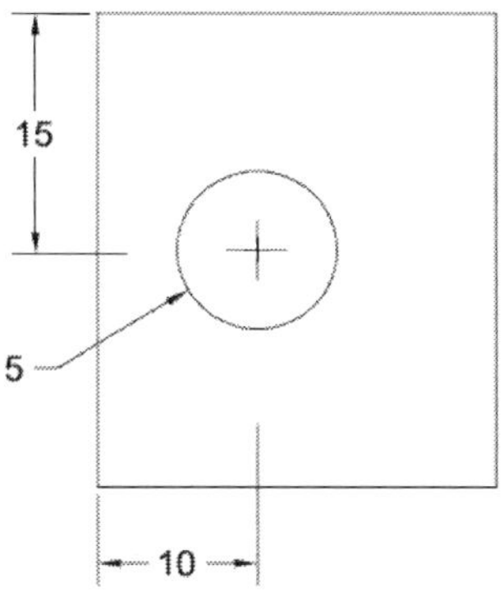

Figure 3-28 The model for Exercise 3

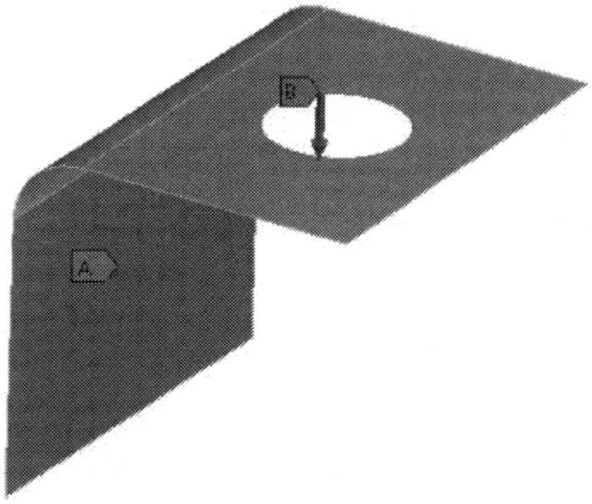

Figure 3-29 Boundary conditions for the model

Answers to Self-Evaluation Test

1. Details, 2. Create Mesh Plot, **3.** Create Mesh, **4.** F, **5.** T

Chapter 4

Linear Static Study

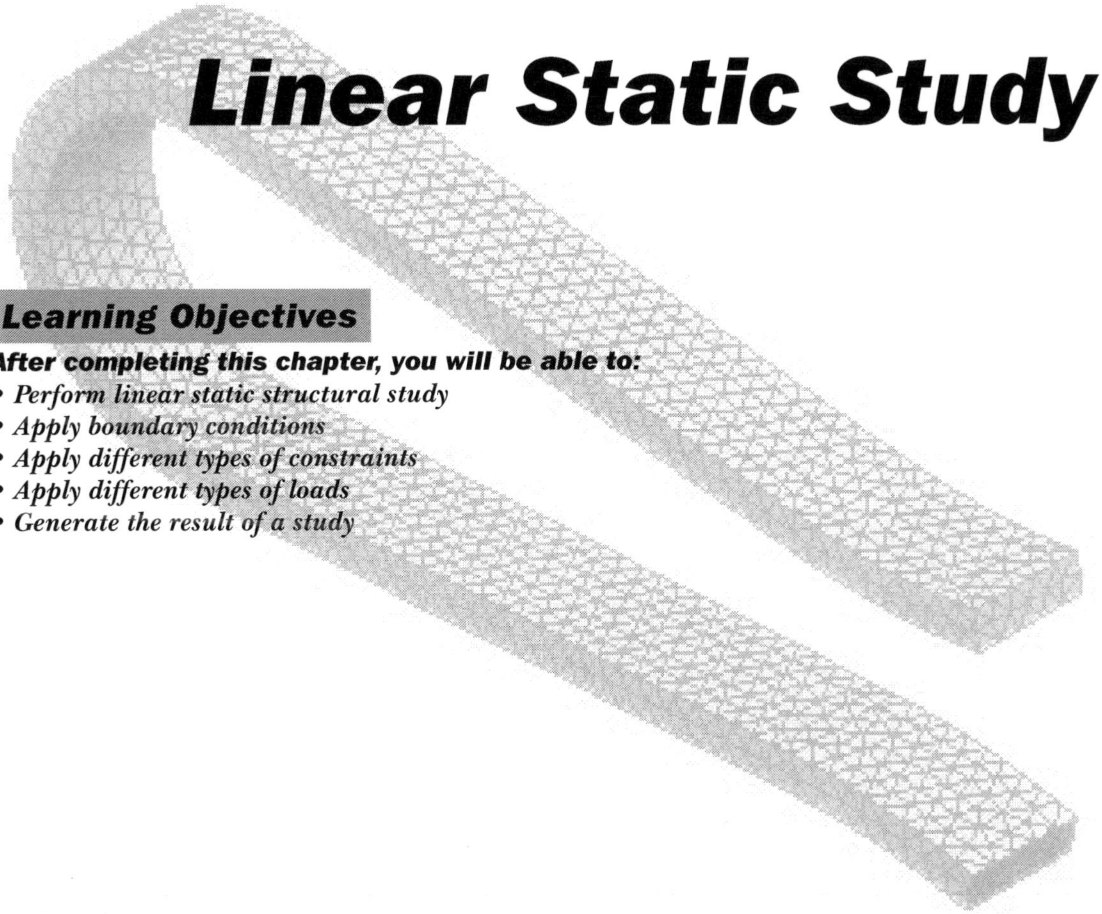

Learning Objectives

After completing this chapter, you will be able to:

- *Perform linear static structural study*
- *Apply boundary conditions*
- *Apply different types of constraints*
- *Apply different types of loads*
- *Generate the result of a study*

INTRODUCTION

In this chapter, you will learn about linear static structural analysis problems. In this type of analysis, load or field conditions do not vary with time. Linear static analysis is a type of structural analysis and is used to determine the displacements, stresses, strains, and forces in structures or components caused by loads.

Linear Static Analysis

When loads are applied to a body, the effect of loads is transmitted throughout the body, causing it to undergo deformation. The external loads induce internal forces and reactions to keep the body in a state of equilibrium. Linear static analysis is used to calculate displacements, strains, stresses, and reaction forces under the effect of applied loads in equilibrium state.

Assumptions of Static Analysis

- Loads are applied slowly and gradually.
- Loads remain constant (time-invariant) after reaching their maximum value.
- Relation between loads and induced responses is linear.
- Hook's law is applicable.
- All material properties such as Young's modulus and Poisson's ratio remain constant throughout the study.
- Loads must be constant in magnitude, direction, and distribution.

TUTORIALS

Tutorial 1

In this tutorial, you will download *c04_tut01* file (bracket model) and analyze the effect of loads. Refer to Figure 4-1 for the given boundary conditions. Consider that the bracket is of Alloy Steel (SS) material and its back face is fixed. You need to determine the von Mises stresses under the load of 5 N. You can download this model from www.cadcim.com. The path of the file is as follows:

Textbooks > CAE Simulation > SOLIDWORKS Simulation > SOLIDWORKS Simulation 2022: A Tutorial Approach > Input Files > c04_simulation_2022_input.zip

(Expected time: 30 min)

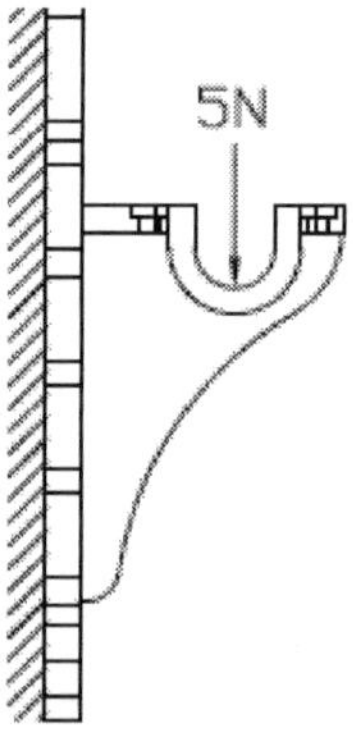

Figure 4-1 *The boundary conditions of the bracket*

The following steps are required to complete this tutorial:

a. Download the model and save it.
b. Start a new static study.
c. Apply material, constraints, and loads.
d. Create mesh and run the analysis.
e. Generate the report.

Downloading the File and Starting a New Study

In this section, you need to download the model *(c04_tut01)* and start a new study.

1. Download the input files of the model from *www.cadcim.com* to perform the analysis.

2. Save the model at the location given below:
   \Documents\SOLIDWORKS\c04

3. Open the model *(c04_tut01)* in SOLIDWORKS.

4. Choose the **Simulation** tab from the **CommandManager**; the **Simulation CommandManager** is displayed.

5. Choose the **New Study** tool from the **New Study** flyout in the **Simulation CommandManager,** refer to Figure 4-2; the **Study PropertyManager** is displayed, as shown in Figure 4-3.

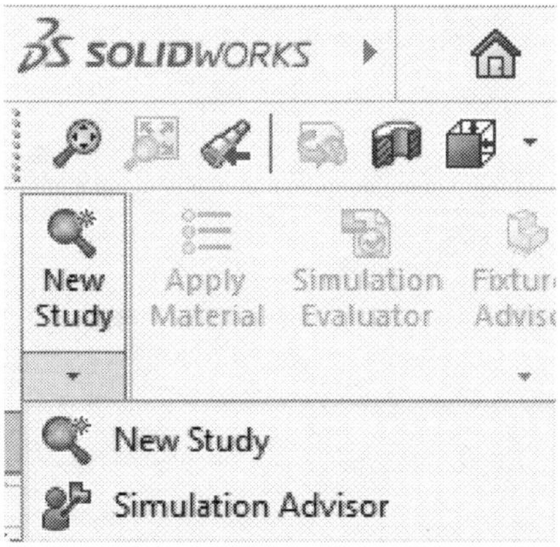

*Figure 4-2 The **New Study** flyout*

6. Choose the **Static** button, refer to Figure 4-3, if it is not chosen, and then enter the name of the study as **Bracket** in the **Study name** edit box of the PropertyManager. Next, choose the **OK** button; a Simulation Tree is added below the **FeatureManager Design Tree.**

Applying Material

Now, you need to apply the alloy steel material to the model.

1. Choose the **Apply Material** button from the **Simulation CommandManager**, refer to Figure 4-4; the **Material** dialog box is displayed.

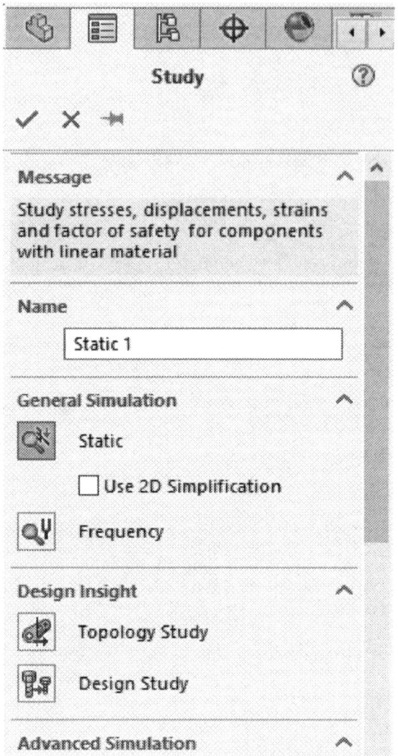

Figure 4-3** Partial view of the **Study PropertyManager

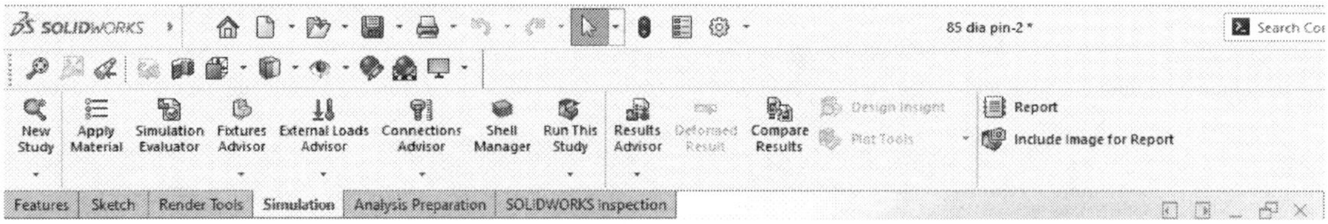

Figure 4-4** The **Simulation CommandManager

2. Expand the **SOLIDWORKS Materials** node in the **Material** dialog box. Next, expand the **Steel** sub-node under it and choose the **Alloy Steel (SS)** material. All the properties of the selected material are displayed on the right in the dialog box.

3. Choose the **Apply** button from the dialog box; the alloy steel material gets applied to the model. Next, close the **Material** dialog box.

Applying Constraints

After applying the material to the model, you need to apply constraints to fix one face of the model.

1. Choose the **Fixed Geometry** tool from the **Fixtures Advisor** flyout in the **Simulation CommandManager**; the **Fixture PropertyManager** is displayed, as shown in Figure 4-5.
2. Make sure that the **Fixed Geometry** button is chosen in the **Standard (Fixed Geometry)** rollout of the PropertyManager.
3. Rotate the model and select the back face of the model; the **Fixed Geometry** callout is attached to the selected face, as shown in Figure 4-6.
4. Choose the **OK** button from the PropertyManager.

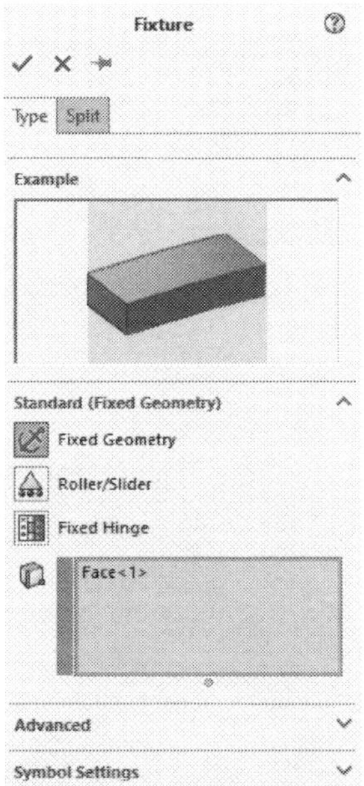

Figure 4-5** The **Fixture PropertyManager

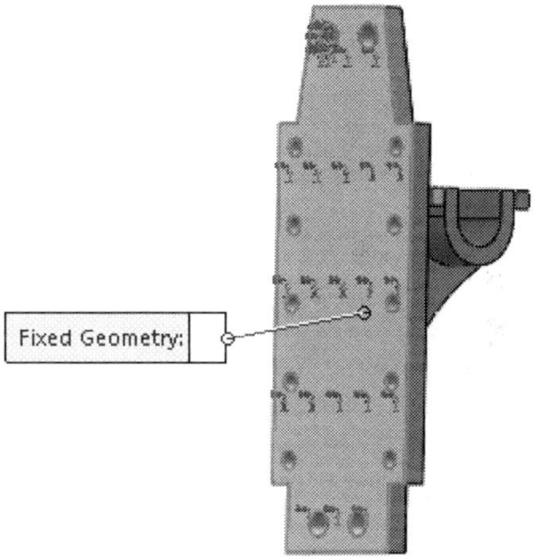

***Figure 4-6** Back face of the model to be selected for applying fixed constraint*

Applying Loads

After applying the material and constraints to the model, you need to apply load.

1. Choose the **Force** tool from the **External Loads Advisor** flyout of the **Simulation CommandManager**; the **Force/Torque PropertyManager** is displayed, as shown in Figure 4-7.

2. Make sure that the **Force** button is chosen in the **Force/Torque** rollout of the PropertyManager.

3. Select the top semi-circular face of the model to apply the force, refer to Figure 4-8; the force callout is attached to the selected face. Now, you need to specify the direction of force.

4. Select the **Selected direction** radio button from the **Force/Torque** rollout in the PropertyManager; the **Face**, **Edge**, **Plane for Direction** selection box is displayed in the **Force/Torque** rollout.

5. Select any vertical edge of the model to apply force in vertical direction.

   If the direction of the applied force is vertically upward, you need to reverse it to vertically downward.

6. Select the **Reverse direction** check box from the **Force** rollout of the PropertyManager to reverse the direction of the force applied if required.

7. Select the **SI** option from the **Unit** drop-down list of the **Units** rollout in the PropertyManager.

8. Enter **5** in the **Along Edge** edit box of the **Force** rollout.

9. Accept other default settings and choose the **OK** button to apply the load. Figure 4-8 shows the model with applied load.

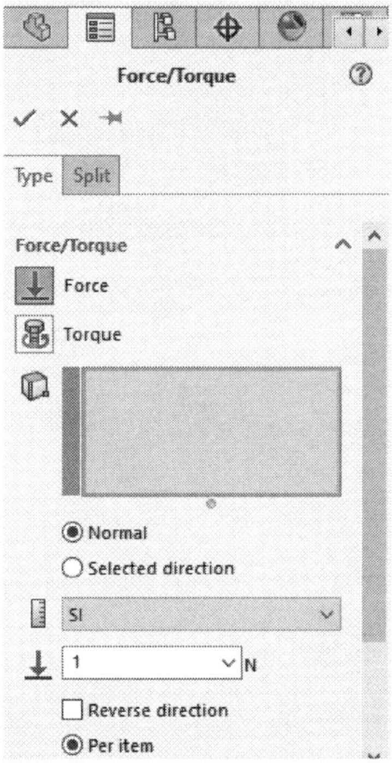

Figure 4-7 *The* ***Force/Torque PropertyManager***

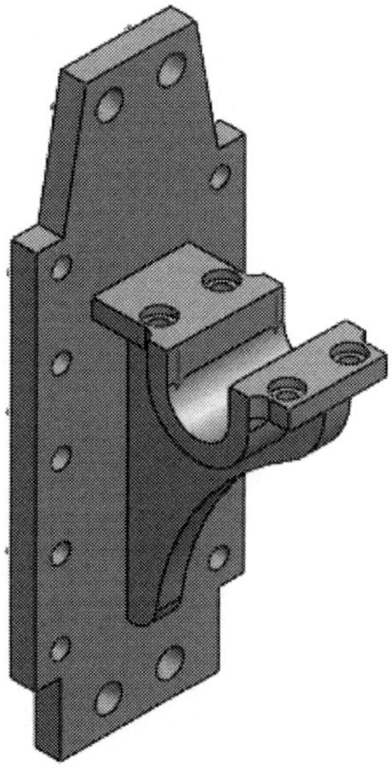

Figure 4-8 *Model after applying boundary condition and load*

Creating Mesh

After applying the boundary condition and load, you need to create mesh.

1. Choose the **Create Mesh** button from the **Run This Study** flyout of the **Simulation CommandManager**; the **Mesh PropertyManager** is displayed, as shown in Figure 4-9.

2. Accept all the default settings and choose the **OK** button from the PropertyManager; the **Mesh Progress** window is displayed. Once the process of meshing is complete, the meshed model is displayed in the drawing area, as shown in Figure 4-10.

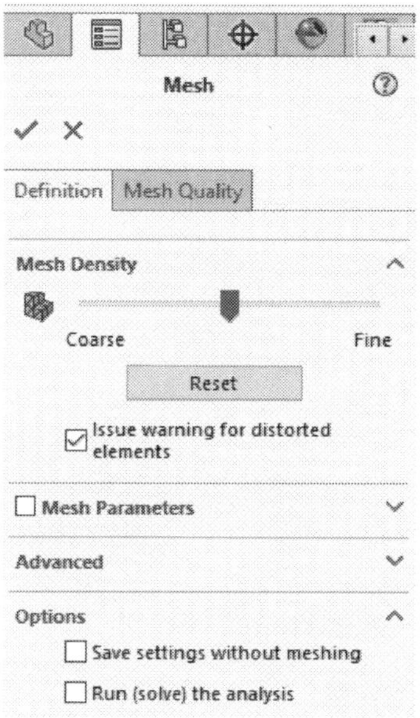

Figure 4-9** The **Mesh PropertyManager

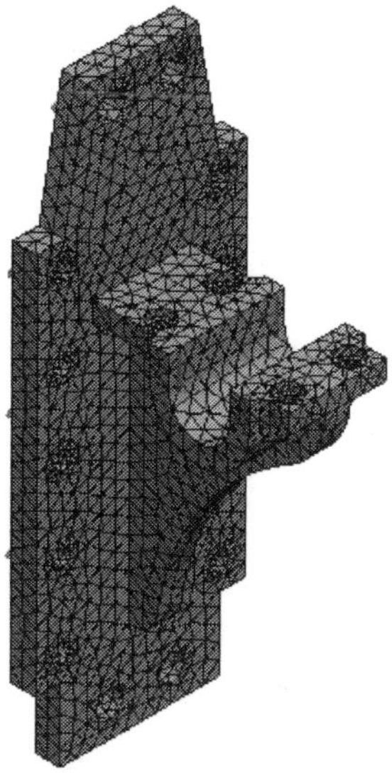

***Figure 4-10** The model after meshing*

Running the Study

Now, you need to run the study on the model.

1. Choose the **Run This Study** button from the **Simulation CommandManager**; the **Bracket** window is displayed and the process of solving the finite element model gets started. Once the solving process is completed, a new node named **Results** is added to the Simulation Tree. Also, the analyzed model is displayed in the drawing area with the von Mises stress contour, as shown in Figure 4-11.

Tip

*To animate the displacement plot, double-click on **Displacement1 (-Res disp-)** under the **Results** node and then right-click on it; a shortcut menu will be displayed. Choose the **Animate** option; the animation starts with the default settings. Also, the **Animation PropertyManager** is displayed on the left in the screen enabling you to change the settings according to your requirement.*

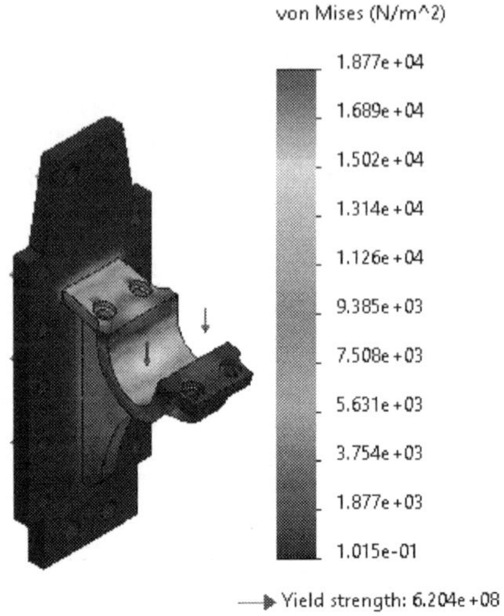

***Figure 4-11** The resultant model with von Mises stress contour*

Generating Report

Now, you need to generate the report.

1. Choose the **Report** button from the **Simulation CommandManager**; the **Report Options** dialog box is displayed.

2. Select all the sections that you want to add to the report by selecting their respective check boxes available in the **Report sections** area of the dialog box.

   As discussed earlier, you can also add information such as designer, company, URL, logo, and so on to the header of the report. To do so, you need to add information in the respective edit boxes of the dialog box that will be enabled after selecting the check box corresponding to them.

3. Enter all the required information in the **Report Options** dialog box and then choose the **Publish** button; the report is generated in Word format and the **Report** node is added to the Simulation Tree. After viewing the report, you can close it.

4. Save and close the file.

Tip
*After closing the report, you can again open it by double-clicking on its name displayed under the **Report** node of the Simulation Tree.*

Tutorial 2

In this tutorial, you will perform a mixed mesh study which depends upon the element type (solid, shell, and beam). Figure 4-12 shows the dimensions of the model used in this tutorial. The model is fixed at one end and a load of 1000 N is applied, as shown in Figure 4-13. The model is made of Plain Carbon Steel material. **(Expected time: 1 hr)**

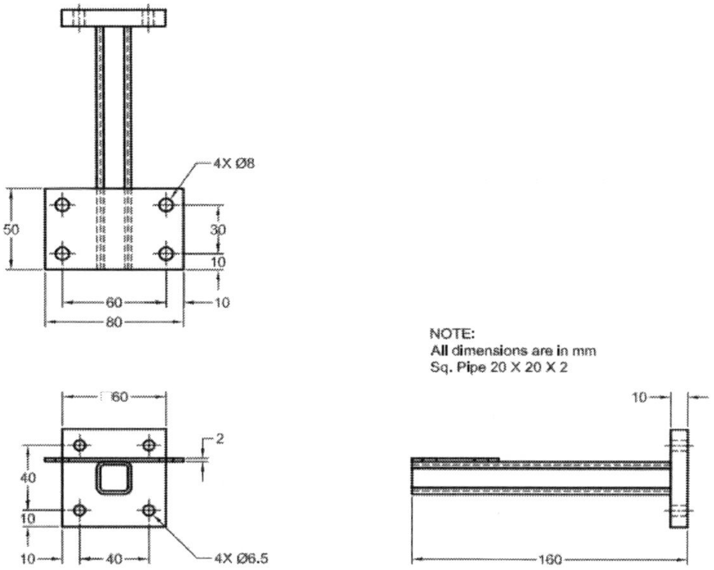

Figure 4-12 *Dimensions of the model for Tutorial 2*

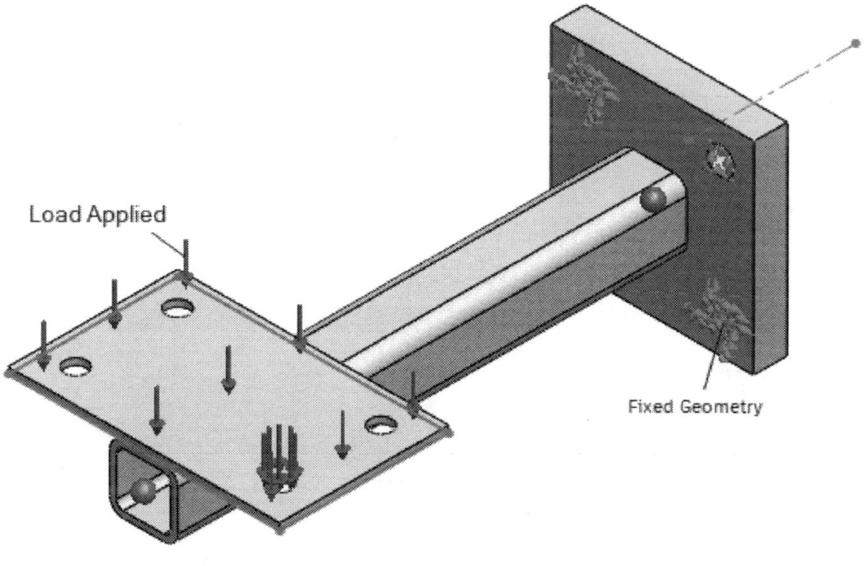

Figure 4-13 *Fixture and load applied on the model*

The following steps are required to complete this tutorial:

a. Create the model.
b. Start a new Static study.
c. Apply material, fixture, and load.
d. Create mesh and run the study.
e. Generate the report.
f. Save the results.

Creating Model for Analysis

First, you need to create the base feature of the model. Next, you need to create the square pipe using weldment structure. Finally, you will create the sheet metal plate on the square pipe, refer to Figure 4-13.

1. Start a new SOLIDWORKS part document, select the front plane, and draw a square of size 60 mm by using the **Center Rectangle** tool.

2. Extrude the sketch upto 10 mm depth.

3. Draw a line of 150 mm length on the right plane whose one end is connected to the extruded square surface, as shown in Figure 4-14.

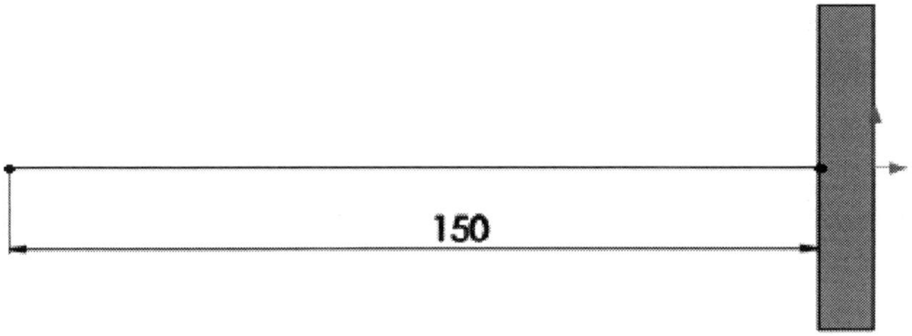

***Figure 4-14** Line drawn on right plane with one end connected to the square plate*

4. Choose the **Structural Member** tool from the **Weldments CommandManager**; the **Structural Member PropertyManager** is displayed.

5. Select iso from the **Standard** drop-down list and **square tube** from the **Type** drop-down list of the **Selections** rollout in the **Structural Member PropertyManager**.

6. Select **20 x 20 x 2** from the **Size** drop-down list in the **Selections** rollout, refer to Figure 4-15.

7. Click on the **Groups** selection box and then select the line from the drawing window; Group1 is added to the **Group** selection box and a preview of the model is displayed, as shown in Figure 4-16.

8. Choose the **OK** button from the **Structure Member PropertyManager.**

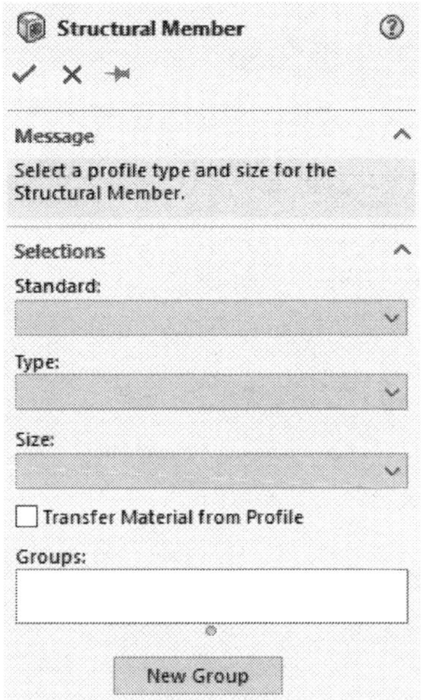

Figure 4-15 *The* ***Structural Member PropertyManager***

9. Select the top surface of the structure as the sketching plane and create a sketch, as shown in Figure 4-17.

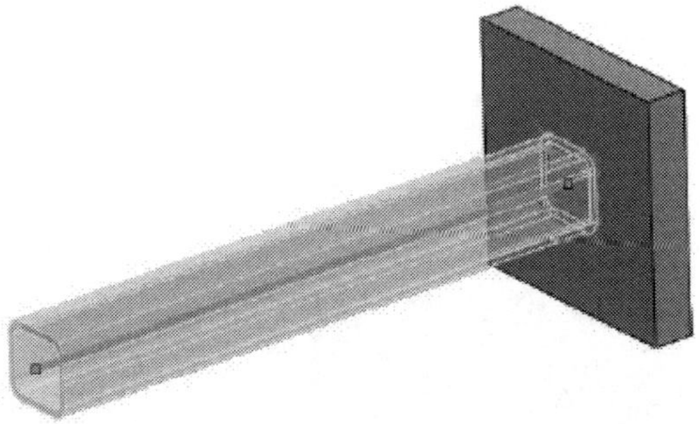

Figure 4-16 *Preview of structure with square plate*

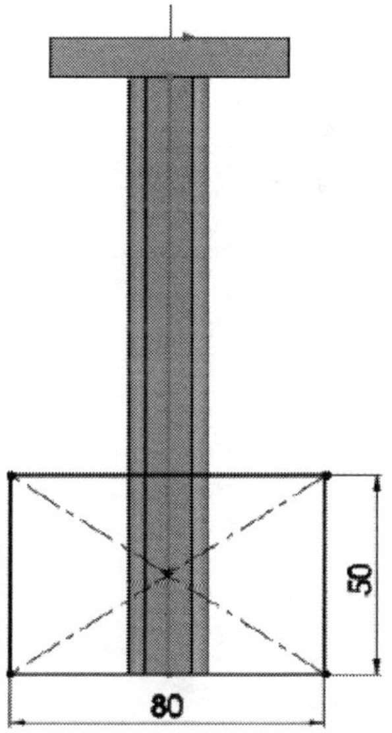

Figure 4-17 *Sketch drawn on the top surface of the structure*

10. Choose the **Base Flange/Tab** tool from the **Sheet Metal CommandManager**; the **Base Flange PropertyManager** is displayed, as shown in Figure 4-18.

11. Set the sheet thickness to 2 mm in the **Thickness** spinner of the **Sheet Metal Parameters** rollout and then set the required direction of the extrusion.

12. Go with other default values and choose the **OK** button from the **Base Flange PropertyManager;** the base flange is added to the model, as shown in Figure 4-19.

13. Next, use the **Extruded Cut** tool and create holes in both the plates, refer to Figure 4-12 for dimensions of the holes. The final model for study is shown in Figure 4-20.

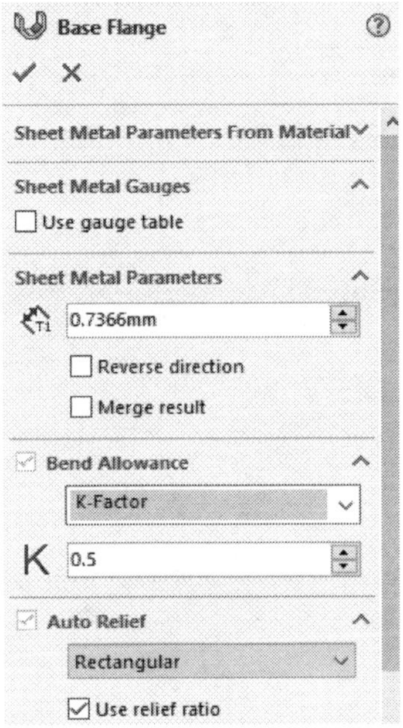

Figure 4-18** The **Base Flange PropertyManager

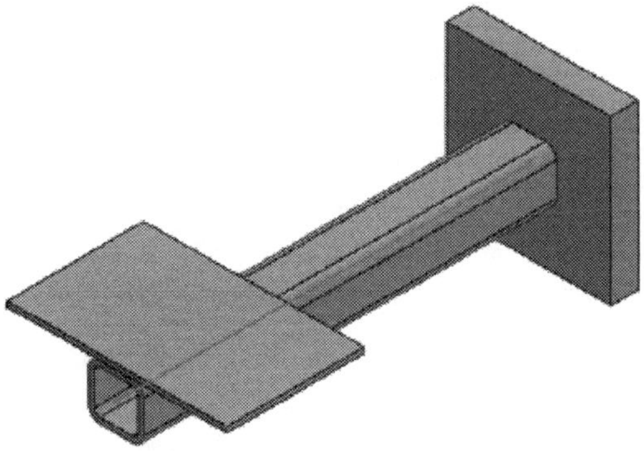

***Figure 4-19** Model after adding base flange*

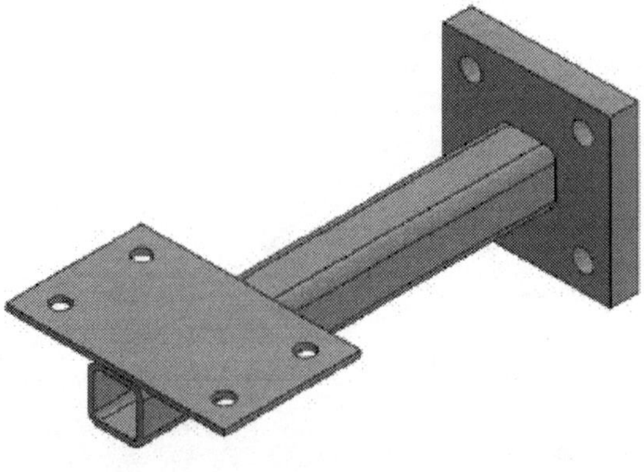

***Figure 4-20** Final model for study*

14. Save the model with the name *c04_tut_02* at the desired location.

Creating Mixed Mesh Study

1. Select the **New Study** option from the **New Study** flyout of the **Simulation CommandManager;** the **Study PropertyManager** is displayed.

2. Choose the **Static** button from the **General Simulation** rollout and enter **Mixed Meshing** as the name of the study in the **Study name** edit box.

3. Choose the **OK** button from the **Study PropertyManager**; a new study is added.

Defining Element Type

SOLIDWORKS Simulation defines the element type for each element on the basis of the shape of the model. But you can change the default element type manually. The time elapsed in meshing for an element depends upon the element type.

1. Expand the *c04_tut_02* node in the Simulation Tree, as shown in Figure 4-21.

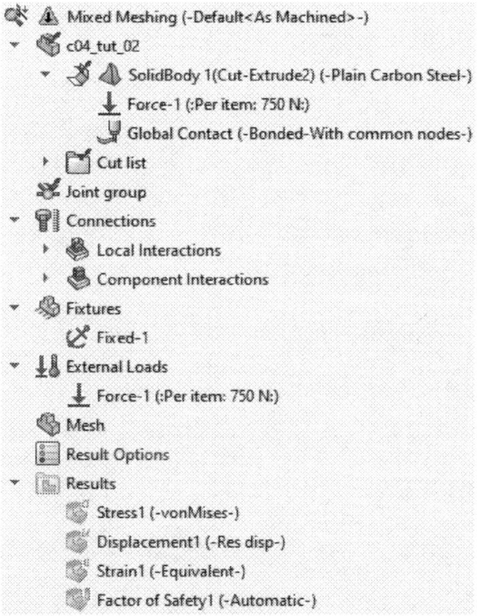

Figure 4-21 *Simulation Tree for Mixed Meshing*

In Figure 4-21, you will notice that the base plate is treated as solid element, weldment structure is treated as beam element, and the base flange created using sheet metal is treated as shell by default. So you do not have to change the element type.

Assigning Material

1. Right-click on the **SolidBody 3** node in the **Mixed Meshing** Simulation Study and choose **Apply/Edit Material** from the shortcut menu; the **Material** dialog box is displayed.

2. Select the **Plain Carbon Steel** from the **Steel** node and then choose the **Apply** button from the **Material** dialog box.

3. Choose the **Close** button to close the **Material** dialog box.

4. Similarly, apply Plain Carbon Steel material to the structure and rectangular plate.

Specifying Interactions

Now, you need to specify the Interactions in between the components. To specify contacts, you need to follow the steps given below:

1. Choose the **Local Interaction** tool from the **Connections Advisor** flyout in the **Simulation CommandManager**; the **Local Interactions PropertyManager** is displayed.

2. Select the **Automatically find Local Interactions** radio button from the **Interaction** rollout. Make sure that the **Touching faces** radio button is selected under the **Options** rollout.

3. Select the components from the drawing area; the names of the selected components are displayed under the **Components** rollout. Next, choose the **Find Local Interactions** button; the available contacts are displayed in the **Results** rollout.

   Make sure the **Contact** option is selected in the **Type** drop-down list. The **Contact** option does not allow bodies to penetrate each other during the simulation process.

4. Choose the **OK** button from the **Local Interactions PropertyManager**; the **Simulation** message box is displayed. Choose the **Yes** button from the message box to create a local interaction.

   You will notice that a new local interaction is created in Simulation Tree. Now, you need to Delete existing local interaction if available for the same components, refer to Figure 4-22.

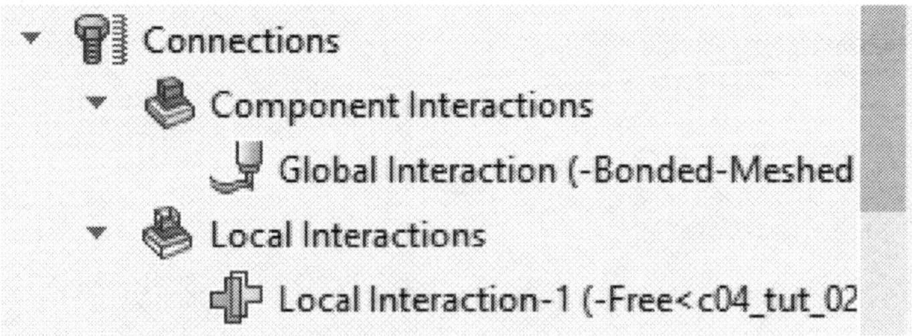

*Figure 4-22 The **Mesh PropertyManager***

Applying Boundary Conditions

After assigning the contacts, you need to apply boundary or loading conditions to perform the simulation study.

1. Choose the **Fixed Geometry** tool from the **Fixtures Advisor** flyout in the **Simulation CommandManager**; the **Fixture PropertyManager** is displayed.

2. Select the holes (faces) from the model, as shown in Figure 4-13. Before selecting the holes to create fixture, make sure that the **Fixed Geometry** button is chosen in the **Standard (Fixed Geometry)** rollout.

3. Choose the **OK** button from the **Fixture PropertyManager**; the Fixed geometry fixture is applied to the selected holes.

   After the fixture is defined for the model, you need to define the load for which simulation is to be carried out.

4. Choose the **Force** tool from the **External Loads Advisor** flyout in **Simulation CommandManager**; the **Force/Torque PropertyManager** is displayed.

5. Select the face, as shown in Figure 4-13; the name of the selected face is displayed in the **Force/Torque** rollout.

6. Specify the magnitude of the load in the **Force Value** edit box as **750**. Make sure the **Normal** radio button and the **SI** option are selected in the **Force/Torque** rollout. After specifying the load value, choose the **OK** button to exit the PropertyManager.

Creating Mesh

In this section, you need to divide the model into elements.

1. To create the mesh, right-click on the **Mesh** node in the Simulation Tree and choose the **Create Mesh** option from the shortcut menu; the **Mesh PropertyManager** is displayed.

2. Accept the default settings and choose the **OK** button from the PropertyManager; the mesh is created on the model, as shown in Figure 4-23.

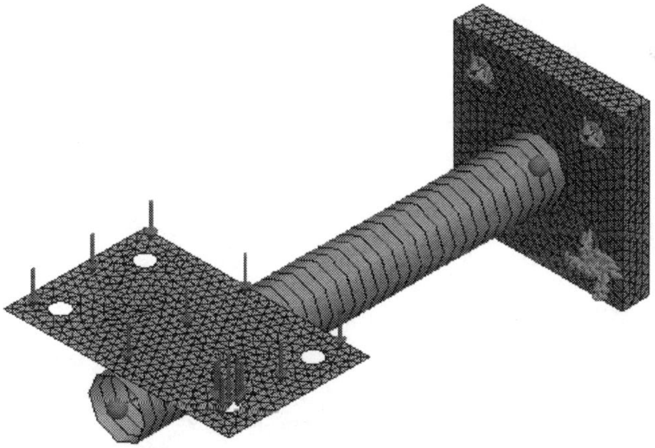

Figure 4-23 *Model after meshing*

Running Study and Interpreting the Results

Now you need to run study to get results based on the input given to the model.

1. Right-click on the **Mixed Meshing** node in the Simulation Tree and choose the **Run** option from the shortcut menu; the **Mixed Meshing** window is displayed and the solution process gets started.

When the analysis is completed, a results node is added to the Simulation Tree. Also, the analyzed model is displayed with von-Mises stress result plot. By default, von-Mises stress is displayed in the plot, as shown in Figure 4-24. To view the other results' plot, double-click on the respective option in the **Results** node of the Simulation Tree.

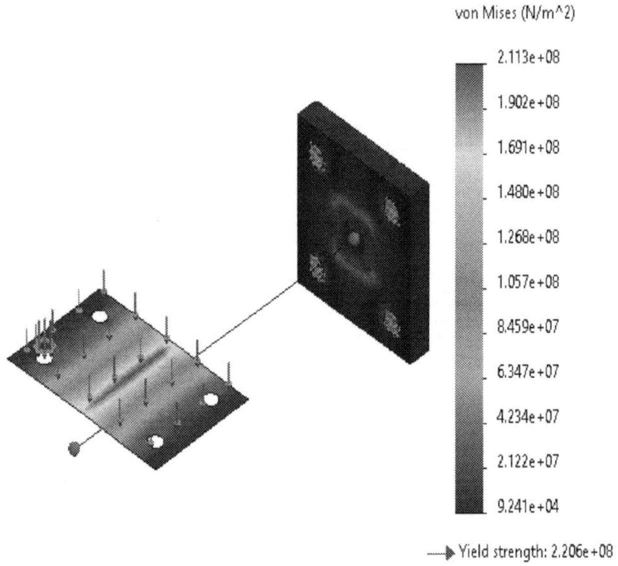

Figure 4-24 *Von-Mises stress plot*

Calculating the Factor of Safety (FOS)

1. Right-click on the **Results** node in the Simulation Tree and choose the **Define Factor Of Safety Plot** from the shortcut menu; the **Factor of Safety PropertyManager** is displayed.

   Now, you need to specify the option to check the safety of your design based on a failure criterion. Select the **Automatic** option under the **Step 1 of 3** rollout. On doing so, the software will be enabled to select the most appropriate failure criterion across all element types.

2. Choose the **Next** button in the **Factor of Safety PropertyManager** to specify the multiplication factor. Enter **1** in the **Multiplication Factor** edit box. It allows you to enter a multiplication factor for calculating the selected stress limit. Now, choose the **Next** button again.

3. Select the **Areas below factor of safety** radio button and then enter **1** in the **Minimum factor of safety** edit box. Next, choose the **OK** button; the factor of safety plot is displayed, as shown in Figure 4-25.

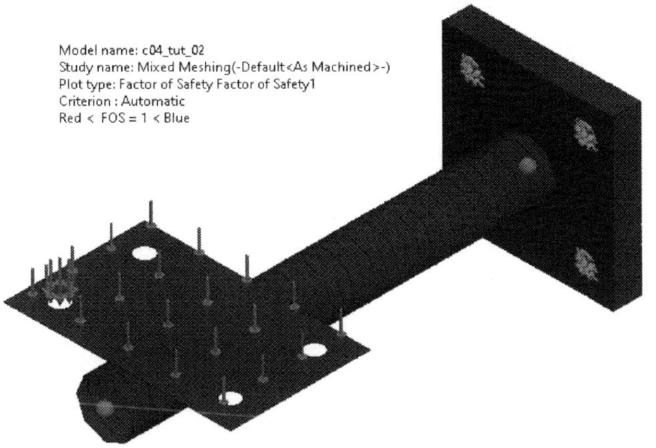

Figure 4-25 *Factor of safety plot*

The **Factor of safety distribution** option is used to plot the distribution of the factor of safety. The areas below the **Factor of Safety** option are used to plot the regions below the specified factor of safety. Note that regions below the specified factor of safety are displayed in red color and above the specified factor of safety are displayed in blue color.

Saving the File

After the desired solution is achieved, you need to save the model and results.

1. Choose **File > Save** from the menu bar to save the simulation files.

Tutorial 3

In this tutorial, you will perform static study in three cases. Assume that an Alloy Steel plate is supported and loaded. In this plate, the support is rigid and a force of 20000 N is applied along the end face of the model, as shown in Figure 4-26. You will reduce the weight of the plate without affecting the working conditions. Next, you will perform three case studies:

Case Study-1: Apply the boundary condition, as shown in Figure 4-26 and run study.

Case Study-2: Create a cut of diameter 65 mm at the centre of the plate and run the study with same boundary condition.

Case Study-3: Create a cut of diameter 50 mm and run the study with same boundary conditions. After completion of all studies, you will compare and analyze the results.

(Expected time: 1 hr)

The following steps are required to complete this tutorial:

a. Create the model.
b. Start a new Static study.

c. Apply material, fixture, and load.
d. Create mesh.
e. Run the study.
f. View the results.
g. Create second case study.
h. Create third case study.
i. Compare the results.

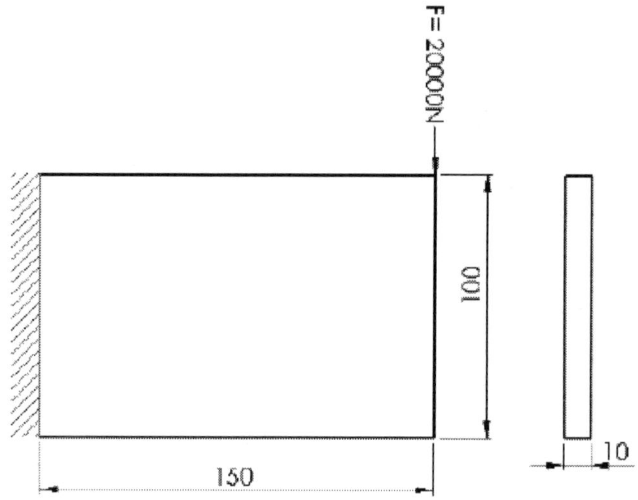

Figure 4-26 *The dimensions and the boundary conditions of the plate*

Creating Model for Analysis

Now, you need to create the model of the rectangular plate for analysis.

1. Create the rectangular plate, as shown in Figure 4-27. The dimensions of model are given in Figure 4-26.

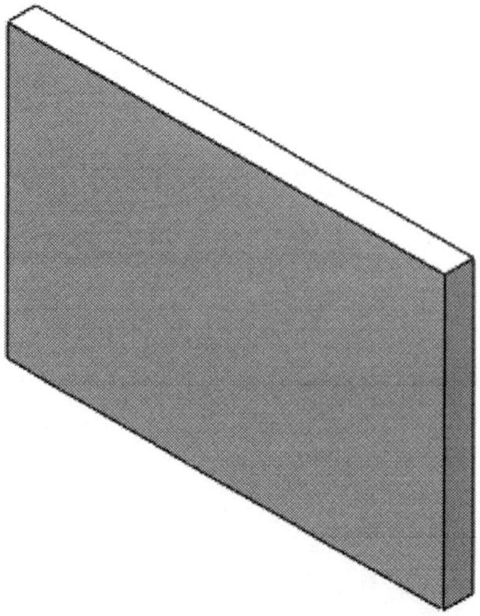

Figure 4-27 *Model for Case study-1*

2. Save the model with the name *c04_tut_03* at the location given below:
\Documents\SOLIDWORKS\c04

Adding Configurations for Cut Feature

Next, you need to create configurations named cut dia. 65 and cut dia. 50 for cut feature of diameter 65 and 50, respectively.

1. Select the **ConfigurationManager** from the Manager Pane; the **ConfigurationManager** along with the default configuration **(Default [c04_tut_03])** will be displayed.

2. Select the name of the part in configurations and right-click; a shortcut menu is displayed, as shown in Figure 4-28.

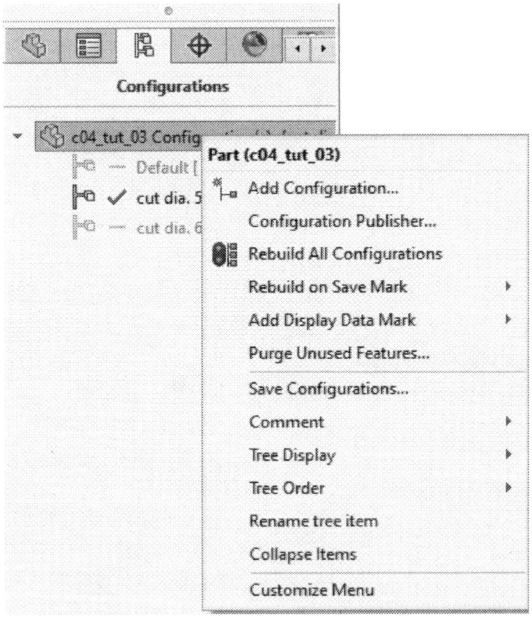

***Figure 4-28** Shortcut menu displayed on right-clicking on the name of the configurations*

3. Select the **Add Configuration** option from this; the **Add Configuration PropertyManager** is displayed.

4. Enter **cut dia. 65** as name of the configuration in the **Configuration name** text box and choose the **OK** button; configuration is created in the **ConfigurationManager** tree. The newly created configuration will be activated automatically.

5. Create a cut of diameter 65 mm at the center of the plate, as shown in Figure 4-29 for Case Study 2. Next, suppress the cut feature created in configuration **cut dia. 65** from the Feature Manager Design Tree.

6. Choose the **Feature Manager Design Tree** button in the **Manager pane**; **Feature Manager Design Tree** is displayed. Select the cut feature and then suppress it from the **Design Tree.**

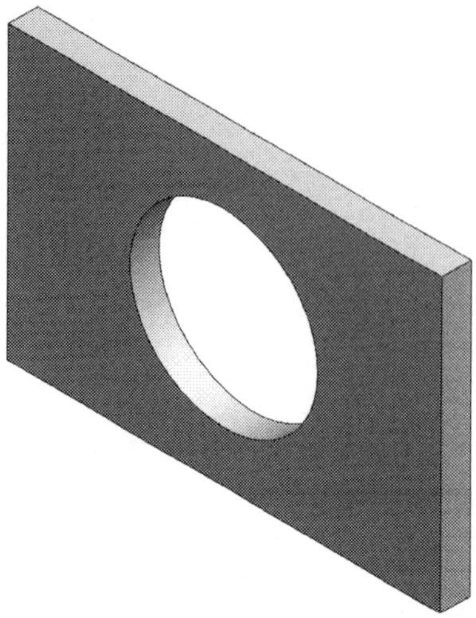

Figure 4-29 Model for Case study-2

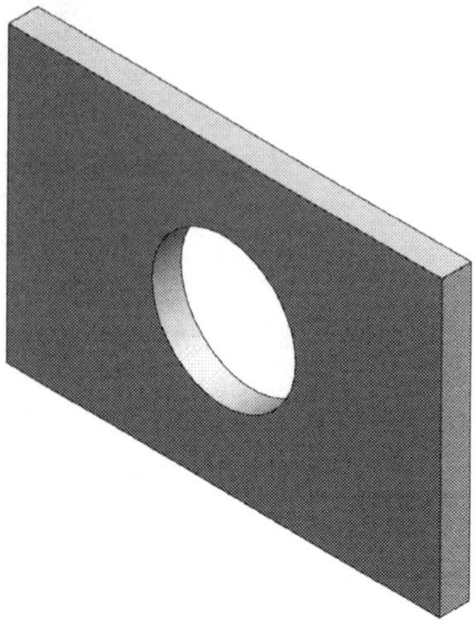

Figure 4-30 Model for Case study-3

7. Now, add the third configuration of diameter 50 mm and name it cut dia 50.

8. Similarly, create configuration for a cut of diameter 50 for case study 3, as shown in Figure 4-30. To activate a particular configuration, double-click on the name of the configuration.

Starting Static Study Case-1

You need to start the study with default configuration of the model.

1. Activate the default configuration of the model for Case Study-1.

2. Select the **New Study** option from the **New Study** flyout of the **Simulation CommandManager;** the **Study PropertyManager** is displayed. Alternatively, you can invoke the **Study PropertyManager** by selecting the **Study** option from the **Simulation** menu.

3. Select the **Static** option from the **Study Property Manager** and enter **Case Study-1** as the name of the study. Next, choose the **OK** button; the Simulation Tree will be displayed, as shown in Figure 4-31.

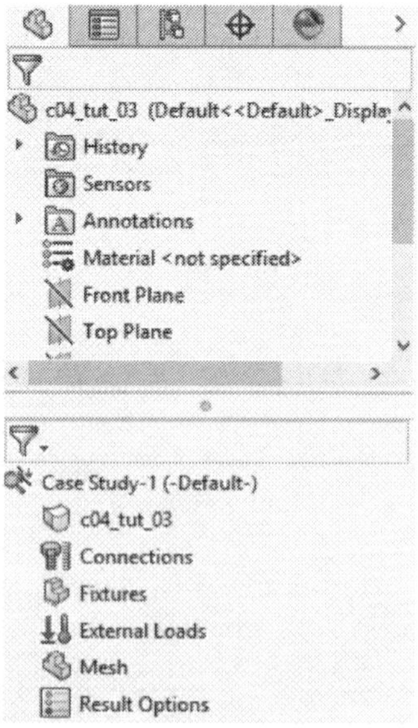

Figure 4-31 The Simulation Tree

Applying Material

Now, you need to apply **Alloy Steel** material to part *c04_tut_03*.

1. Right-click on the **c04_tut_03** and select the **Apply/Edit Material** option from the shortcut menu; the **Material** dialog box is displayed.

2. Expand the **SOLIDWORKS Materials** node in the **Material** dialog box**.** Next, expand the **Steel** sub-node under it and choose the **Alloy Steel** material; all the properties of the selected material are displayed on the right in the dialog box.

3. Choose the **Apply** button from the dialog box; the alloy steel material is applied to the model.

4. Choose the **Close** button to close the **Material** dialog box.

Applying Restraints

You need to define restraints to make the model stable.

1. Right-click on **Fixtures** node in the Simulation Tree and choose the **Fixed Geometry** option from the shortcut menu; the **Fixture PropertyManager** is displayed, as shown in Figure 4-32.

2. Choose the **Fixed Geometry** button from the **Standard(Fixed Geometry)** rollout, if it is not chosen by default.

3. Rotate the model and select the left face; the fixed geometry callout gets attached to the selected face of the model, as shown in Figure 4-33.

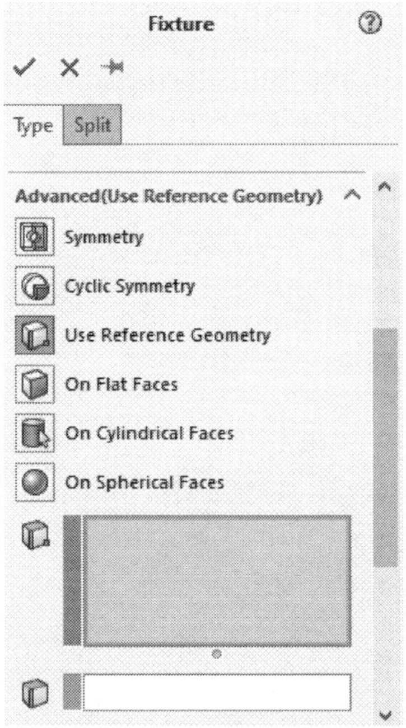

Figure 4-32 *The* ***Fixture PropertyManager***

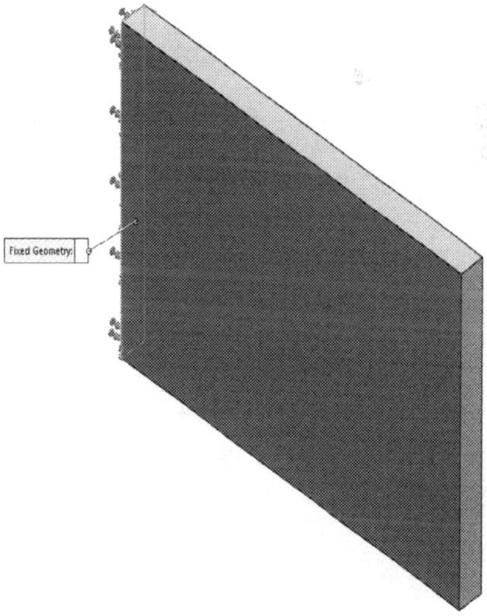

Figure 4-33 *Selected face of the model to apply the Fixed Geometry restraint*

4. Choose the **OK** button from the PropertyManager; the selected face gets fixed.

Applying Loads

You need to apply the load on the right face of the model in vertically downward direction.

1. Right-click on the **External Loads** node in the Simulation Tree and choose the **Force** option from the shortcut menu; the **Force/Torque PropertyManager** is displayed, as shown in Figure 4-34.

2. Choose the **Force** button from the **Force/Torque** rollout, if it is not chosen by default; the **Faces and Shell Edges for Normal Force** selection box is activated.

3. Select the right face of the model; the preview of the applied force is displayed.

4. Select the **Selected direction** radio button; the **Face, Edge, Plane for Direction** selection box is displayed in the **Force/Torque** rollout.

5. Select any vertical edge of the model for specifying the direction of the force applied on the right face.

6. Select the **SI** unit system from the **Unit** drop-down list of the **Units** rollout.

7. Enter **20000** in the **Along Edge** edit box of the **Force** rollout and select the **Reverse direction** check box if the direction of the applied force is upward.

8. Choose the **OK** button from the PropertyManager; the model after applying the restraint and force is shown in Figure 4-35.

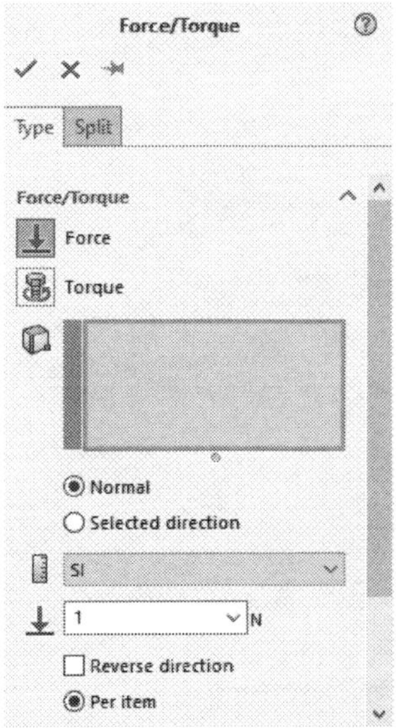

*Figure 4-34 The **Force/Torque PropertyManager***

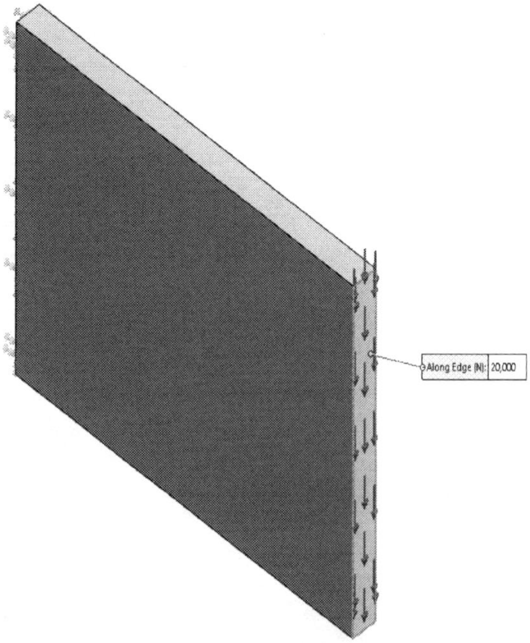

Figure 4-35 Model after applying restraint and load

Creating the Mesh

You need to create mesh to divide the model into small pieces.

1. Right-click on the **Mesh** node in the Simulation Tree and choose the **Create Mesh** option from the shortcut menu; the **Mesh PropertyManager** is displayed, as shown in Figure 4-36.

2. Accept the default settings and choose the **OK** button from the PropertyManager; the mesh is created in the model, refer to Figure 4-37.

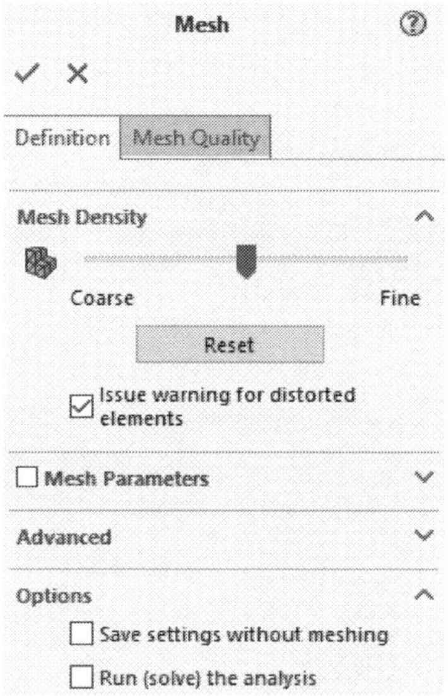

Figure 4-36 The Mesh PropertyManager

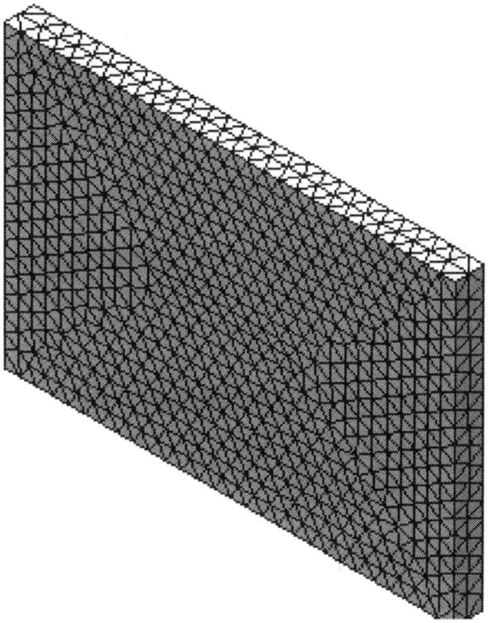

Figure 4-37 Model after meshing

Running the Study

You need to run the study to get results based on the input given to the model.

1. Right-click on **Case Study-1** node in the Simulation Tree and choose the **Run** option from the shortcut menu displayed; the Case Study-1 window is displayed and the solution process starts.

   When the analysis is completed, a **Results** node is added in the Simulation Tree. Also, the analyzed model is displayed with von Mises stress plot, as shown in Figure 4-38. By default, von Mises stress is displayed in the plot. To view the other result plot, double-click on the respective option in the **Results** node of the Simulation Tree.

Note

Since in this study, you have to reduce weight of the material without affecting the working conditions, therefore you need to calculate Factor of Safety (FOS) as FOS depends on the working stress of the model.

Finding Factor of Safety (FOS)

1. Right-click on the **Results** node in the Simulation Tree and select the **Define Factor Of Safety Plot** option from the shortcut menu; the **Factor of Safety PropertyManager** is displayed, as shown in Figure 4-39.

2. Choose the **Next** button in the **Factor of Safety PropertyManager** with other default options selected; the **Factor of Safety PropertyManager** is modified, as shown in Figure 4-40.

3. Choose the **Next** button again with default value **1** in the **Multiplication factor** edit box of the **Step 2 of 3** rollout; the **Factor of Safety PropertyManager** is modified again, as shown in Figure 4-41.

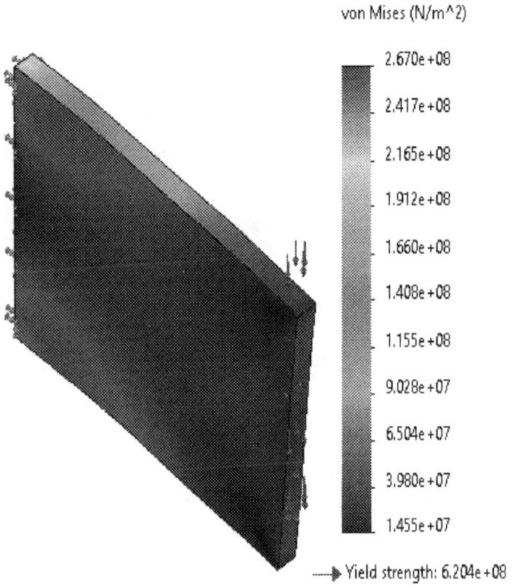

Figure 4-38 Visualization of model with von Mises stress plot

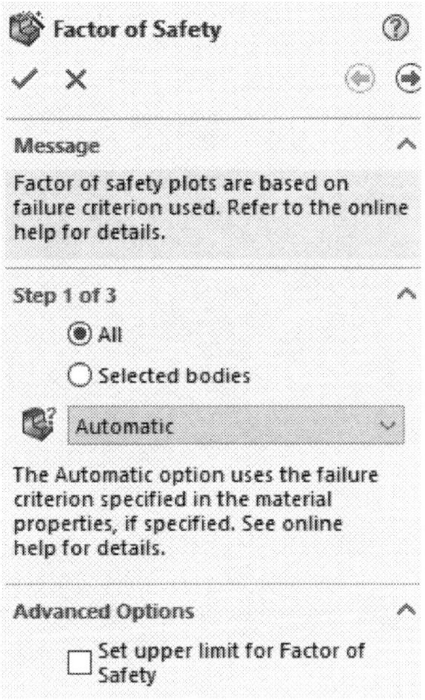

Figure 4-39** The **Factor of Safety PropertyManager

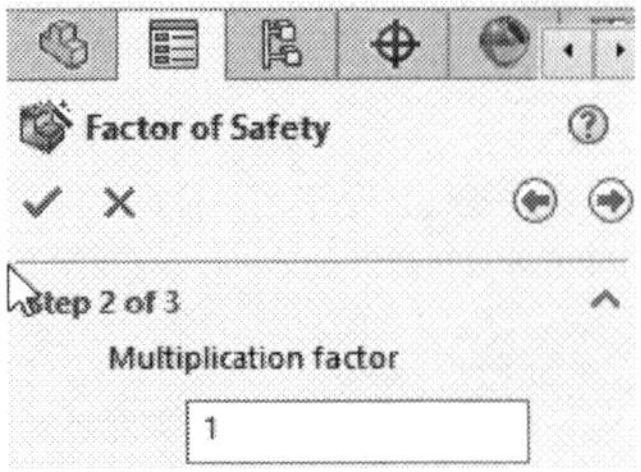

Figure 4-40** The modified **FOS PropertyManager

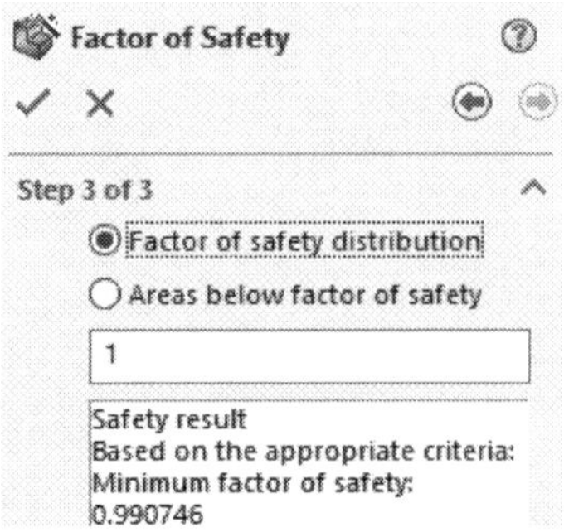

***Figure 4-41** The **FOS PropertyManager** with further modification*

4. Choose the **OK** button with default value **1** in the **Minimum factor of safety** edit box; the factor of safety distribution plot is displayed, as shown in Figure 4-42.

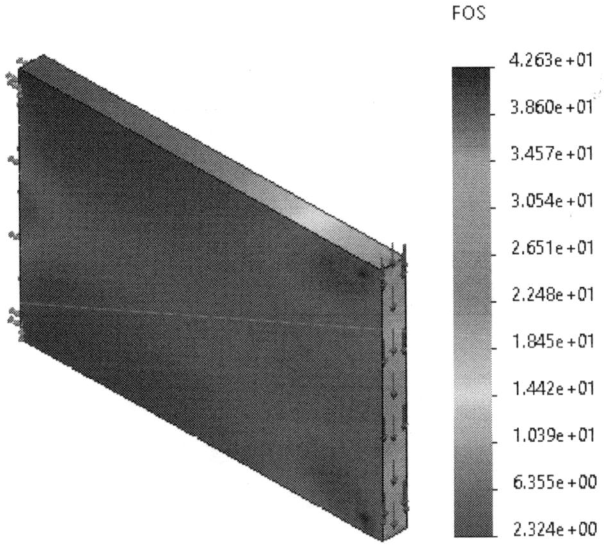

Figure 4-42 *The FOS distribution plot*

The plot results show that the minimum FOS value for this study is 2.3. So you need to try and keep the next study FOS equal to or greater than 2.3.

Creating Case Study-2

In SOLIDWORKS Simulation, you can create multiple studies. The results of the existing study can be copied to the new study. In this session, you need to create duplicate of Case Study-1 to create a new case study.

1. Right-click on the **Case Study-1** tab below the drawing area and choose the **Copy Study** option from the shortcut menu; the **Copy Study PropertyManager** is displayed, as shown in Figure 4-43.

2. Enter **Case Study-2** in the **Study name** edit box and select **Cut Dia. 65** in the **Configuration to use** drop-down list.

3. Select the **Static** button from the **Target Study** rollout and choose the **OK** button from the **Copy Study PropertyManager**; Case Study-2 tab is created.

4. Right-click on the **Case Study-2** node in the Simulation Tree and choose **Activate SW Configuration** from the shortcut menu displayed.

5. Next, un-suppress the cut feature from the Design Tree, created earlier for cut dia 65.

The configuration Cut Dia 65 is activated but some of the nodes contain exclamation symbols indicating errors. These errors are due to the change in the model. To remove these errors, you need to remesh and run the study again.

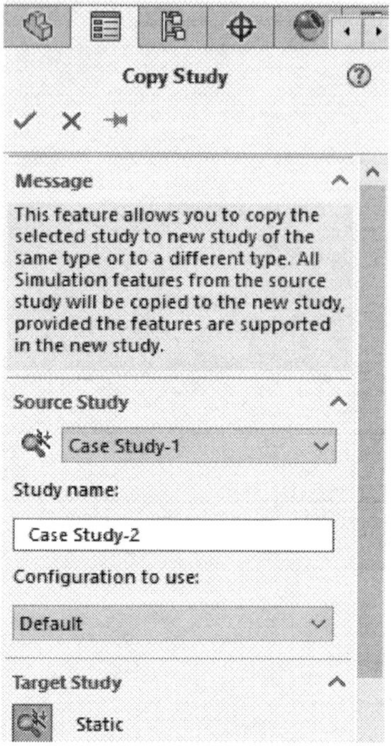

*Figure 4-43 The **Copy Study PropertyManager***

6. Right-click on the **Mesh** node and choose **Create Mesh** from the shortcut menu; the **Mesh PropertyManager** is displayed.

7. Choose the **OK** button and go with the default values of **Mesh PropertyManager**; the **Mesh Progress** dialog box is displayed and meshing gets completed.

8. Right-click on the Case Study-2 node and choose the **Run** option from the shortcut menu displayed; the **Case Study-2** window is displayed and the solution process starts .

   When the analysis is complete, a results node is added in the Simulation Tree. Also, the analyzed model is displayed with von Mises stress plot, as shown in Figure 4-44.

Finding Factor of Safety

Now, you need to find FOS.

1. Follow the same steps as mentioned in the Case Study-1; the FOS plot result is displayed, as shown in Figure 4-45.

   Since in Case Study-2, FOS value is 1.8 which is less than the FOS value of Case Study-1, so you need to try one more study with cut dia of 50 mm in the center of the plate.

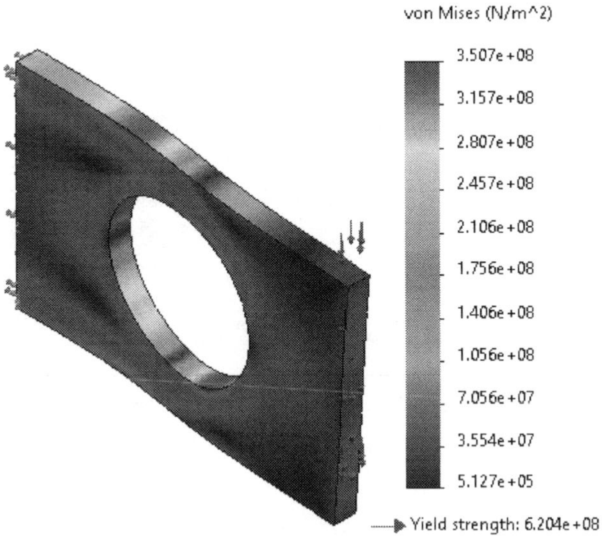

Figure 4-44 *Visualization of model with von Mises stress plot*

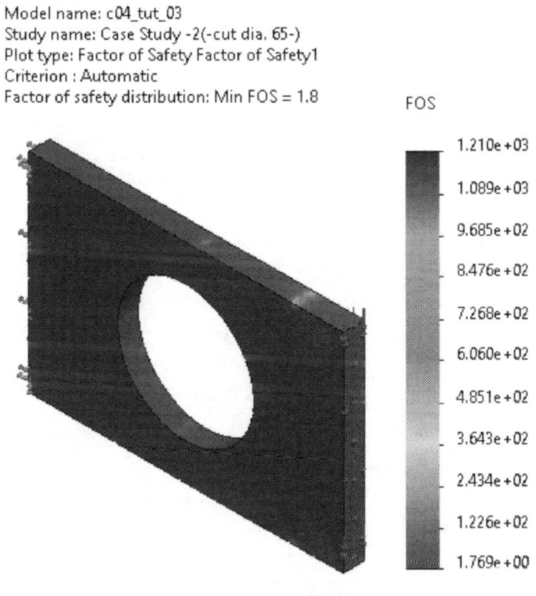

Figure 4-45 *The FOS distribution plot*

Creating Case Study-3

Now, you need to perform static study as Case Study 3 for cut dia. 50.

1. Follow the steps mentioned in Case Study-2 and then find Stress and FOS plots; the stress and FOS plots obtained in Case Study-3 are displayed in Figure 4-46 and Figure 4-47, respectively. In Case Study-3, FOS value is 2.58 which is close to the FOS value of Case Study-1.

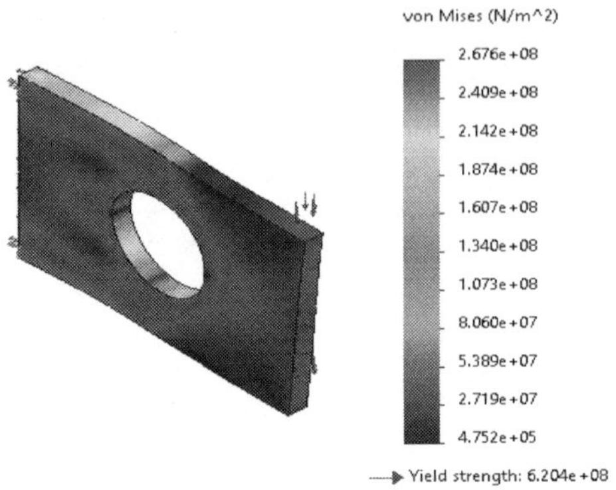

Figure 4-46 *Visualization of model with Mises stress plot*

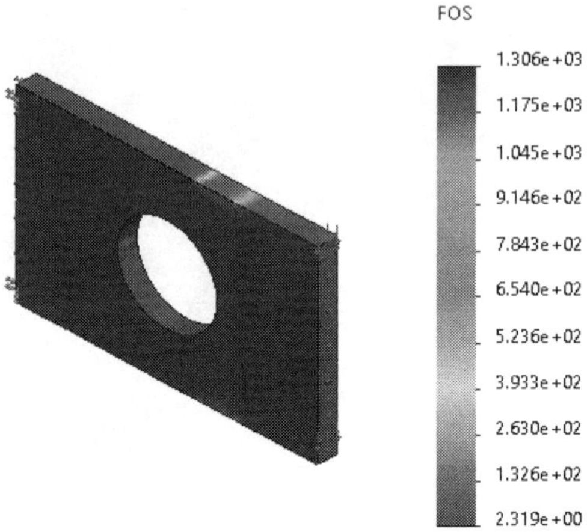

Figure 4-47 *The FOS distribution plot*

Saving the Studies

1. Choose the **Save** button in the menu bar or press **Ctrl+S** to save all the studies.

Comparing Results of FOS

Next, you need to compare all these three studies and find out which model will be better for this working condition.

1. Choose the **Compare Results** button in the **Simulation CommandManager**; the **Compare Results PropertyManager** is displayed, as shown in Figure 4-48.

2. Select the **All studies in all configurations(slower)** radio button; results of all the studies in all configurations is displayed in the **Compare Results PropertyManager.**

3. Select only the **Factor of Safety** check box in the results of all the three case studies and choose the **OK** button; the results of all the studies are displayed, as shown in Figure 4-49.

   Note that the minimum FOS for Case Study 1 and Case Study 3 are closer to each other. However, the weight of the Alloy Steel Plate in Case Study 3 is less than Case Study 1. Therefore, Case Study 3 is more acceptable based on the design optimization point of view.

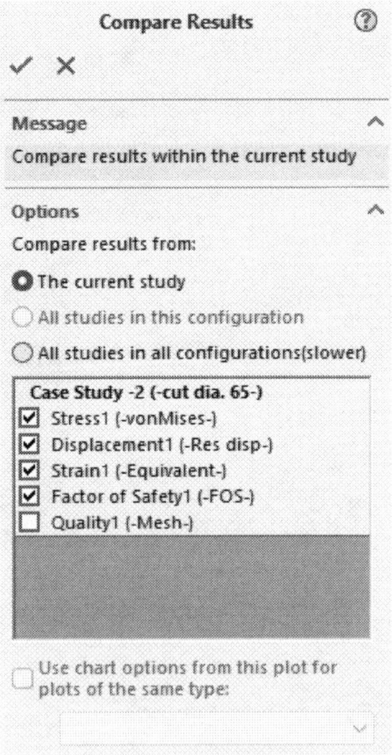

Figure 4-48 *The* ***Compare Results PropertyManager***

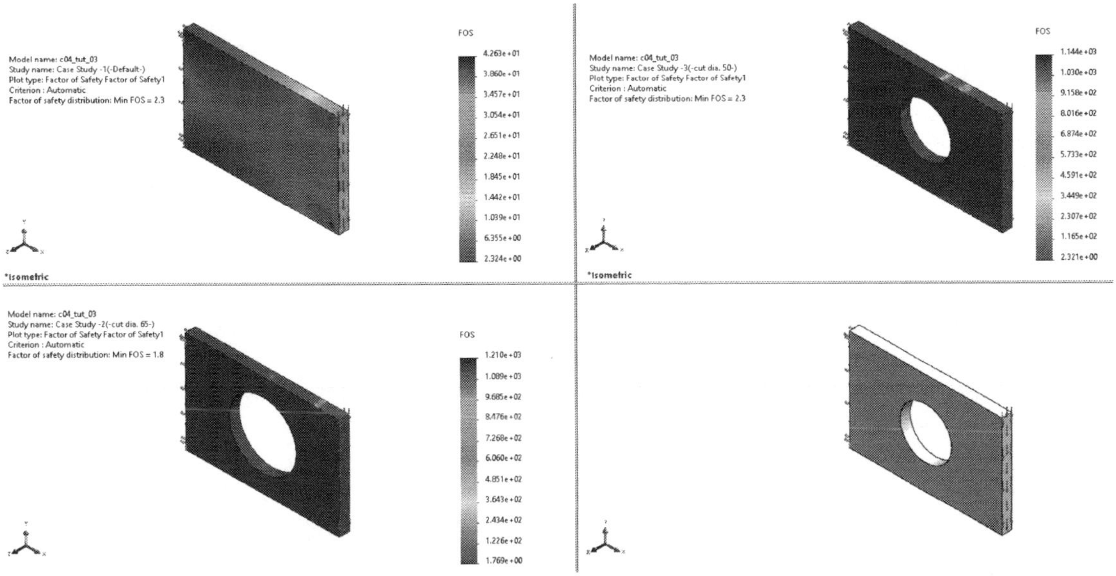

Figure 4-49 The FOS results plot for all studies

Tutorial 4

In this tutorial, you will perform static analysis on a frame structure. The dimensions, boundary conditions, and load applied on the frame are shown in Figure 4-50. The frame to be built is made up of hollow aluminum (5052-H38) tubing of 0.5 inch and 40 schedule.

(Expected time: 1 hr)

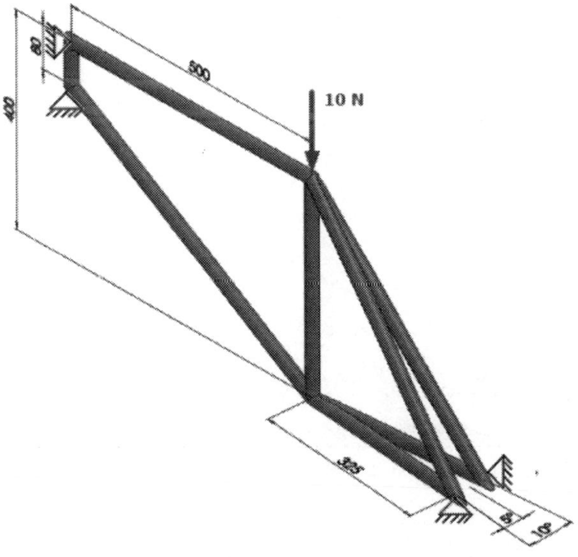

Figure 4-50 Model of the frame

The following steps are required to complete this tutorial:

a. Create the model for analysis.
b. Start a new static study.
c. Apply material, constraints, and loads.
d. Create mesh and run the analysis.
e. View the results.

Creating Model for Analysis

1. Start a new SOLIDWORKS part document and then invoke the Sketch flyout **3D Sketch** tool from the **Sketch** flyout of the **Sketch CommandManager**, refer to Figure 4-51; the 3D sketch environment is invoked.

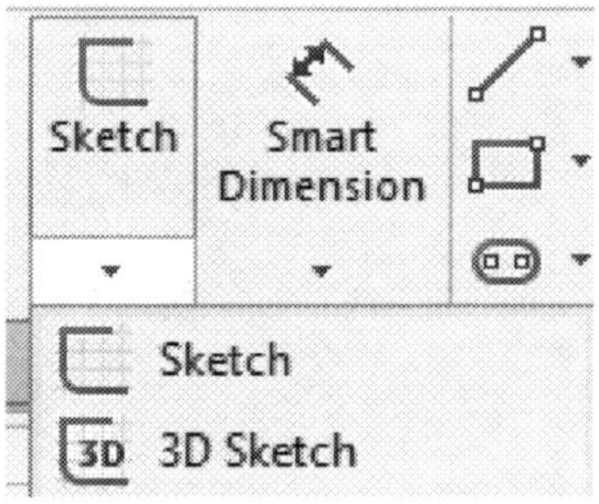

Figure 4-51 *The* ***Sketch*** *flyout*

2. Create the 3D sketch frame of the bicycle model, as shown in Figure 4-52. Next, exit the 3D sketch environment. You can also download the 3D sketch frame of the bicycle model from *www.cadcim.com.*

Note
You can create 3D sketch by using the tools available in the modeling environment of SOLIDWORKS.

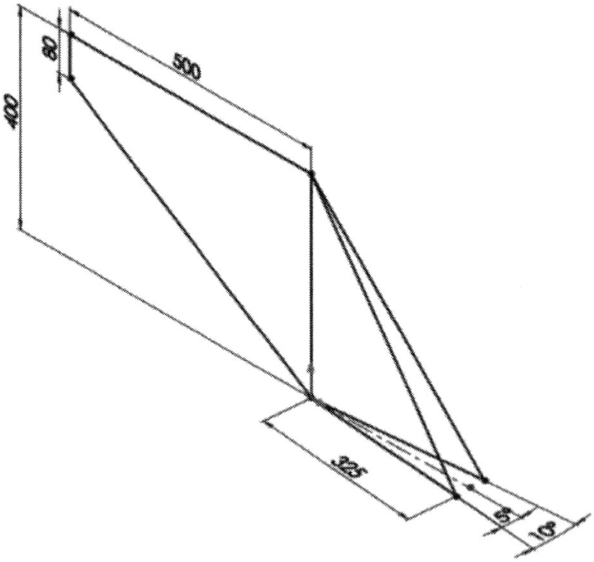

Figure 4-52 *The 3D sketch frame*

3. Invoke the **Weldments CommandManager** by choosing the **Weldments** tab from the **CommandManager.**

Note
If the ***Weldments*** *tab is not added to the* ***CommandManager*** *by default, you can customize the CommandManager.*

SOLIDWORKS provides you with the options to create a weldment structure in which the members are grouped together to form a complex structure. In SOLIDWORKS, the weldment structures are created by using the tools available in the **Weldments CommandManager**. You can use the 2D and 3D sketches to define the basic framework for creating the weldment structure.

After creating the 3D sketch frame of the model, you need to create the weldment structure by using the **Structural Member** tool of the **Weldments CommandManager.**

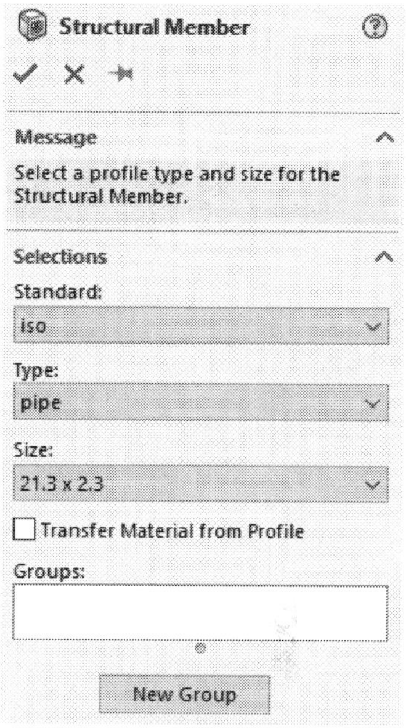

Figure 4-53** The **Structural Member PropertyManager

4. Choose the **Structural Member** tool from the **Weldments CommandManager**; the **Structural Member Property Manager** is displayed, as shown in Figure 4-53.

   The **Structural Member** tool is used to create structural members by sweeping a pre-defined profile along the user-defined path.

5. Select **ansi inch** from the **Standard** drop-down list of the **Selections** rollout in the PropertyManager.

6. Select **pipe** from the **Type** drop-down list and **0.5 sch 40** from the **Size** drop-down list of the **Selections** rollout of the PropertyManager.

7. Click in the **Groups** selection box of the PropertyManager to activate it. Next, select the first set of entities as the entities of the first group, refer to Figure 4-54; a preview of the structural members is displayed in the drawing area, as shown in Figure 4-55. Also, a group is created in the **Groups** selection box of the PropertyManager.

8. Make sure that the **Apply corner treatment** check box is selected in the **Settings** rollout of the **Structural Member PropertyManager**. Next, choose the **End Miter** button available below the **Apply corner treatment** check box, if not selected.

9. Clear the **Merge miter trimmed bodies** check box of the **Settings** rollout. Next, set the value in the **Gap between Connected Segments** in **Same Group** edit box to **0.**

Note

In SOLIDWORKS, you can create a weldment structure that contains one or more groups. All the members of a group are treated as a single unit. You can select parallel or continuous line segments in a group. Also, the weldment profile of all the groups within a structural member feature remains same.

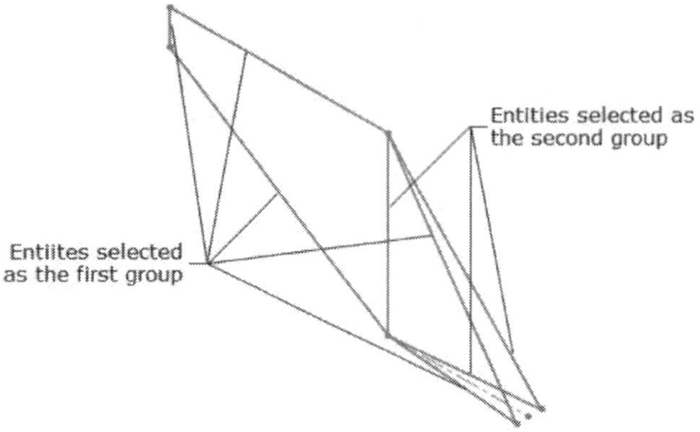

Figure 4-54 *Different sets of entities to be selected*

Figure 4-55 *Preview of the structural members*

10. Choose the **New Group** button from the **Structural Member PropertyManager** to create a new group of structural members; the new group is added to the **Groups** selection box of the PropertyManager.

11. Select the other set of entities as the entities of the second group, refer to Figure 4-54; a preview of the structural members is displayed in the drawing area.

12. Make sure that the **Apply corner treatment** check box is selected in the **Settings** rollout of the **Structural Member PropertyManager**. Next, choose the **End Miter** button available below the **Apply corner treatment** check box.

13. Clear the **Merge miter trimmed bodies** check box in the **Settings** rollout. Next, enter **0** as the value in the **Gap between Connected Segments in Same Group** edit box.

14. Choose the **OK** button from the **Structural Member PropertyManager**; the weldment structure is created, as shown in Figure 4-56.

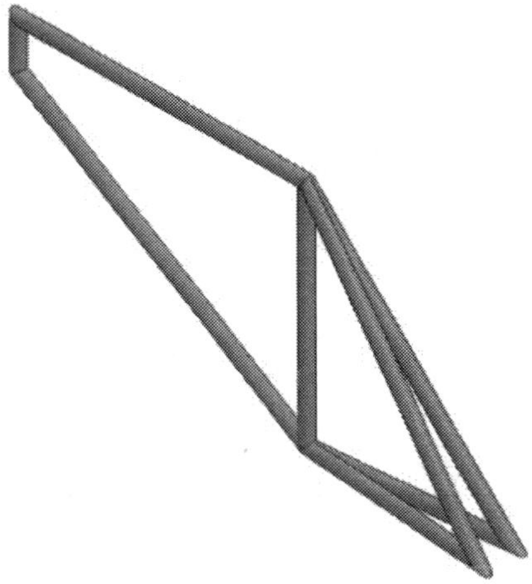

Figure 4-56 *The weldment structure of the bicycle frame*

Starting a New Study

Now, you need to start a new analysis study.

1. Choose the **Simulation** tab from the **CommandManager**; the **Simulation CommandManager** is displayed.

2. Choose the **New Study** tool from the **New Study** flyout in the **Simulation CommandManager;** the **Study PropertyManager** is displayed.

3. Choose the **Static** button if it is not chosen by default and then enter the name of the study as **Bicycle Frame** in the **Study name** edit box of the PropertyManager. Next, choose **OK**; a new Simulation Tree named **Bicycle Frame** is added below the **FeatureManager Design Tree**. Also, joints are displayed in the drawing area, as shown in Figure 4-57.

   A joint is displayed at the free end of a structural member or at the intersection of two or more structural members.

Note

Pink color joints will be displayed at the intersection of two or more structural members and light green color joints will be displayed in a structural member at its end and mid points.

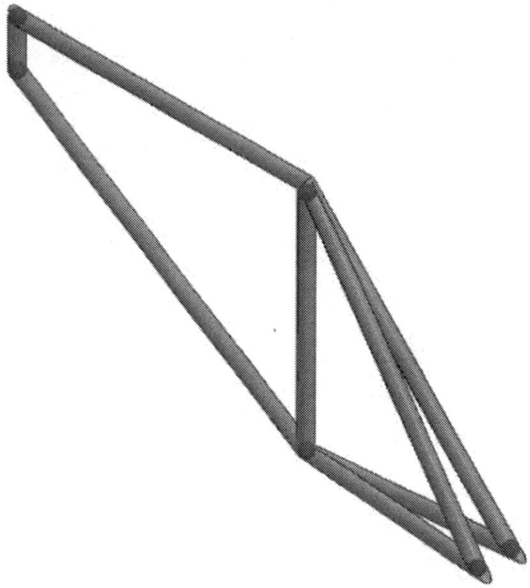

Figure 4-57 *The weldment structure with joints*

Applying the Material

Now, you need to apply the aluminium material to the frame.

1. Choose the **Apply Material** tool from the **Simulation CommandManager**; the **Material** dialog box is displayed.

2. Expand the **SOLIDWORKS Materials** node in the **Material** dialog box. Next, expand the **Aluminium Alloys** node under the **SOLIDWORKS Materials** node.

3. Select the **5052-H38** material; all the properties of the selected material are displayed on the right in the dialog box.

4. Choose the **Apply** button from the dialog box; the selected material is applied to the model. Next, close the **Materia**l dialog box.

Applying Constraints

1. Choose the **Fixed Geometry** tool from the **Fixtures Advisor** flyout of the **Simulation CommandManager**; the **Fixture PropertyManager** is displayed.

2. Make sure that the **Fixed Geometry** button is chosen in the **Standard (Fixed Geometry)** rollout of PropertyManager.

3. Select four joints one by one from the drawing area, refer to Figure 4-58; a symbol of fixed geometry is displayed on the joints. Next, choose the **OK** button from the PropertyManager; the degrees of freedom of the selected joints is restrained. Figure 4-59 shows the weldment structure after applying the constraints.

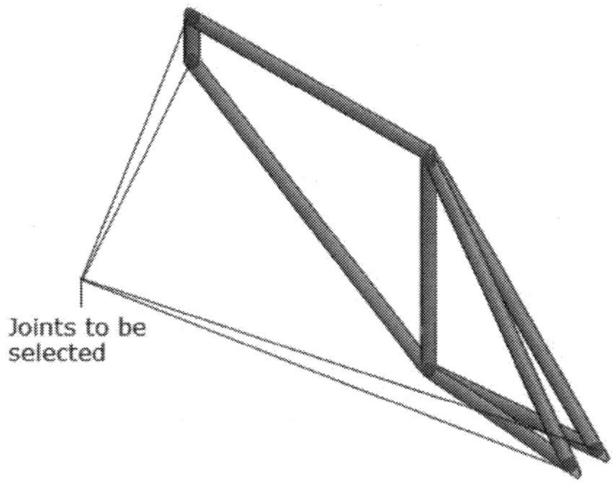

Figure 4-58 Joints to be selected

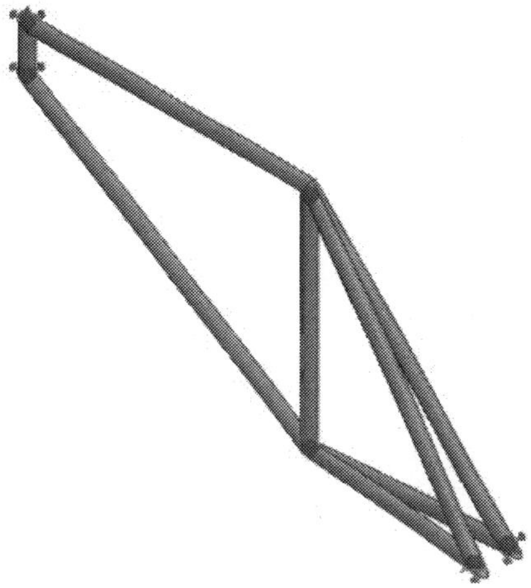

Figure 4-59 The weldment structure after applying the fixed constraint

Applying Loads

After applying the material and constraints to the model, you need to apply loads.

1. Choose the **Force** button from the **External Loads Advisor** flyout of the **Simulation CommandManager**; the **Force/Torque PropertyManager** is displayed.
2. Choose the **Joints** button available at the left of the **Joint** selection box of the **Selection** rollout of the PropertyManager.

3. Select the joint displayed on the top of the middle vertical member of the frame from the drawing area to apply the force.
4. Click on the **Face, Edge, Plane for Direction** selection box in the **Selection** rollout of the PropertyManager to activate it.

   Now, you need to specify the direction of force.

5. Select the **Top Plane** from the **FeatureManager Design Tree** and then select **SI** from the **Unit** drop-down list in the **Units** rollout in the PropertyManager. You can also select the vertical member to define the direction of the load.
6. Choose the **Normal to Plane** button from the **Force** rollout of the PropertyManager; the **Normal to Plane** edit box is enabled.

7. Enter **10** in the **Normal to Plane** edit box of the **Force** rollout. Next, select the **Reverse Direction** check box to reverse the direction of force to vertically downward.
8. Choose the **OK** button from the PropertyManager. Figure 4-60 shows the structural frame after applying the load.

Creating Mesh

After applying the boundary condition and load, you need to create mesh.

1. Choose the **Create Mesh** tool from the **Run This Study** flyout of the **Simulation CommandManager**; the **Mesh PropertyManager** is displayed. 

2. Accept all the default settings and choose the **OK** button from the PropertyManager; the **Mesh Progress** window is displayed. Once the process of meshing is done, the meshed model is displayed in the drawing area, as shown in Figure 4-61.

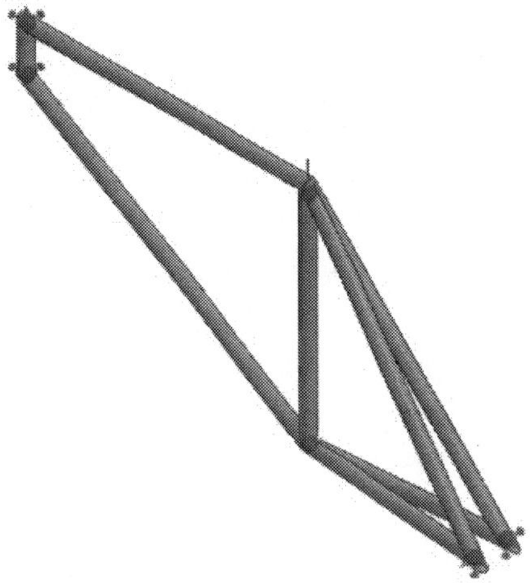

Figure 4-60 *Structure after applying the load*

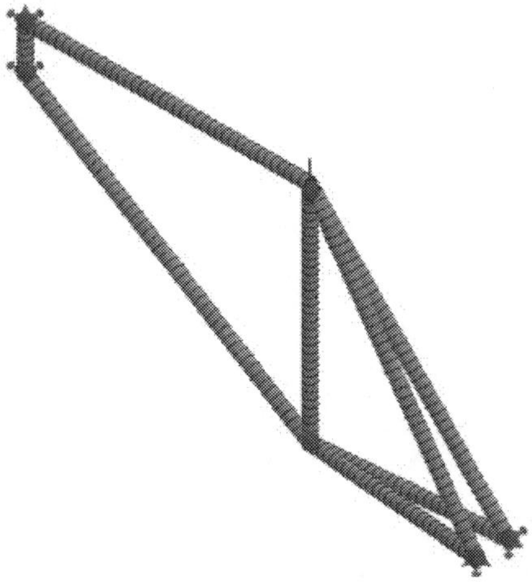

Figure 4-61 *Structure after meshing*

Running the Study

Now, you need to run the analysis on the model.

1. Choose the **Run this Study** button from the **Simulation CommandManager**; the **Bicycle Frame** window is displayed and the solution process is started. Once the analysis is completed, a new node named **Results** is added to the **FeatureManager Design Tree**. Also, the resultant structure is displayed in the drawing area with the stress contour, as shown in Figure 4-62.

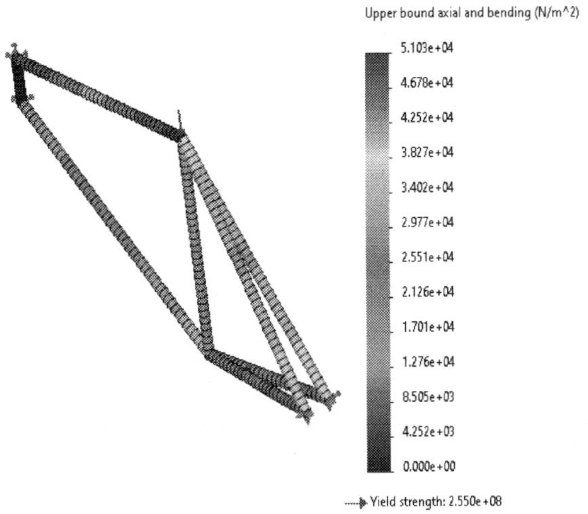

***Figure 4-62** The weldment structure with stress contour*

Viewing the Results

Viewing the stress and displacement contours is same as discussed in the previous tutorials. In this tutorial, you need to find out the safety factor of the design.

1. Right-click on the **Results** node; a shortcut menu is displayed. Next, choose the **Define Factor Of Safety Plot** option from it; the **Factor of Safety PropertyManager** is displayed, as shown in Figure 4-63. By default, the **Automatic** option is selected in the **Criterion** drop-down list of the **Step 1 of 3** rollout in the PropertyManager.

   The **Automatic** option as the failure criteria for safety is selected in the **Criterion** drop-down list. The **Automatic** option is the most appropriate failure criterion across all element types.

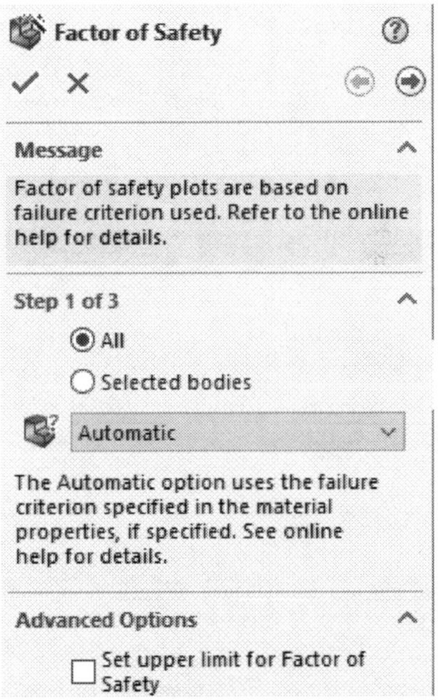

Figure 4-63** The **Factor of Safety PropertyManager

2. Accept the default settings and choose the **Next** button from the PropertyManager; the **Step 1 of 3** rollout is replaced by the **Step 2 of 3** rollout in the PropertyManager.

   By default, the value in the **Multiplication factor** edit box in the **Step 2 of 3** rollout is set to 1.

   The **Multiplication factor** edit box allows you to specify a multiplication factor for calculating the selected stress limit. For example, if you apply a factor of 0.6 to the yield stress limit of 2000 psi, the factor of safety will be 0.6 x 2000 = 1200 psi for the stress limit.

3. Accept the default settings and then choose the **Next** button to go to step 3. Choose **OK** to accept and exit the PropertyManager; the factor of safety plot is displayed, as shown in Figure 4-64.

Saving the File

After all the actions are performed and the desired solution is achieved, you need to save the model.

1. Choose **File > Save** from the menu bar to save the simulation files.

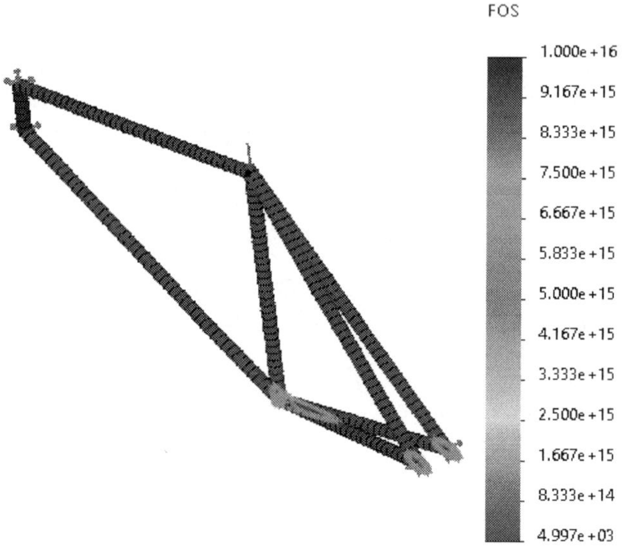

***Figure 4-64** The weldment structure with FOS*

Self-Evaluation Test

Answer the following questions and then compare them to those given at the end of this chapter:

1. The __________ tool is used to determine the Equivalent Stress.

2. You can specify the magnitude of the force load in the __________ edit box.

3. When you choose the **New Study** tool from the **Study Advisor** flyout in the **Simulation CommandManager**, the __________ is displayed.

4. In SOLIDWORKS Simulation, you cannot apply Fixed support on the cylindrical surface. (T/F)

5. In SOLIDWORKS Simulation, you can apply supports on multiple surfaces. (T/F)

6. Static Study calculates stresses, strains, displacements, and reaction forces in your model. (T/F)
7. If the value of Factor of Safety comes out as 0.5, the design is safe. (T/F)
8. An analysis study represents a scenario of analysis type, materials, loads, and fixtures. (T/F)
9. Analysis is a process in which the field performance of design is simulated. (T/F)
10. The main steps involved in performing static study are: create a study, assign materials, apply fixtures, apply loads, mesh the model, run the analysis, and visualize the results. (T/F)

Review Questions

Answer the following questions:

1. When you choose the _________ button from the **Run This Study** flyout of the **Simulation CommandManager**, the **Mesh PropertyManager** is displayed.
2. When you choose the _________ button from the **Simulation CommandManager**, a Simulation Solver window is displayed and the solution process is started. Once the study is completed, a new node named **Results** is added to the Simulation Tree.
3. The von-mises Stress is the resultant of stresses along all axes. (T/F)
4. By applying Fixed support to a part, you can restrict its few degrees of freedom. (T/F)
5. The options available in the **Joint selection** area of the **Selection r**ollout of the **PropertyManager** are used to apply joints in the analysis. (T/F)

EXERCISES

To perform the exercises of this chapter, download the input file from *www.cadcim.com*. The complete path for downloading the file is:

Textbooks > CAE Simulation > SOLIDWORKS Simulation > SOLIDWORKS Simulation 2022: A Tutorial Approach > Input Files > c04_simulation_2022_input.zip

Extract and save the downloaded **c04_simulation_2022_input** zipped file at the location *C:\SOLIDWORKS Simulation\c04.*

Exercise 1

In this exercise, you will perform the linear static analysis. To perform the analysis, open the file *c04_exr01* that you have downloaded earlier and then assign AISI 1045 Steel cold drawn material from the SOLIDWORKS material library to the model. Figure 4-65 and Figure 4-66 show the model, dimensions and boundary conditions, respectively.

(Expected time: 45 min)

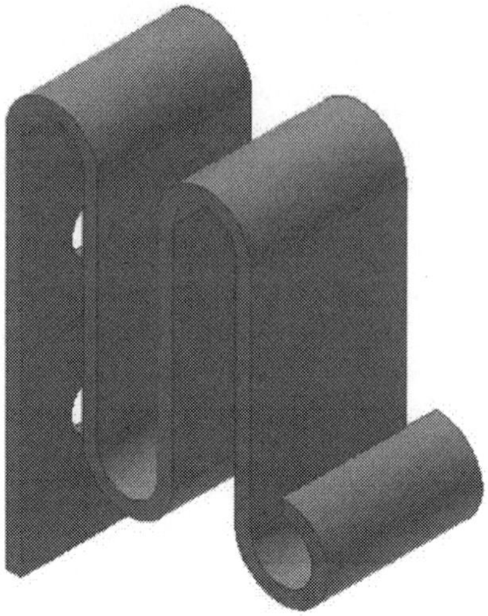

Figure 4-65 *The model for Exercise 1*

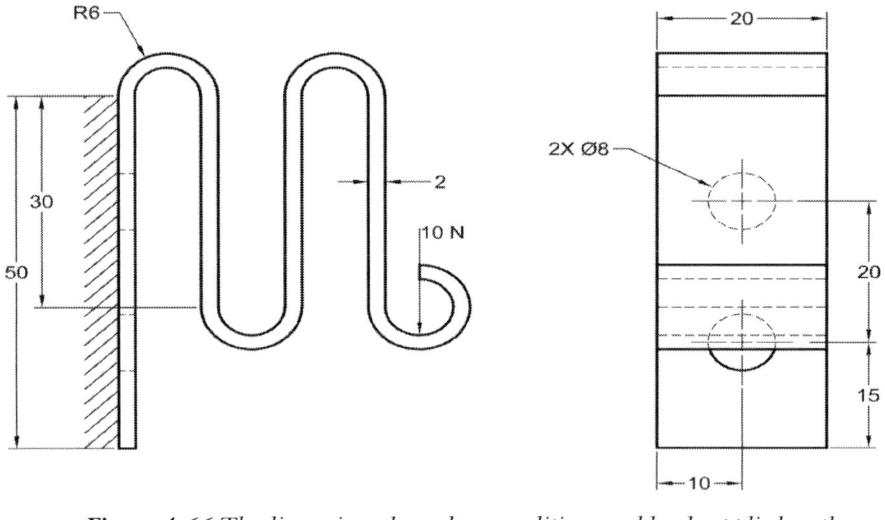

Figure 4-66 *The dimensions, boundary condition, and loads applied on the model*

Exercise 2

In this exercise, you will use *c04_exr02* file to perform the static analysis on the model shown in Figure 4-67. The boundary conditions and load applied on the model are also shown in Figure 4-67. The model is made up of alloy steel material. The load applied on the split face of the model is 50 N.

(Expected time: 45 min)

Figure 4-67 *The model for Exercise 2*

Answers to Self-Evaluation Test

1. **Stress**, **2.** **Force Value**, **3.** **Study PropertyManager**, **4.** F, **5.** T, **6.** T, **7.** T, **8.** T, **9.** T, **10.** T

Chapter 5

Advanced Structural Analysis

Learning Objectives

After completing this chapter, you will be able to:

- *Perform multi body static structural analysis*
- *Perform fatigue analysis*
- *Perform buckling analysis*
- *Generate the result of an analysis*

INTRODUCTION

Structural analysis is a process to analyze a structural system to predict its responses and behaviors by using physical laws or conditions and mathematical equations. The main objective of structural analysis is to determine the internal forces, stresses, and deformations of structures under various load effects. In other words, in structural analysis, mathematical processes are used to extract a structure's response under service and seismic loads. Various types of structural analysis are discussed in this chapter such as fatigue analysis and buckling analysis.

FATIGUE ANALYSIS

Fatigue analysis is used to evaluate the life expectancy of a model under repeating stress conditions where the actual load on the part is less than the yield strength. In such structures, however, the oscillating stresses at lower magnitudes may cause shear stress in the micro-structure of the part. This results in planes of the structure slipping over each other, thereby creating micro-fissures which grow into cracks. Therefore, this analysis is a crucial step of any production process because it predicts failure in the structure for the long term.

SOLIDWORKS Simulation provides tools to simulate different types of loading stresses which can be constant (Constant Amplitude Stress Event) or varying in load (Variable Amplitude Stress Event). These events are discussed next.

Constant Amplitude Stress Event

All cycles of a constant amplitude event have the same alternating and mean stresses. A constant amplitude fatigue event is fully defined by an alternating stress, by a mean stress (or stress ratio), and by the number of cycles.

Variable Amplitude Stress Event

A variable amplitude fatigue event is a load record that defines the fluctuation history of the load. In case of single variable amplitude event, the time does not play any role for fatigue studies and it is adequate to define the amplitude sequence. For studies with more than one variable amplitude fatigue events, time value is required for each record to correlate with other loading events.

BUCKLING ANALYSIS

Buckling refers to sudden large displacements due to axial loads. Slender structures that are subject to axial loads can fail due to buckling at load levels lower than those required to cause material failure. Buckling can occur in different modes under the effect of different load levels. Buckling studies can help to avoid failure due to buckling. Usually, the lowest buckling load is of interest, and it is lower than those required to cause material failure.

When a force F is applied axially on a structure and the structure is slender, it might bend laterally with respect to the force applied. The measurement of slenderness is defined by l/r which is the structure or column's length divided by its cross-sectional radius. This sudden sideways failure is called buckling and is caused at loads less than the ultimate strength of the material. The load at which buckling occurs depends on the material's stiffness. This analysis type is conducted for slender-walled members like steel trusses, cranes, cantilevers, and so on.

In Finite Element Method (FEM), Euler's formula is utilized for column analysis. The formula defines different load factors for several support conditions. It further implies that the effective load applied varies greatly due to the nature of fixtures applied, which can be fixed supports, pin supports, or simple/free supports.

$$F = \frac{\pi^2 EI}{(KL)^2}$$

In the above Euler's formula, F indicates the critical force, E indicates the modulus of elasticity, I indicates the area moment of inertia, L refers to the unsupported length of column, and K refers to the column effective length factor which depends upon the type of supports.

Support 1	Support 2	K Value
Fixed	Fixed	0.5
Fixed	Free	2.0
Fixed	Pinned	0.7071
Pinned	Pinned	1.0

TUTORIALS

To perform the tutorials, you need to download the input files from *www.cadcim.com*. The complete path for downloading the files is as follows:

Textbooks > CAE Simulation > SOLIDWORKS Simulation > SOLIDWORKS Simulation 2022: A Tutorial Approach > Input Files > c05_simulation_2022_input.zip

Extract the downloaded zipped file and save it.

Tutorial 1

In this tutorial, you will open the *c05_tut01* file that was downloaded earlier and perform the static structural study on a rim model, refer to Figure 5-1. The boundary conditions are given in Figure 5-2 and the model is made up of **Zinc AC 41A Alloy, as Cast material**. Also, to understand the importance of geometry optimization, you will optimize the model, and then evaluate the results. **(Expected time: 1.5 hr)**

Figure 5-1 Full model of the rim

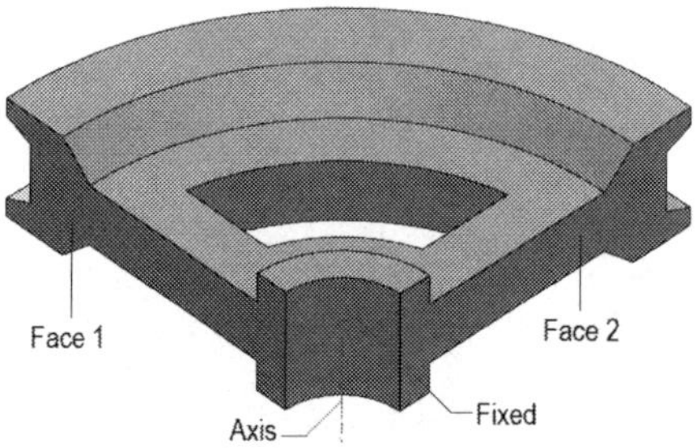

Figure 5-2 1/4th model of the rim

The following steps are required to complete this tutorial:

a. Start SOLIDWORKS Simulation and import the model.
b. Start a new static study.
c. Apply material, fixture, and load.
d. Create mesh.
e. Run the study.
f. View the results.
g. Create second case study.
h. Create third case study.
i. Compare the results.

Starting SOLIDWORKS Simulation and Importing the Model

Now, you need to import the model of pully for analysis.

1. Choose **Open** from the **File** menu in the **SOLIDWORKS** menus; the **Open** dialog box is displayed.

2. Select the **c05_tut01** (Pully model) and choose the **Open** button from the dialog box; the selected model is opened in SOLIDWORKS.

Starting Simulation Study

Start the study with default configuration of the model.

1. Select the **New Study** option from the **New Study** flyout of the **Simulation CommandManager**, refer to Figure 5-3; the **Study PropertyManager** is displayed. Alternatively, you can invoke the **Study PropertyManager** by selecting the **Study** option from the **Simulation > Study** in the **SOLIDWORKS** menu.

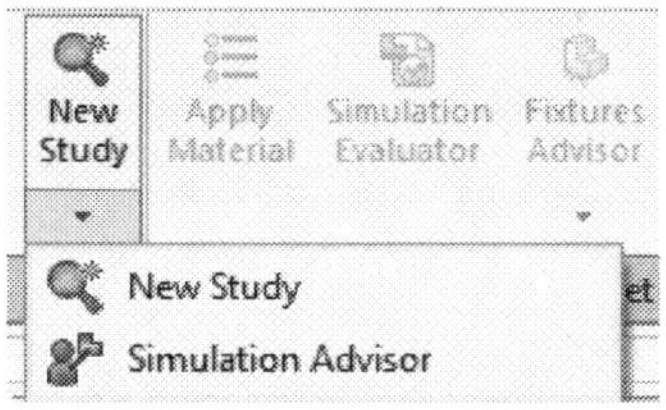

Figure 5-3 The ***New Study*** *fllyout*

2. Select the **Static** button from the **Study PropertyManager** and enter study name as **Static Study.** Next, choose OK; the Simulation Tree is displayed, refer to Figures 5-4 and 5-5.

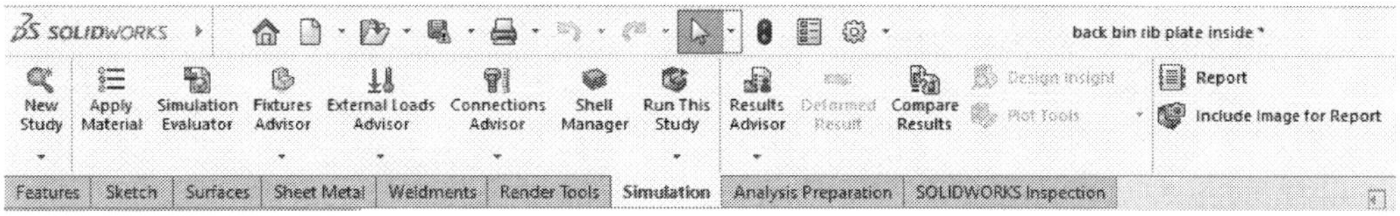

Figure 5-4 The ***Simulation CommandManager***

Applying Material

Now, you need to apply Zinc Alloy material to model.

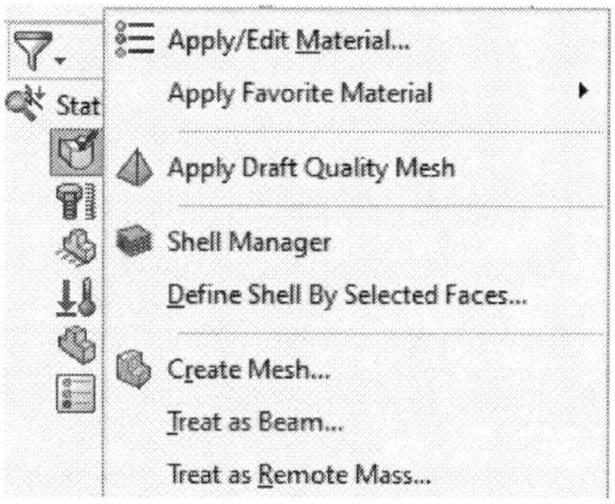

Figure 5-5 The Simulation Tree

1. Right-click on the **c05_tut01** part node in Simulation Tree, refer to Figure 5-5, and choose the **Apply/Edit Material** option from the shortcut menu; the **Material** dialog box is displayed.

2. Expand the **SOLIDWORKS Materials** node in the **Material** dialog box. Next, expand the **Zinc Alloys** sub-node under it and choose the **Zinc AC41A Alloy, As Cast** material; all the properties of the selected material are displayed under the **Properties** tab of the dialog box.

3. Choose the **Apply** button from the dialog box; the zinc alloy material is applied to the model.

4. Next, choose the **Close** button to close the **Material** dialog box.

Applying Restraints

The types of loads and restraints available depend on the type of the study. A load or restraint is applied to make the model stable by using the corresponding PropertyManager. You can use symmetry condition to 1/4th portion of the model instead of the full model. The results on the un-modeled portions are deducted from the modeled portion. If required, you can take advantage of symmetrical approach to help you reduce the size of a problem and obtain more accurate results.

1. Right-click on the **Fixtures** node in the Simulation Tree, refer to Figure 5-6, and choose the **Advanced Fixtures** option from the shortcut menu; the **Fixture PropertyManage**r is displayed, as shown in Figure 5-7.

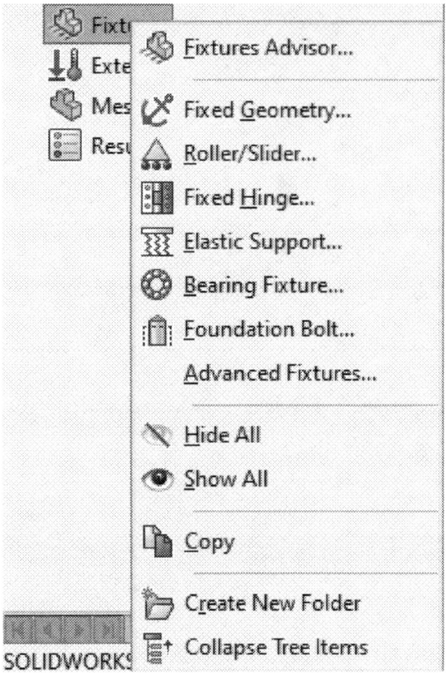

Figure 5-6 *The shortcut menu displayed*

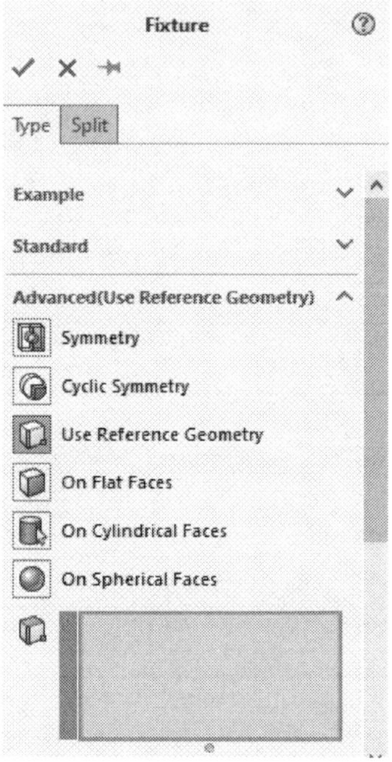

Figure 5-7 *The* ***Fixture PropertyManager***

2. Choose the **Cyclic Symmetry** button from the **Advanced** (Use Reference Geometry) rollout.

3. Select the faces, as shown in Figure 5-8, in the **Selection(Face)** boxes. Note that Face 1 should be in the first Selection (Face) box and the Face 2 should be in the second Selection (Face) box.

4. Next, select the axis of the model under the **Axis** selection. If you are not be able to select the axis from the graphics window, select it from the Design Tree. Next, choose the **OK** button from the **Fixture PropertyManager** to close the PropertyManager.

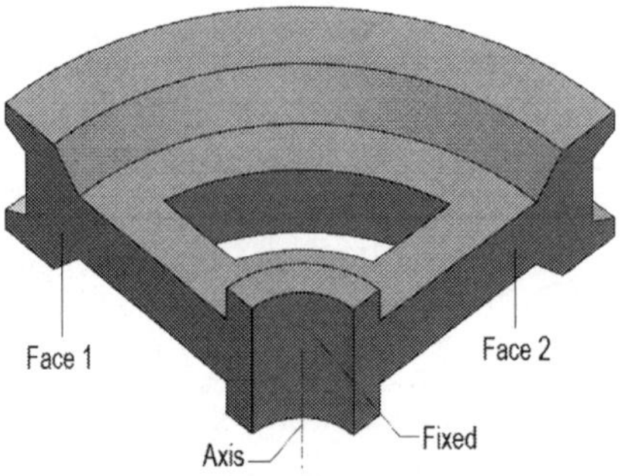

Figure 5-8 *The faces and axis to be selected*

Now, you need to apply the Fixed Geometry restraints to the circular face of the pully.

5. Right-click on the **Fixture** node and then choose the **Fixed Geometry** option from the shortcut menu; the **Fixture PropertyManager** is displayed.

6. Select the mentioned circular face, as shown in Figure 5-8 (circular surface of the hole) in the **Faces, Edges, Vertices** for Fixture selection box and then choose the **OK** button from the PropertyManager.

Applying Loads

Now, you need to apply 100 Nm torque on the circular face of the model about the axis.

1. Right-click on the **External Loads** node in the Simulation Tree and choose the **Torque** option from the shortcut menu, refer Figure 5-9; the **Force/Torque PropertyManager** is displayed.

2. Choose the **Torque** button from the **Force/Torque** rollout if it is not chosen; the selection boxes corresponding to **Faces for Torque** and **Axis Cylindrical Face for Direction** are displayed.

3. Select the mentioned face under the **Face for Torque** selection box and then select the axis under the **Axis, Cylindrical face for direction** selection box, refer to Figures 5-10. Figure 5-11 shows the face and axis to be selected.

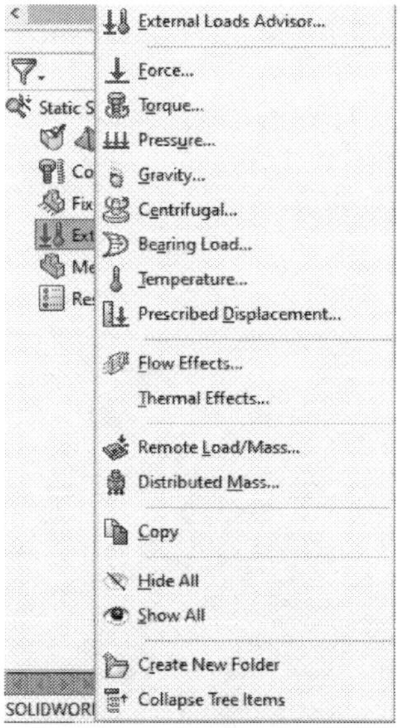

Figure 5-9 The shortcut menu

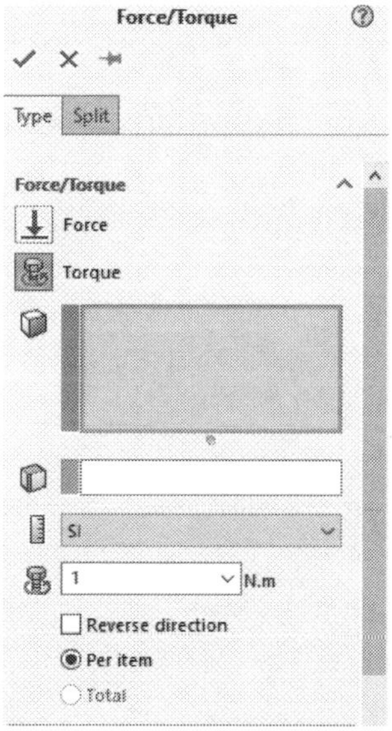

Figure 5-10 The Force/Torque PropertyManager

4. Select the **SI** unit system from the **Unit** drop-down list and then enter **100** in the **Torque Value** edit box of the **Force/Torque PropertyManager**. If required, you can change the direction of the applied torque by using the **Reverse direction** check box.

5. Choose the **OK** button to close the PropertyManager.

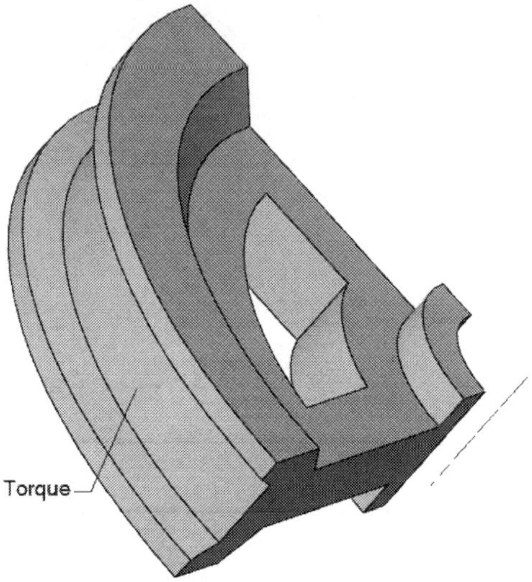

***Figure 5-11** The face and axis to be selected for applying torque*

Creating Mesh

Now, you need to create mesh to divide the model into small pieces.

1. Right-click on the **Mesh** node in the Simulation Tree and choose the **Create Mesh** option from the shortcut menu; the **Mesh PropertyManager** is displayed, as shown in Figure 5-12.

2. Accept the default settings and choose the **OK** button from the PropertyManager; a progress window is displayed and mesh is created on the model, refer to Figure 5-13. Note that if you change the mesh density, the result you get will change.

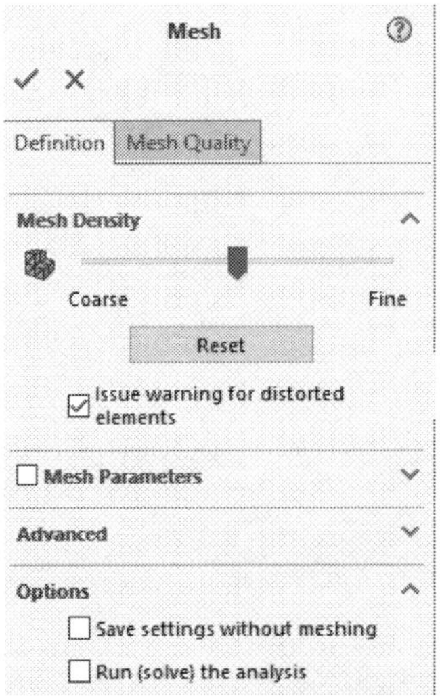

Figure 5-12** The **Mesh PropertyManager

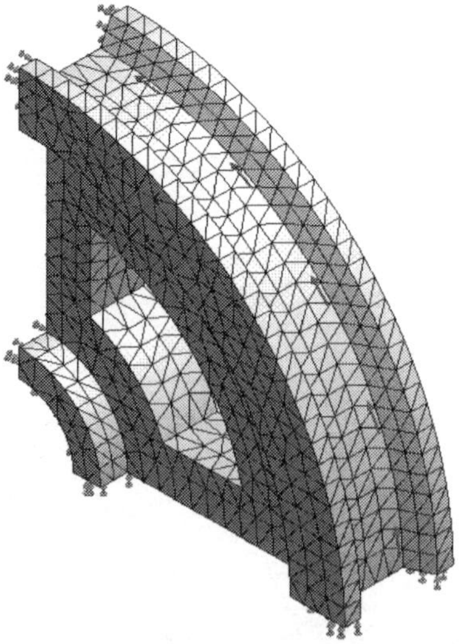

***Figure 5-13** Model after meshing*

Running the Study

Now, you need to run the simulation study to get results based on the input given to the model.

1. Right-click on the **Static Study** node in the Simulation Tree and choose the **Run** option from the shortcut menu; the **Static Study** Process window is displayed and the solution process starts.

   When the solution process is complete, a node **Results** is added in the Simulation Tree. Also, the resultant model is displayed with von-Mises stress plot, as shown in Figure 5-14. You can also plot the von-Mises stress result for the full model.

2. To do so, select the **Stress1(-vonMises-)** sub-node under the **Results** node in the Simulation Tree. Next, right-click on it; a shortcut menu is displayed.

3. Choose the **Edit Definition** option from the shortcut menu; the **Stress plot PropertyManager** is displayed.

4. Click on the down arrow of the **Advanced Options** rollout to expand it.

5. Now, select the **Display symmetric results** check box under this rollout and then choose the **OK** button from the **Stress plot PropertyManager;** von-Mises stress plot is displayed for full model. Figure 5-15 shows the von-Mises stress result for full model.

   By default, von-Mises stress is displayed in the plot. To view other results, double-click on the respective option in the **Results** node of the Simulation Tree.

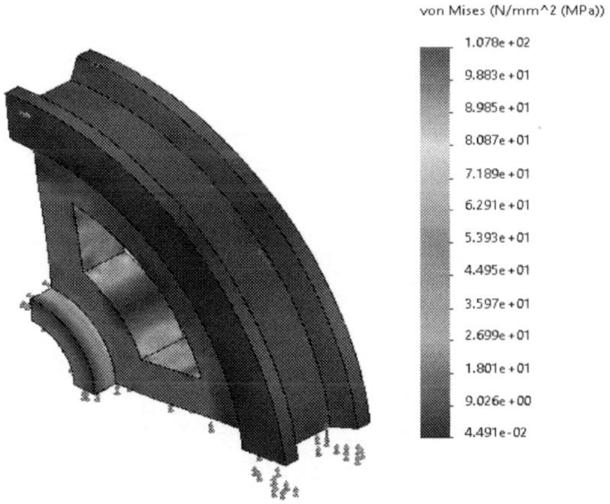

Figure 5-14 *von-Mises stress plot for the Study*

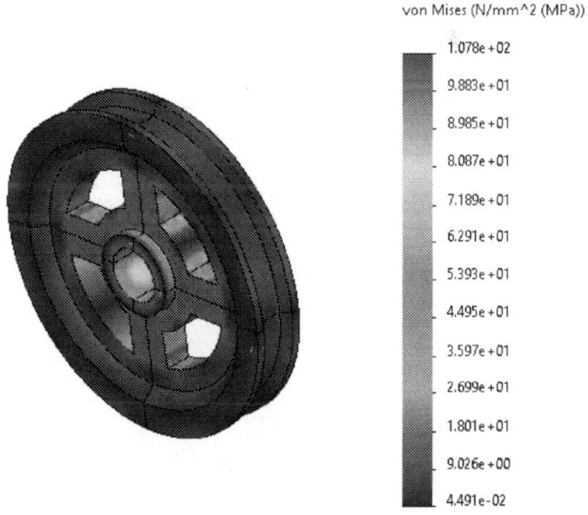

Figure 5-15 *von-Mises stress plot for full model*

Note

*You can also change the units of the results. To do so, right-click on von Mises legend to display a menu. Next, choose **Edit Definition**; the **Stress plot PropertyManager** is displayed. Next, choose **Definition** tab > **Display** rollout and then select the desired unit from the **Unit** drop-down list. Make sure you have selected the desired unit system. You can also change the unit system. To do so, choose **Options** from the **Simulation** menu; the **System Options - General dialog box** is displayed. Next, choose the **Default Options** tab and then select the desired unit system from the **Unit System** area.*

Finding FOS

1. Right-click on the **Results** node in the Simulation Tree and select the **Define Factor Of Safety** Plot from the shortcut menu displayed; the **Factor of Safety PropertyManager** is displayed, as shown in Figure 5-16.

2. Keep all the options intact in the **Factor of Safety PropertyManager** and choose the Next button; the **Factor of Safety PropertyManager** is modified, as shown in Figure 5-17.

3. In the **Multiplication factor** edit box of the **Step 2 of 3** rollout, retain the default value 1 and then choose the **Next** button; the **Factor of Safety PropertyManager** is changed again, as shown in Figure 5-18.

4. Enter 2 in the **Minimum factor of safety** edit box and choose the **OK** button; the Factor of Safety distribution plot is displayed, as shown in Figure 5-19.

   The plot result shows that minimum FOS for this study is 2.7. So, the next study FOS should be equal to 2.7 or greater than 2.7.

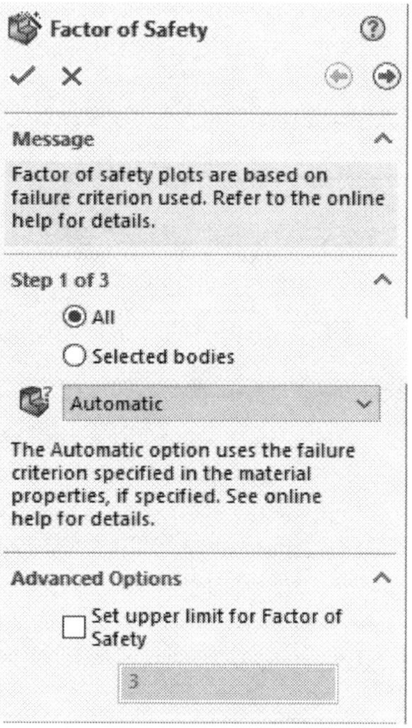

Figure 5-16** The **Factor of Safety PropertyManager

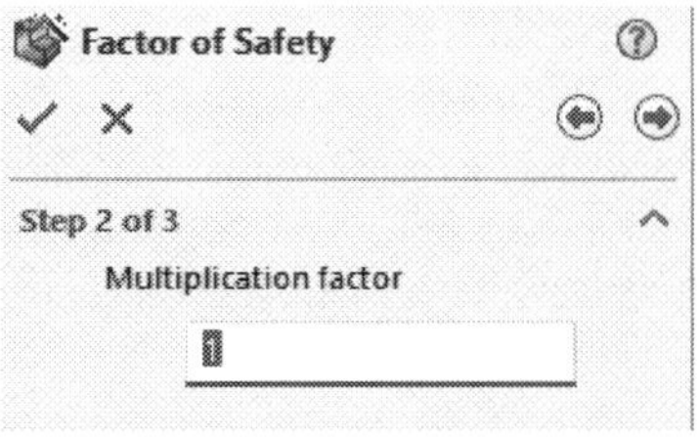

***Figure 5-17** The **Factor of Safety PropertyManager** in second step*

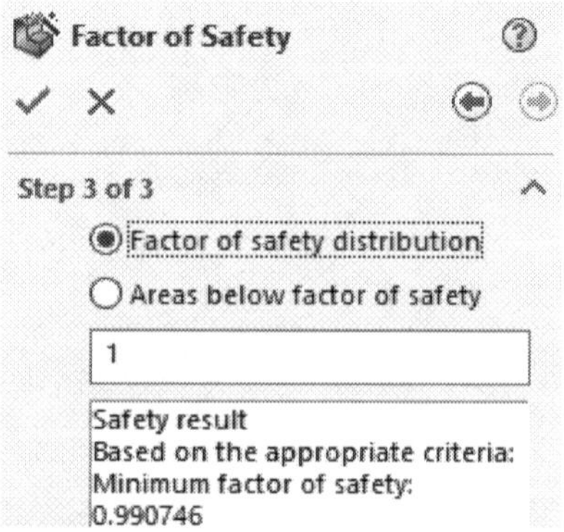

***Figure 5-18** The **Factor of Safety PropertyManager** in third step*

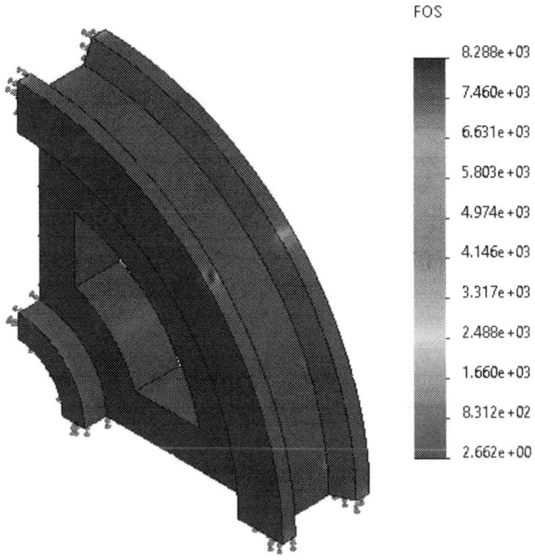

Figure 5-19 *The FOS distribution plot*

Saving the Study

After performing the study on the model, you need to save the results and files in the specified folder and then exit the SOLIDWORKS Simulation.

1. Choose the Save button in the menu bar to save all the studies.

2. Exit SOLIDWORKS Simulation.

Tutorial 2

In this tutorial, you will open *c05_tut02* file that was downloaded earlier and create the contacts, connections and perform static structural study. The model is made up of the materials such as Alloy steel material to the shoulder screw, Alloy of Aluminium to the wheel, and Cast Iron material to the other parts, as shown in Figure 5-20. Also, you will optimize the model and then evaluate the results to understand the importance of optimization. The Base of the assembly is fixed and 1 Nm torque is applied on the wheel of the assembly in the clockwise direction.

(Expected time: 1 hr)

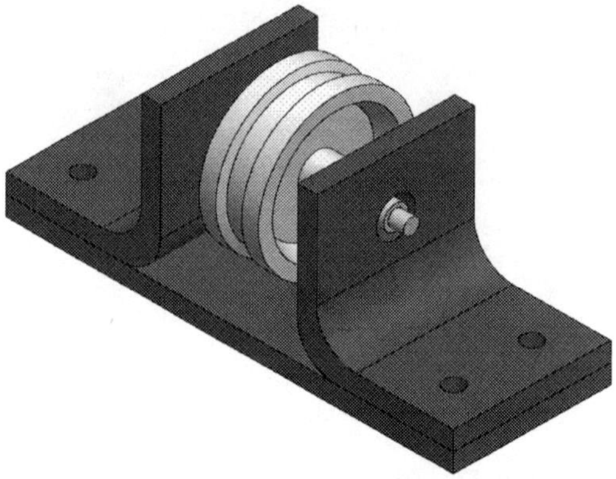

Figure 5-20 *Wheel Support Assembly model*

The following steps are required to complete this tutorial:

a. Start SOLIDWORKS Simulation and import the model.
b. Start a new Static study.
c. Apply material, fixture, contacts, and load.
d. Create mesh.
e Run the study.
f. View the results.

Starting SOLIDWORKS Simulation and Importing the Model

Now, you need to import the model of the Wheel Support Assembly for study.

1. Choose the **Open** option from **File** menu in the SOLIDWORKS menus to import the model (Wheel Support Assembly); the **Open** dialog box is displayed.

2. Select the model and choose the **Open** button in the dialog box; the selected model is opened in SOLIDWORKS and is ready for study.

Starting Simulation Study

1. Start the study with default configuration and position of the model.

2. Select the **New Study** option from the **New Study** flyout of the **Simulation CommandManager**, refer to Figure 5-21; the **Study PropertyManager** is displayed.

   Alternatively, you can get the **Study PropertyManager** by selecting the **Study** option from the **Simulation** menu.

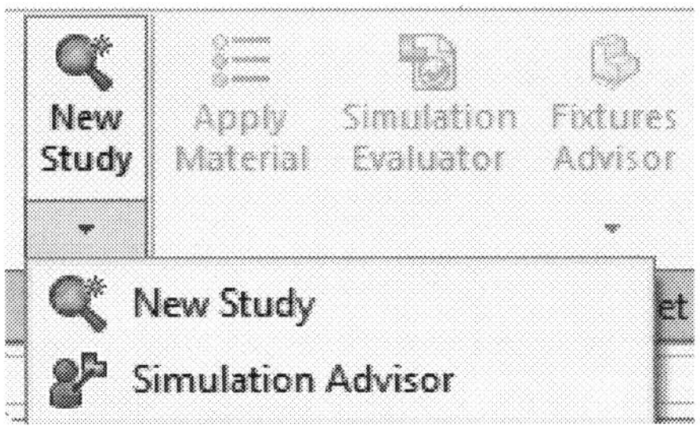

Figure 5-21 *The* ***New Study*** *flyout*

3. Select the **Static** option from the **Study PropertyManager** and enter study name as **Support Assembly**, and then choose the **OK** button; the **Simulation CommandManager** and Simulation Tree are displayed, as shown in Figures 5-22 and 5-23, respectively.

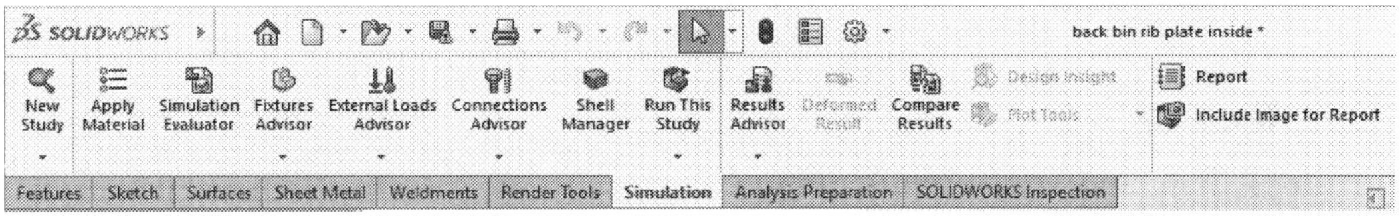

Figure 5-22 *The* ***Simulation CommandManager***

Applying Material

Now, you need to apply Alloy steel material to the shoulder screw, Alloy of Aluminium to the wheel, and Cast Iron material to other parts of the model.

1. Expand the **Parts** node and right-click on the shoulder screw sub-node in the Simulation Tree; a shortcut menu is displayed, refer to Figure 5-23. Next, choose the **Apply/Edit Material** option from the shortcut menu; the **Material** dialog box is displayed.

2. Expand the **SOLIDWORKS Materials** node in the **Material** dialog box. Next, expand the **Steel** sub-node under it and choose **Alloy Steel**; all properties of the selected material are displayed under the **Properties** tab of the dialog box.

3. Choose the **Apply** button from the dialog box; the Alloy steel material is applied to the shoulder screw.

4. Next, choose the **Close** button to close the **Material** dialog box.

5. Similarly, apply Alloy of Aluminium (1060 Alloy) to Wheel and Gray Cast Iron material to other parts of the assembly model (Support 1, Support 2, and Base). After applying the materials to the model, the Simulation Tree is displayed, as shown in Figure 5-24.

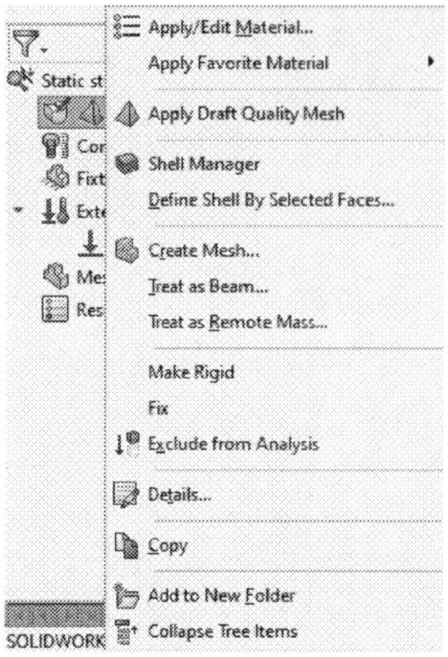

Figure 5-23 The Simulation Tree

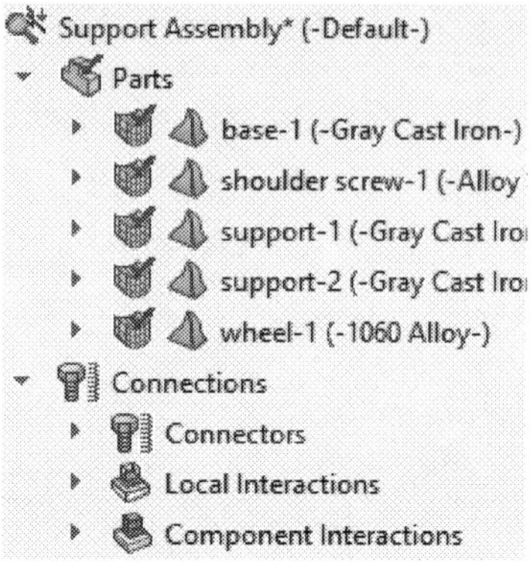

Figure 5-24 *The Simulation Tree after applying material*

Applying Interactions

A interaction describes the contact between part boundaries that come into contact during loading. The contact functionality is available in assembly and multi-body part study or analysis. The Contact prevents interference between the entities but allows gaps to form. This is the most time-consuming option to solve. In case of Bonded contact, bonded entities behave as if they were welded. This type of contact is available for static, nonlinear, frequency, buckling, and thermal studies.

1. Right-click on the **Connections** node in the Simulation Tree; a shortcut menu is displayed, refer to Figure 5-25. Choose the **Local Interaction** option from this menu; the **Local Interactions PropertyManager** is displayed, refer to Figure 5-26.

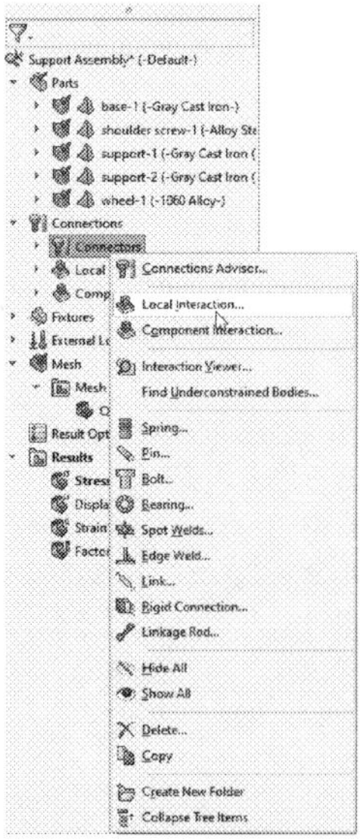

Figure 5-25 *The shortcut menu*

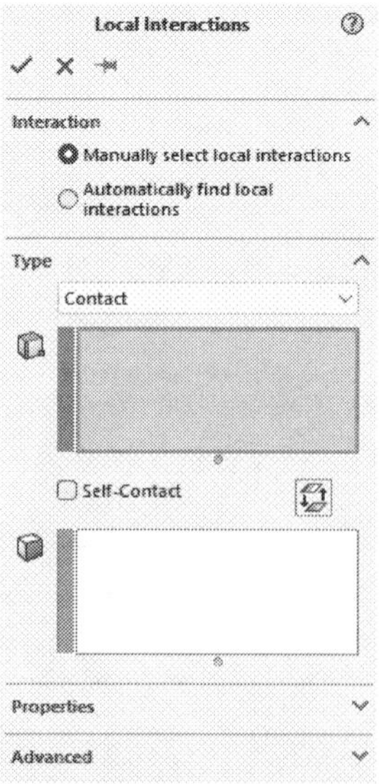

Figure 5-26 *The* ***Local Interactions PropertyManager***

2. Select the **Manually select Local Interactions** option in the **Interaction** rollout and the **Contact** option in the **Type** rollout. Next, you need to select contact faces to apply contacts manually.

3. Select the outer surface of the Shoulder Screw in the **Faces**, **Edges**, **Vertices for Set 1** selection box, as shown in Figure 5-27. Next, select the cylindrical surface of the hole of Wheel in the **Faces for Set 2** selection box, as shown in Figure 5-28.

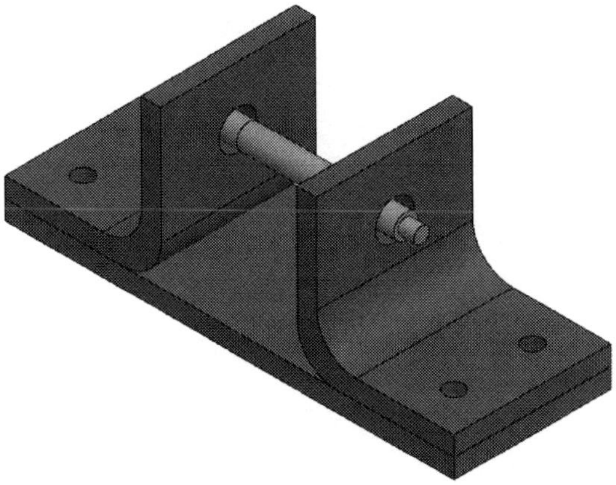

Figure 5-27 *The faces and axis to be selected*

Figure 5-28 *The faces and axis to be selected*

4. Next, select the **Surface to surface** radio button from the **Advanced** rollout and then choose the **OK** button from the PropertyManager.

   Now, you need to apply contacts between the other remaining parts such as Base and Support1, Base and Support2, and Support2 and Shoulder Screw.

5. Right-click on the **Connections** node in the Simulation Tree; a shortcut menu is displayed.

6. Choose the **Local Interaction** option from the shortcut menu; the **Local Interactions PropertyManager** is displayed.

7. Select the **Automatically find Local Interactions** radio button in the **Interaction** rollout and then select the **Touching faces** radio button in the **Options** rollout.

   Next, you need the select all remaining parts to find the Local Interactions under the **Select Components or Bodie**s selection box in the **Components** rollout.

8. Select the **Base, Support1, Support2**, and **Shoulder Screw** under this selection box.

9. Next, choose the **Find Local Interactions** button; you will notice that all the possible contacts are displayed in the **Results** rollout.

10. Select the **Bonded** option in the **Type** rollout of the PropertyManager. Now, choose the **OK** button; the **Simulation** message box is displayed. Choose the **Yes** button from the message

box. You will notice that another Local Interaction is created under the **Connections** > **Local Interactions** node in the Simulation Tree.

Applying Bearing Connector

A bearing connector is a combination of spring elements that join two concentric cylindrical parts and mimics the behavior of a real bearing, allowing rotation between the two parts. Now, you need to apply bearing connectors to the assembly.

1. Right-click on the **Connections** node in the Simulation Tree; a shortcut menu is displayed.

2. Choose the **Bearing** option from the shortcut menu, refer to Figure 5-29; the **Connectors PropertyManager** is displayed, refer to Figure 5-30.

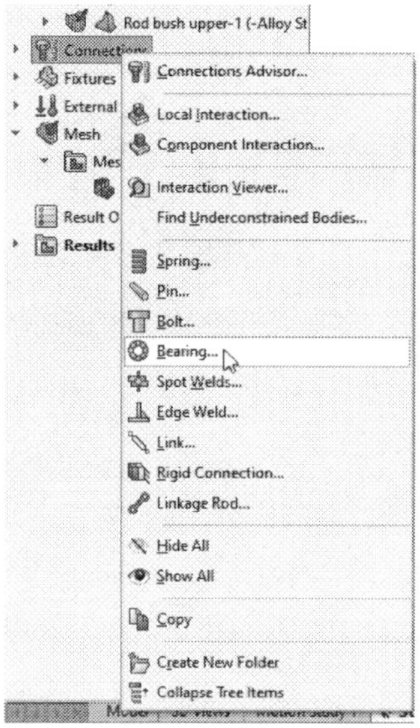

Figure 5-29 *The shortcut menu displayed by clicking on the* ***Connections*** *node*

Figure 5-30 *The* ***Connectors PropertyManager***

3. Next, select the cylindrical faces of the shaft, mentioned in Figure 5-31 in the **Cylindrical face or circular edge of shell** in the **Type** rollout, and then select the circular surface of the support, as shown in Figure 5-32, in the **Cylindrical face or circular edge of shell** selection box. Make sure the **Allow self-alignment** check box is cleared in this rollout.

   Now, you need to set the units for stiffness and other parameters of the study.

4. Select the **SI** option in the **Unit** drop-down list and then select the **Rigid** radio button in the **Stiffness** rollout to apply the rigid behavior of the study. Make sure the **Stabilize shaft rotation** check box is selected in this rollout.

5. Next, choose the **OK** button from the PropertyManager.

6. Similarly, apply bearing connector on the other side. Figure 5-33 shows the model after applying the Bearing connector.

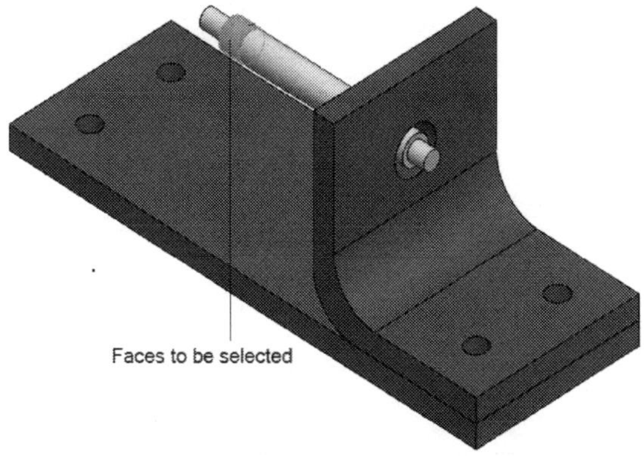

Figure 5-31 The face to be selected from the Shoulder Screw

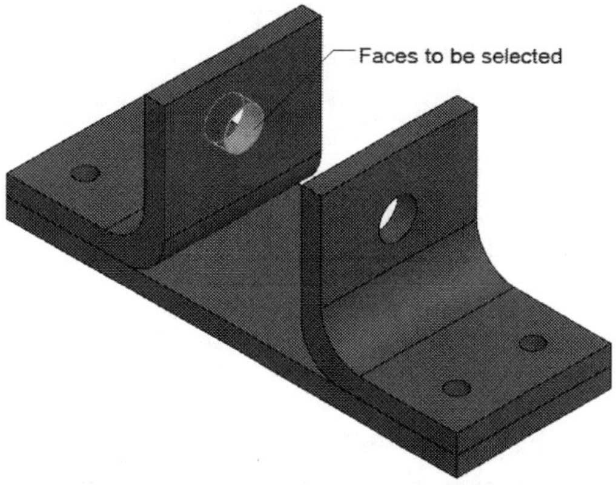

Figure 5-32 The face to be selected from the Support

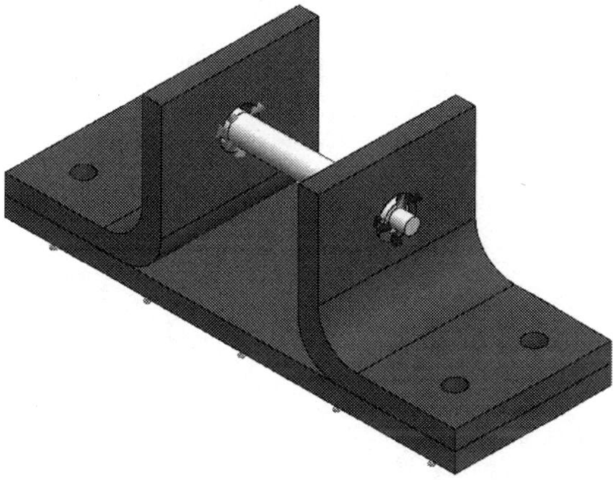

Figure 5-33 The model after applying the Bearing Connector

Applying Restraints

The types of loads and restraints available depend on the type of the study. A load or restraint is applied by using the corresponding PropertyManager and to make the model stable.

1. Right-click on **Fixtures** node in the **Simulation Tree** and choose the **Fixed Geometry** option from the shortcut menu; the **Fixture PropertyManager** is displayed. Select the bottom face of the assembly model to make it stable in the **Faces**, **Edges**, **Vertices for Fixture** selection box under the **Standard (Fixed Geometry)** rollout of the PropertyManager.

2. Next, choose the **OK** button from the PropertyManager; the assembly model is fixed with respect to its bottom face.

Applying Loads

In this section, you need to apply 1 Nm torque on the circular face of the wheel about its axis.

1. Right-click on the **External Loads** folder in the Simulation Tree and choose the **Torque** option from the shortcut menu, refer to Figure 5-34; the **Force/Torque PropertyManager** is displayed, as shown in Figure 5-35.

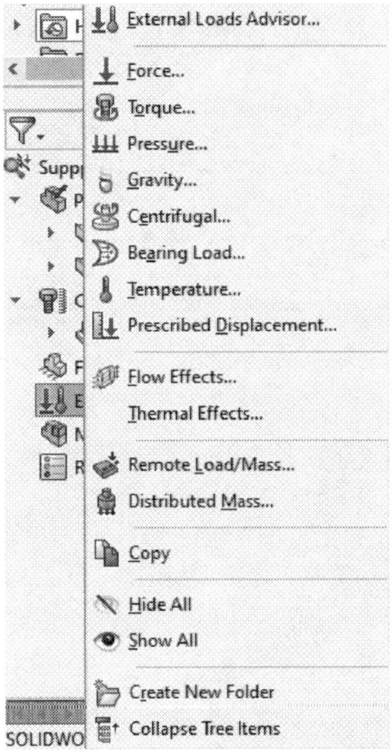

*Figure 5-34 The shortcut menu displayed on clicking on the **External Load** node*

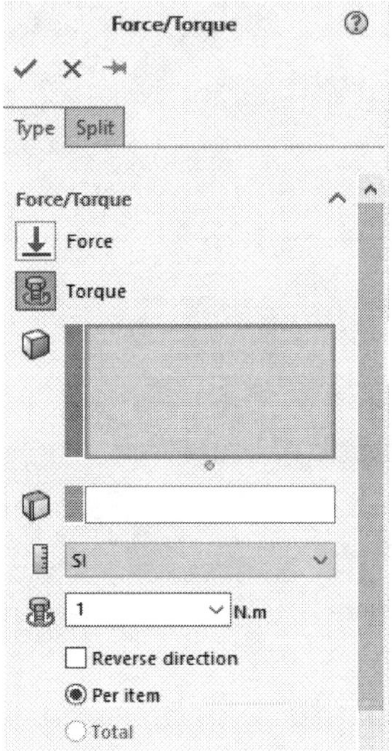

Figure 5-35** The **Force/ Torque PropertyManager

2. Choose the **Torque** button from the **Force/Torque** rollout if it is not chosen; the **Faces for Torque and Axis**, **Cylindrical Face for Direction** selection boxes are activated.

3. Select the face to apply torque under the **Face for Torque** selection box and then select the axis under the **Axis, Cylindrical Face for Direction** selection box.

Note

*You might have to create the axis if it has not been created in the model. To create axis, choose **Axis** from **Reference Geometry** of the **Insert** menu in the menu bar.*

4. Select the **SI** unit system from the **Unit** drop-down list and then enter **1** in the **Torque Value** edit box of the **Force/Torque** rollout. If required, you can change the direction of the applied torque by using the **Reverse direction** check box.

5. Choose the **OK** button to close the PorpertyManager. Figure 5-36 shows the model after applying the torque.

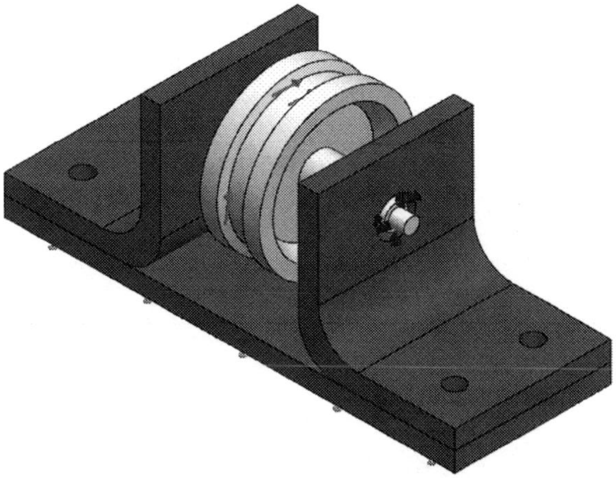

Figure 5-36 *The model after applying the torque*

Creating Mesh

A mesh consists of only one type of element unless the mixed mesh type is specified. Now, you need to create mesh to divide the model into small pieces.

1. Right-click on the **Mesh** node in the Simulation Tree and choose the **Create Mesh** option from the shortcut menu; the **Mesh PropertyManager** is displayed.

2. Set the mesh density slider bar to most right position, as shown in Figure 5-37 and then choose the **OK** button from the PropertyManager; the **Mesh Progress** window is displayed and mesh is created on the model, refer to Figure 5-38.

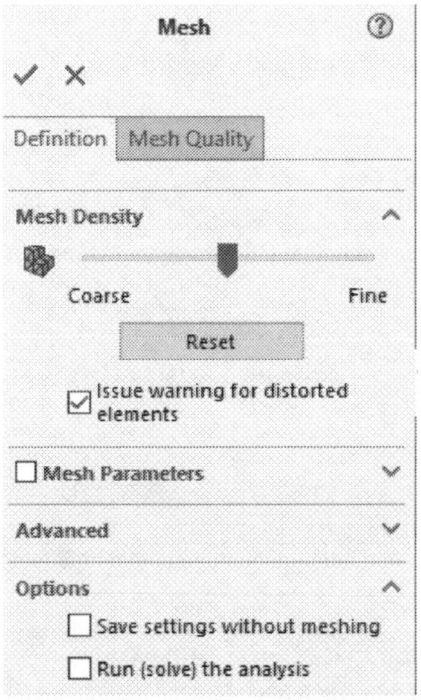

Figure 5-37 *The* ***Mesh PropertyManager***

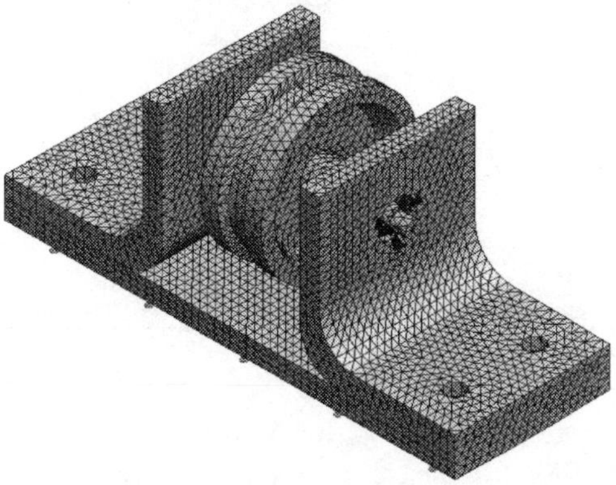

Figure 5-38 *Model after meshing*

Running the Study

Now, you need to perform the study to get results based on the input given to the model.

1. Right-click on the **Support Assembly** node in the Simulation Tree and choose the **Run** option from the shortcut menu; the **Support Assembly** window is displayed and the solution process is started.

   Note that if the **Automatic Solver Selection** check box in the **Solver** area of the **Static** dialog box is not selected then you may get an error message. To set the solver settings, click on the down arrow under **Study Advisor** and select the **Study Properties** option; the **Static** dialog box is displayed. Choose the **Options** tab in this dialog box, and then select **Automatic Solver Selection** check box in the **Solver** area of the dialog box.

2. When the study is completed, the **Results** node is added to the Simulation Tree. Also, the resultant model is displayed with von-Mises stress plot, as shown in Figure 5-39.

   By default, von-Mises stress is displayed in the plot. To view the other plot results, double-click on the respective option in the **Results** node of the Simulation Tree.

Finding FOS

1. Right-click on the **Results** node in the Simulation Tree and choose the **Define Factor Of Safety Plot** from the shortcut menu; the **Factor of Safety PropertyManager** is displayed, as shown in Figure 5-40.

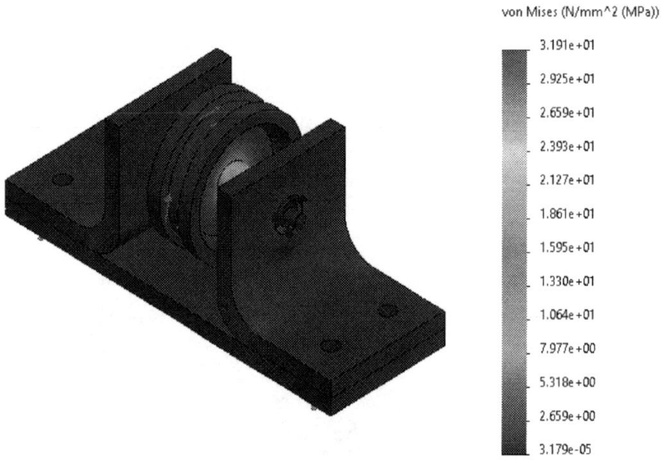

Figure 5-39 Visualization of model with von-Mises stress plot

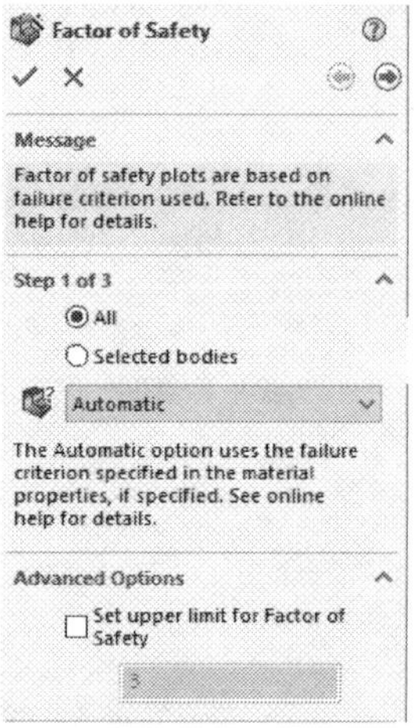

Figure 5-40** The **Factor of Safety PropertyManager

2. Retain the default settings. Next, choose the **Next** button in the **Factor of Safety PropertyManager**; the **Factor of Safety PropertyManager** is modified, as shown in Figure 5-41.

3. Choose the Next button with default value **1** in the **Multiplication factor** edit box of the Step 2 of 3 rollout; the **Factor of Safety PropertyManager** is modified again, as shown in Figure 5-42.

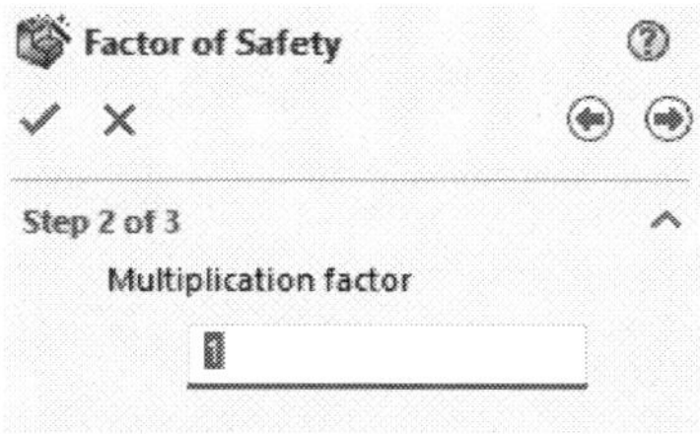

Figure 5-41 ***The Factor of Safety PropertyManager*** *in second step*

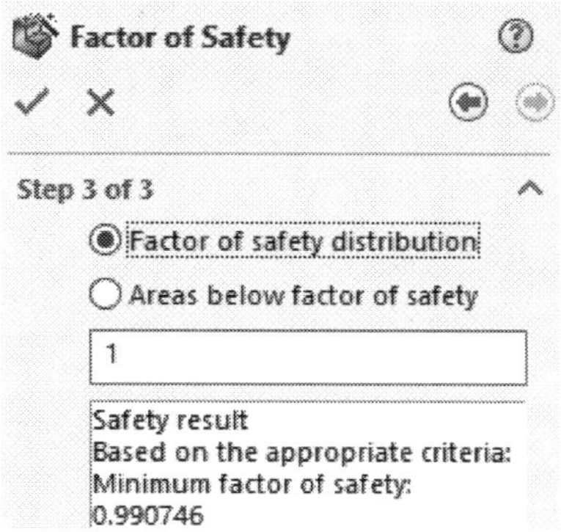

Figure 5-42 *The* ***Factor of Safety PropertyManager*** *in third step*

4. Enter 2 in the **Minimum factor of safety** edit box and choose the **OK** button; the factor of safety distribution plot is displayed, as shown in Figure 5-43.

   This is evident from Figure 5-43 that the minimum FOS is 1.6. Based on the design and applications, the minimum FOS should be 1.5. This proves that the given condition is fine for the safety purpose with applied loading conditions. However, if the FOS is less than 1.5, then you need to change the design, and then check the FOS again till its value is 1.5 or greater than.

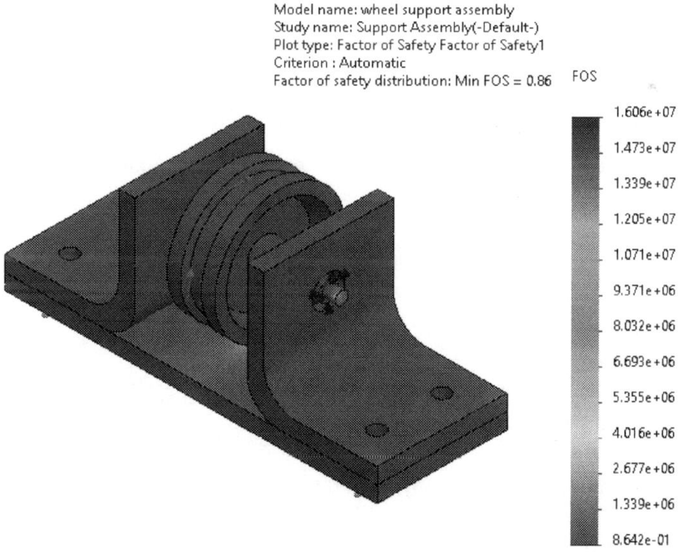

Figure 5-43 *The factor of safety distribution plot*

Saving the Study

After performing the study on the model, you need to save the results and files in the specified folder and then exit SOLIDWORKS Simulation.

1. Choose the **Save** button from the **File** menu and save all the studies.

2. Exit SOLIDWORKS Simulation.

Tutorial 3

In this tutorial, you will open *c05_tut03* file that was downloaded earlier and perform a fatigue analysis of constant amplitude event type on a model. In order to run the fatigue study, you will first create a static study of the model given in Figure 5-44. The model is made up of Alloy Steel SS and you need to apply boundary conditions as given below:

(Expected time: 45min)

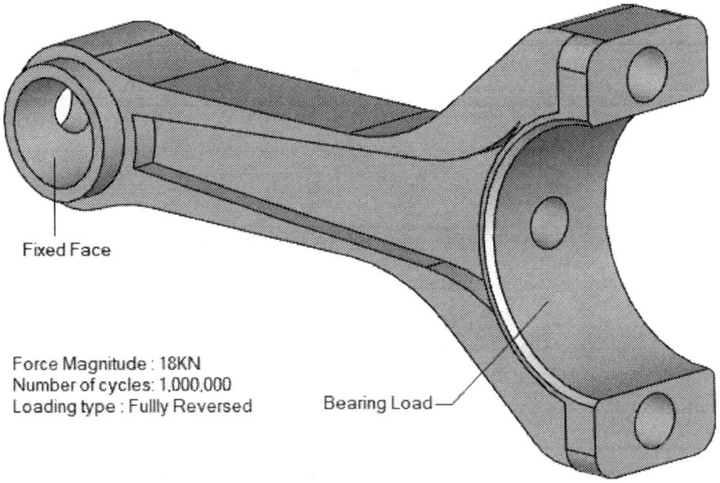

Figure 5-44 *Model for the simulation and the constraints to be applied*

The following steps are required to complete this tutorial:

a. Start SOLIDWORKS Simulation and import the model.
b. Start a new static study.
c. Apply material, fixture, contacts, and load.
d. Create mesh and run the study.
e. Start a fatigue study.
f. Apply event properties in reference to static study.
g. Run the fatigue study and analyze the results.
h. Save the file.

Starting SOLIDWORKS Simulation and Importing the Model

Now, you need to start SOLIDWORKS Simulation and import the model (c05_tut03) for study.

1. Choose **Open** from the **File** menu in the SOLIDWORKS menus; the **Open** dialog box is displayed.

2. Select the model and choose the **Open** button in the dialog box; the model is opened in SOLIDWORKS and you are ready to start the simulation study.

Starting the Static Study

In this section, you need to start the study with default configuration and position of the model.

1. Select the **New Study** option from the **New Study** fllyout in the **Simulation CommandManager,** refer to Figure 5-45 and 5-46; the Study PropertyManager is displayed.

   Alternatively, you can invoke the **Study PropertyManager** by choosing the **Study** option from the **Simulation** menu.

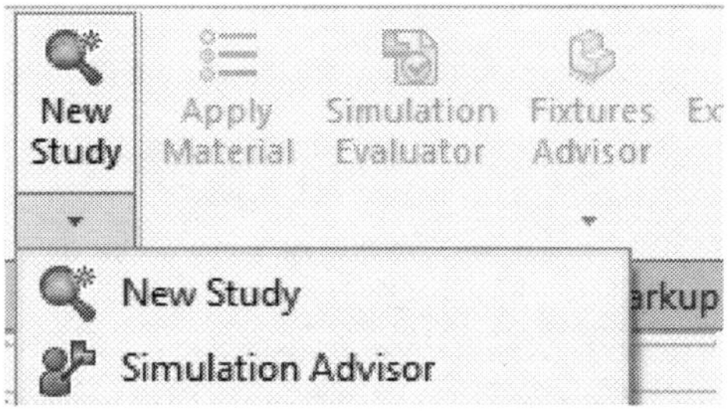

***Figure 5-45** The **Study Advisor** drop-down list*

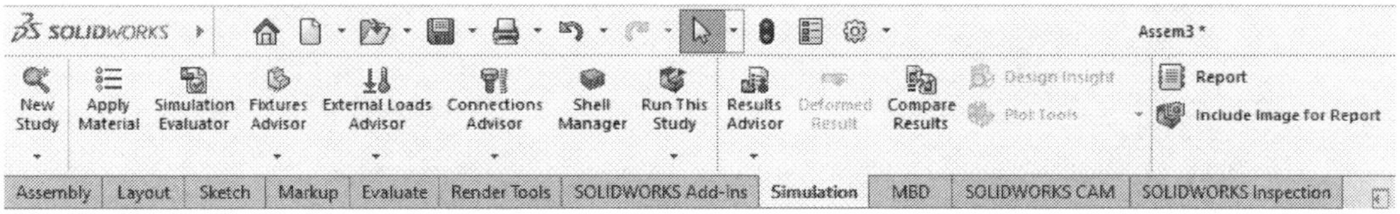

Figure 5-46 The Simulation CommandManager

2. Select the **Static** option from the **Study PropertyManager** and enter study name as **Connecting Rod Static**, and then choose the **OK** button; Connecting Rod Static node gets added in the Simulation tree.

Applying Material

Now, you need to apply Alloy Steel SS material to the model.

1. Right-click on the **c05_tut03** part node and choose the **Apply/Edit Material** option from the shortcut menu, refer to Figure 5-47; the **Material** dialog box is displayed.

2. Expand the **SOLIDWORKS Materials** node in node under it and choose **Alloy Steel**; all properties of the selected material are displayed under the Properties tab of the dialog box.

3. Choose the **Apply** button from the dialog box; the Alloy Steel SS material is applied to the connecting rod. Next, choose the **Close** button to close the **Material** dialog box.

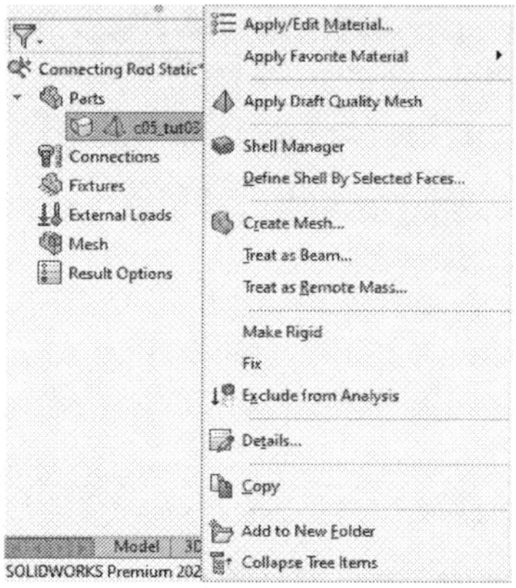

Figure 5-47 Adding material to the model

Applying Boundary Conditions

In this section, you need to prepare the model for simulation by defining all the constraints in terms of fixtures and loads.

1. Right-click on **Fixtures** in the **Simulation Tree**; a shortcut menu is displayed. Choose the **Fixed Geometry** option from this menu; the **Fixture PropertyManager** is displayed. Select the top inner-cylindrical face in the **Faces, Edges, Vertices for Fixture** selection box of the **Standard (Fixed Geometry)** rollout, refer to Figure 5-48, to assign it as a fixture.

2. Choose the **OK** button to accept the fixture.

   Next, you need to apply bearing load. Before applying the load, you need to create a coordinate system that aligns with the lower cylindrical face, refer to Figure 5-49. Make sure the Z-axis aligns with the axis of the cylindrical face.

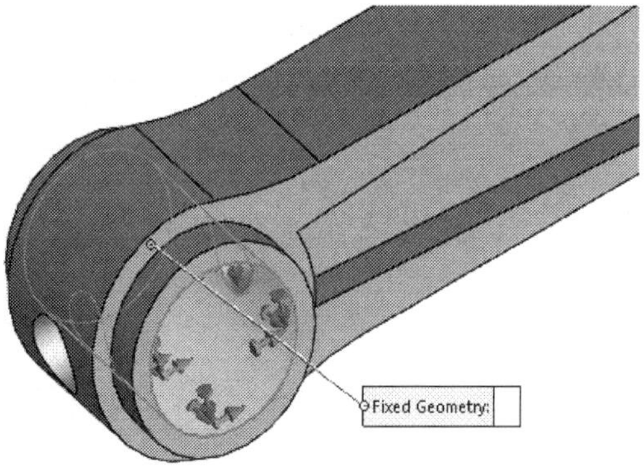

***Figure 5-48** The cylindrical face to be selected for the fixed geometry*

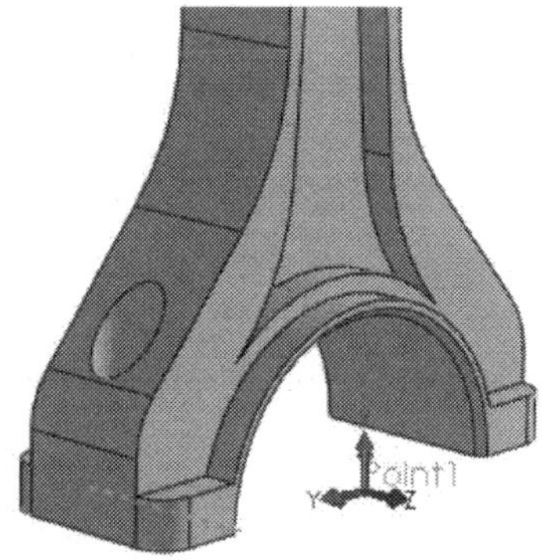

***Figure 5-49** Coordinate axis*

3. Right-click on **External Loads** in the Simulation Tree; a shortcut menu is displayed. Choose the **Bearing Load** option from the shortcut menu; the **Bearing Load PropertyManager** is displayed, as shown in Figure 5-50.

4. Select the bottom inner cylindrical face, refer to Figure 5-51, in the **Cylindrical Faces or Shell Circular Edges for Bearing Load** selection box of the Selected Entities rollout.

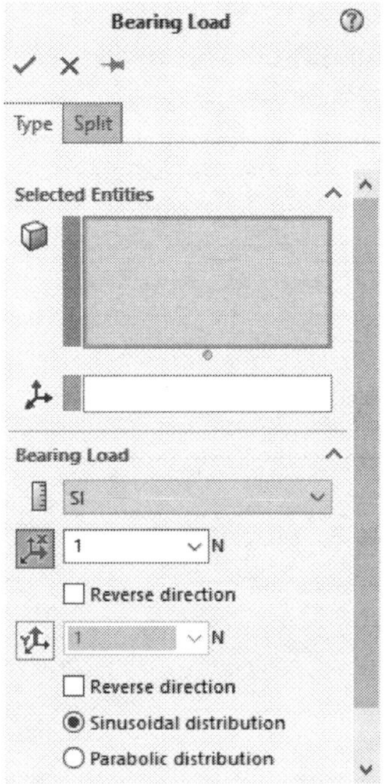

Figure 5-50** The **Bearing Load PropertyManager

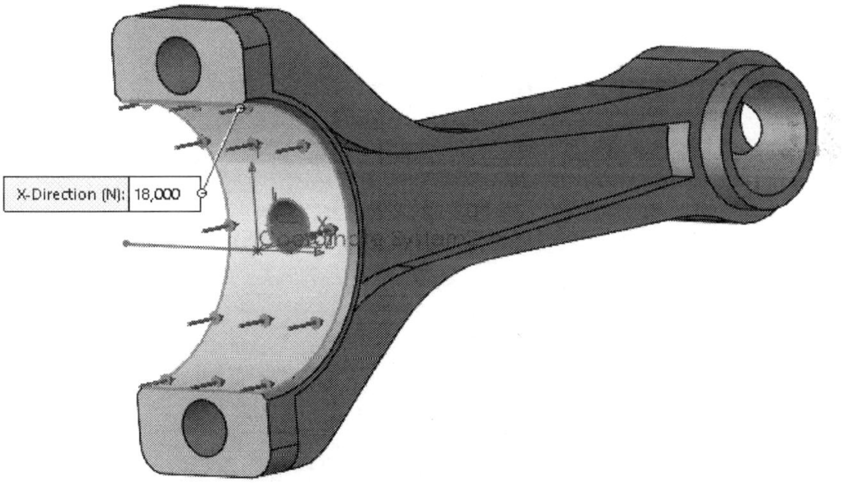

***Figure 5-51** The cylindrical face to be selected*

5. Select the previously constructed coordinate axis in the **Select a Coordinate System** selection box.

6. Enter **18000** in the **X-Direction** edit box in the **Bearing Load** rollout and make sure SI units are selected. Reverse the direction if necessary.

7. Choose the **OK** button to exit the **Bearing Load PropertyManager.**

Creating the Mesh and Running the Study

1. Right-click on **Mesh** in the **Simulation Tree** and choose the **Create Mesh** option from the shortcut menu; the **Mesh PropertyManager** is displayed.

2. Move the slider to the right in the **Mesh Density** rollout to make the mesh fine. Next, retain other default settings and then choose the **OK** button; the **Mesh Progress** window displayed, showing the progress of the mesh creation. Refer to Figure 5-52 for the model after meshing.

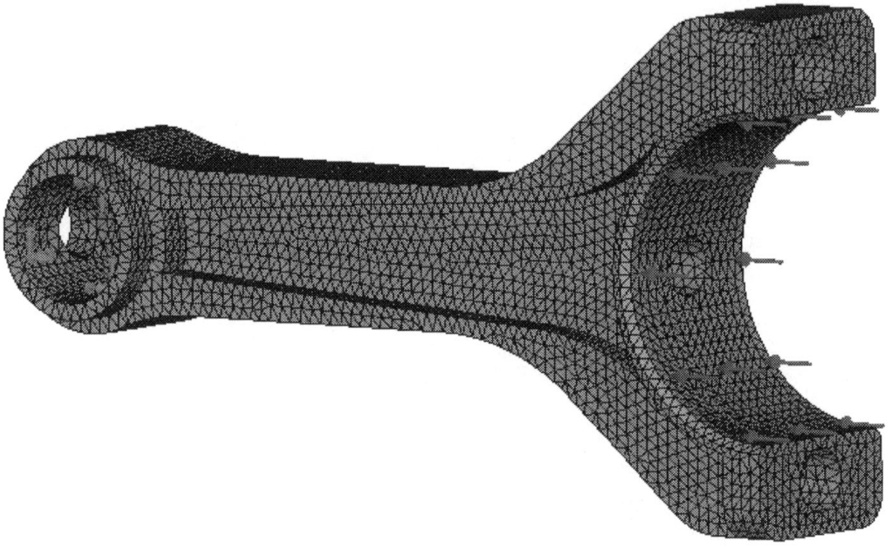

Figure 5-52 *The meshed model*

3. Right-click on the **Connecting Rod Static** node in the Simulation Tree and choose the **Run** option from the shortcut menu; the **Connecting Rod Static Process** window is displayed and the solution process is started.

   When the solution process is completed , a new node with the name **Results** is added in the Simulation Tree. Also, the model is displayed with von-Mises stress plot, as shown in Figure 5-53.

   The stress plot shows that the stress is less than the yield strength in the model (Alloy Steel, Yield Strength = 6.2 x 10 ^ 8 N/m ^ 2). This can be understood by using the color scale and the color distribution across the model. This proves that the given condition is fine for the safety purpose in the static study. Now, you will perform fatigue study investigate the long term results of this stress plot.

Setting Up a New Fatigue Study

1. Select the **New Study** option from the **New Study** fllyout of the **Simulation CommandManager;** the **Study PropertyManager** is displayed.

   Alternatively, you can invoke the **Study PropertyManager** by choosing the **Study** option from the Simulation menu.

2. Select the **Fatigue** option from the **Study PropertyManager** and enter **Connecting Rod Fatigue** as study name, and then choose the **OK** button.

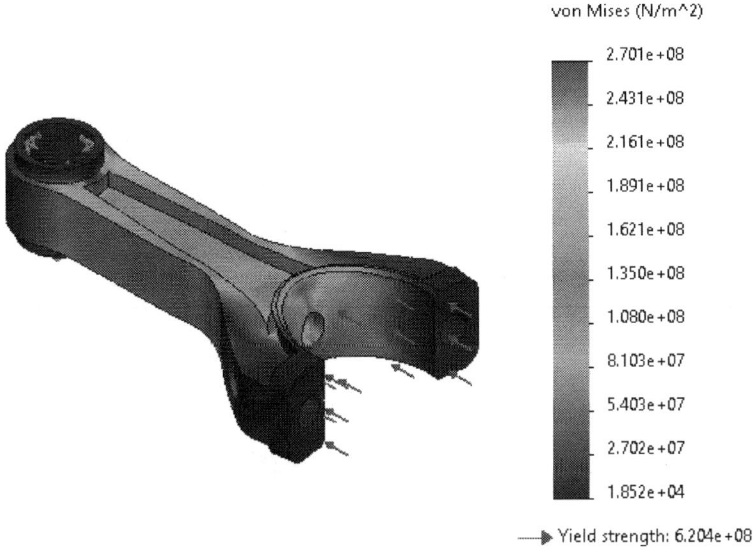

Figure 5-53 *Von-Mises stress plot*

Loading the S-N Curve and Defining the Input

S-N curve is used for evaluating fatigue results by defining alternating stress values versus the number of cycles at a given stress ratio. Before applying an S-N curve to the material, first you need to define the loading event for the analysis. Loading allows you to specify the number of cycles of a particular event you estimate. Two types of events are available in fatigue study such as constant amplitude event and variable amplitude event. A constant amplitude event is fully defined by an alternating stress, mean stress, and the number of cycles. A variable amplitude event is a load history record that defines the fluctuation history of a load. Variable amplitude study options are used to define the number of bins needed for rainflow counting as well as for defining the alternating stress and mean stress. The results of a fatigue analysis are highly dependent on the quality of the input S-N curve(s).

1. Right-click on the **Loading (-Constant Amplitude-)** node in the Simulation Tree, refer to Figure 5-54; a shortcut menu is displayed. Select the **Add Event** option; the **Add Event (Constant) PropertyManager** is displayed, as shown in Figure 5-55.

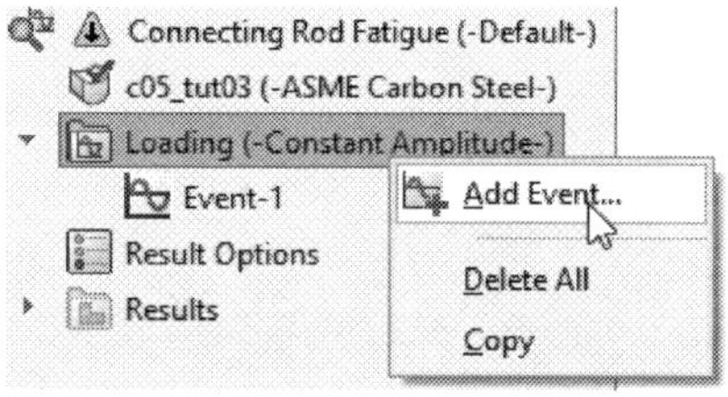

Figure 5-54 Adding a loading event

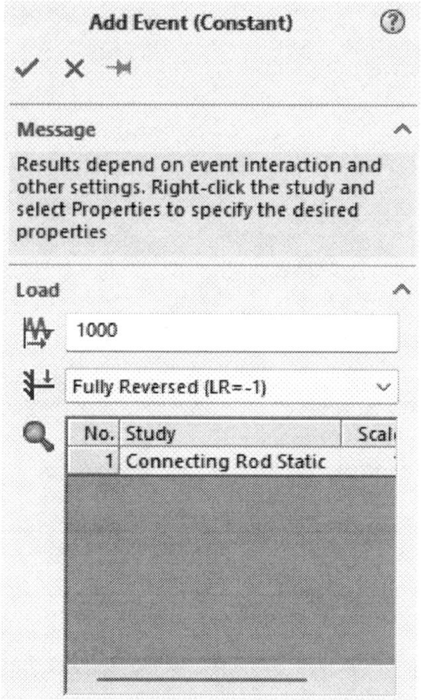

Figure 5-55** The **Add Event (Constant) PropertyManager

2. Enter **1000000** in the Cycles edit box in the **Load** rollout.

   Whenever the number of cycles is greater than 1000, the part is considered as undergoing high-cycle fatigue. The number of cycles can be raised up to 10 million cycles. Maximum stress applied in high-cycle fatigue must be kept much lower than the yield strength of the material. Low-cycle fatigue is considered when the number of cycles is less than 1000.

3. In the **Loading Type** drop-down list, select the **Fully Reversed (LR=-1)** option. Next, make sure the **Connecting Rod Static** is selected in the drop-down list of the **Study Association** table. Choose the **OK** button to exit the PropertyManager.

   A fully reversed load means that the ratio of the applied maximum stress to the minimum stress will be -1. The stress cycle will oscillate between positive and negative values. In real-world application, it means that the part will be periodically under maximum compressional stress and maximum tensile stress. You will now see a new part node called **Connecting Rod Fatigue** added to the Simulation Tree, refer to Figure 5-56.

4. Right-click on the **c05-tut03** node and choose the **Apply/Edit Fatigue Data** option from the shortcut menu; the **Material** dialog box with the **Fatigue SN Curves** tab chosen is displayed.

5. In the **Source** area, select the **Derive from material Elastic Modulus** and the **Based on ASME Carbon Steel curves** radio buttons. Next, choose the **Apply** and **Close** buttons to close the dialog box, refer to Figure 5-57.

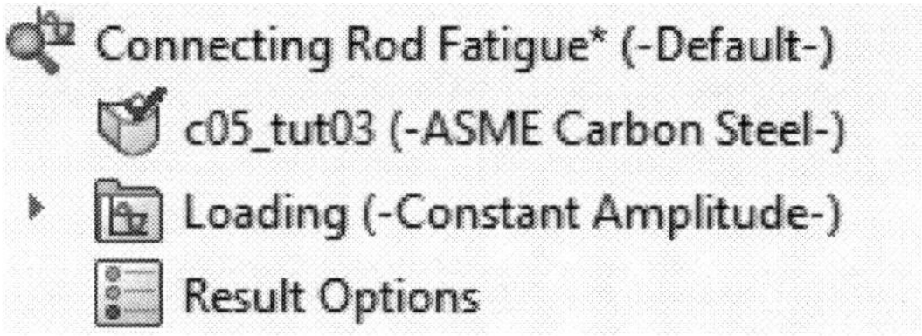

Figure 5-56 A new part node created

An S-N curve contains data of the stress against number of cycles to failure, which is different for each material. S-N curves are in-built with most of the materials in SOLIDWORKS. In case the S-N curve of a particular material does not exist, it can be created by using other data sources.

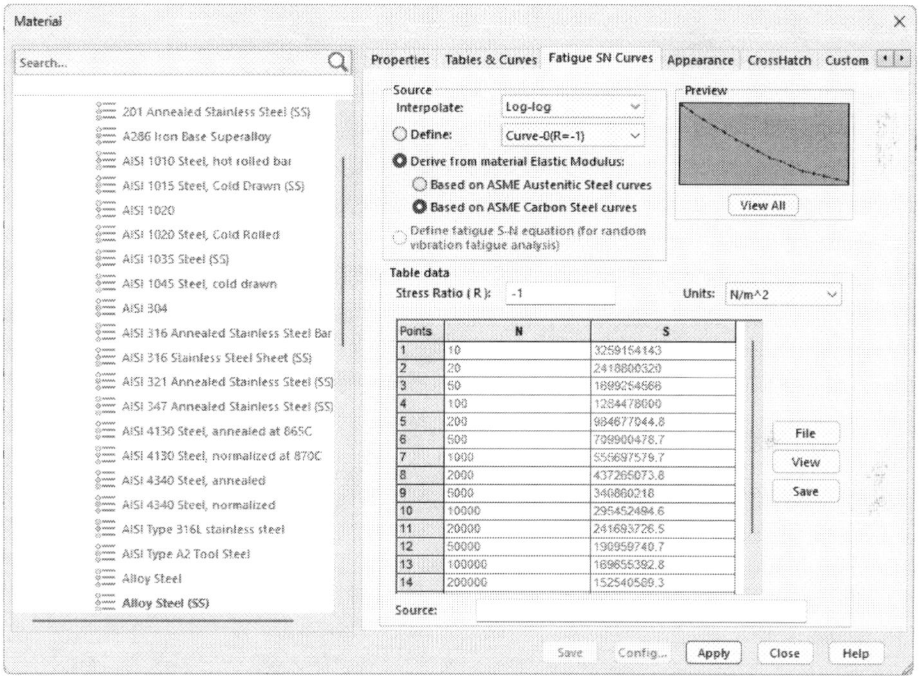

*Figure 5-57 The **Material** dialog box with the **Fatigue SN Curves** tab chosen*

Running the Fatigue Study and Analyze Results

1. Choose the **Run This Study** button from the **Simulation CommandManager**; the **Connecting Rod Fatigue** solver window is displayed and the solution process gets started. Wait for the process to finish.

2. By default, the fatigue plots are added under the Results node in the Simulation Tree. If it is not so then right-click on the **Results node** and select the **Define Fatigue Plot** option, the **Fatigue Plot PropertyManager** will be displayed. Select the required radio button in the **Plot Type** rollout and then choose the **OK** button to view the fatigue plot.

Analyzing the Results

Now, you will analyze the fatigue results. You will use three types of fatigue plots: Life, Damage, and Load Factor.

1. Double-click on the **Results1 (-Damage-)** under the **Results** node to make it visible, if it is not displayed by default. Refer to Figure 5-58 for the damage plot.

   A damage plot shows the percentage of the life of the structure consumed by the defined fatigue events. So, the value of 100 implies that the structure will be completely used up by the times the desired number of cycles are completed.

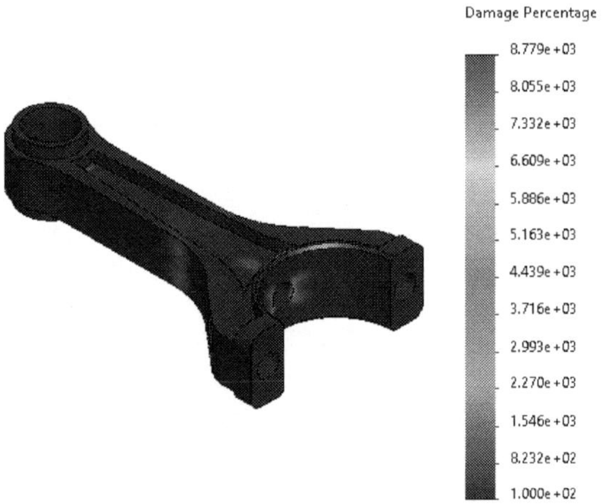

Figure 5-58 *The model with the damaged fatigue plot*

   According to the damaged plot, most parts of the model will last a million cycles. Only a part of the fillet near the load-bearing cylindrical face shows more damage and will get consumed 10-16 times more than the rest of the model.

2. Now, double-click on the **Results2 (-Life-)** sub-node under the **Results** node in the Simulation Tree to make it visible.

   A life plot shows the number of cycles (for constant amplitude event studies) or the number of blocks (for variable amplitude studies) that causes fatigue failure at each location. So, a part that has a value of 5000, will only last upto 5000 cycles before succumbing to failure.

   The life plot displays the longer lasting parts in red by default. To make the plot more intuitive, you can change the colors of the plot by right-clicking on the **Results2 (-Life-)** plot folder.

3. Right-Click on the plot and choose the **Chart Options** option from the shortcut menu to open the **Fatigue Plot PropertyManager**.

4. In the **Fatigue Plot PropertyManager**, expand the Color Options rollout and select the Flip check box to flip the color scale. Refer to Figure 5-59 for the **Color Options** rollout in the **Fatigue Plot PropertyManager**.

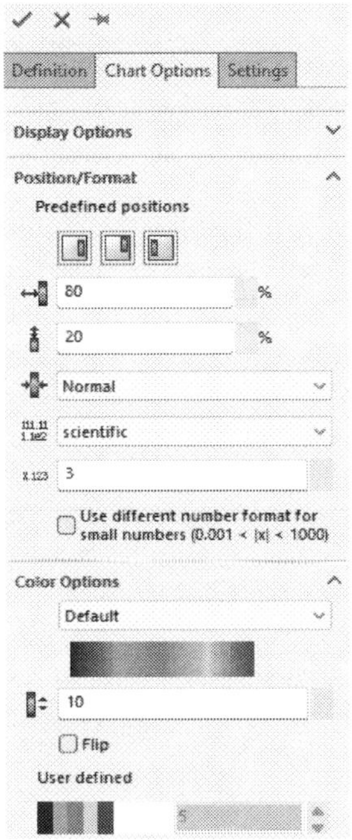

Figure 5-59 *The* ***Color Options*** *rollout*

Now, the areas where the life expectancies are low are displayed in red and where they are high are displayed in blue color in plot, refer to Figure 5-60.

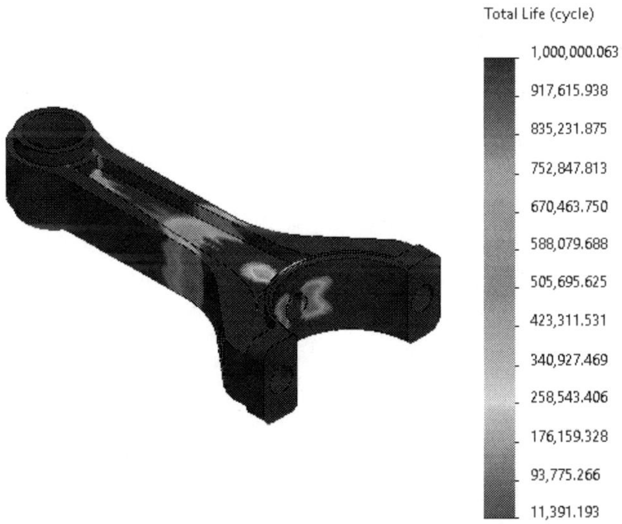

Figure 5-60 *The model with the updated life fatigue plot*

5. Choose the **OK** button to exit the **Fatigue Plot PropertyManager.**

6. Double-click on the **Results3 (-Load Factor-)** node plot to view it. If it is not added by default, you can add it.

Note

*To add a result node (-Load Factor-), right-click on the **Results** node and then choose **Define Fatigue Plot** from the shortcut menu displayed. Next, select **Load Factor** from the PropertyManager and then choose **OK**.*

The Load Factor plot shows the factor of fatigue safety across a model. A value of 4.5, for instance, means that if the load is multiplied by 4.5, the model will fail due to fatigue at that particular part.

The plot shows the model mostly in red, which corresponds to a very low factor of safety. You need to further investigate it.

7. Right-click on the **Results 3(-Load Factor-)** plot; a shortcut menu is displayed. Select the **Chart Options** option from the menu; the **Fatiuge plot PropertyManager** is displayed.

8. From the **Display Options** rollout, select the **Show min annotation** and **Show max annotation** check boxes. On doing so, the callouts for the respective nodes get attached to the model in the drawing area, refer to Figure 5-61. Next, choose **OK.**

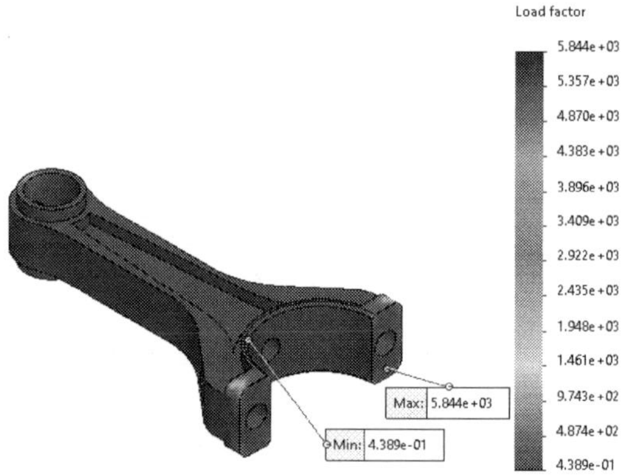

***Figure 5-61** The load factor plot with minimum and maximum annotations*

9. Right-click on the **Results3 (-Load Factor-)** plot option and choose the **Probe** option from the shortcut menu displayed.

10. Make sure the **At location** radio button is selected in the **Options** rollout. Now, select 5 points in the red regions of the model to probe. Refer to Figure 5-62 to get an idea of the probe locations.

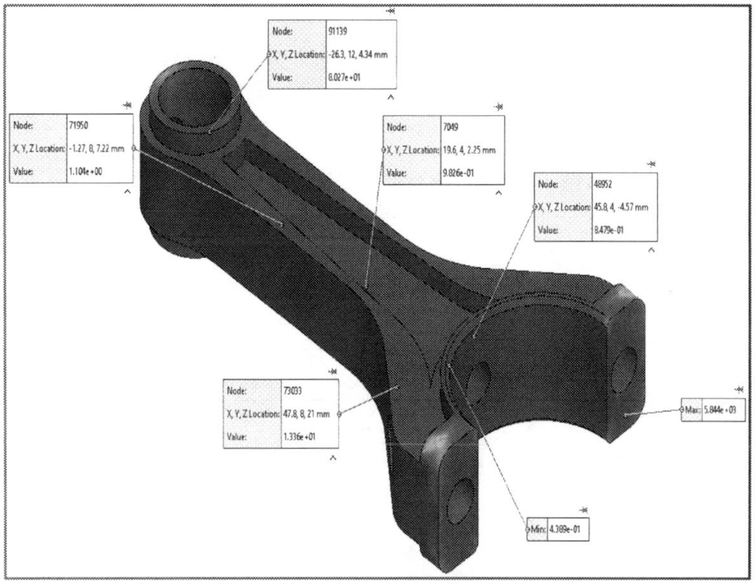

Figure 5-62 *Probe locations*

11. Scroll down and choose the **Plot** button in the **Report Options** rollout to view the load factor vs. node graph, refer to Figure 5-63. You can assess the load factor for other nodes by repeating this procedure and picking different nodal points.

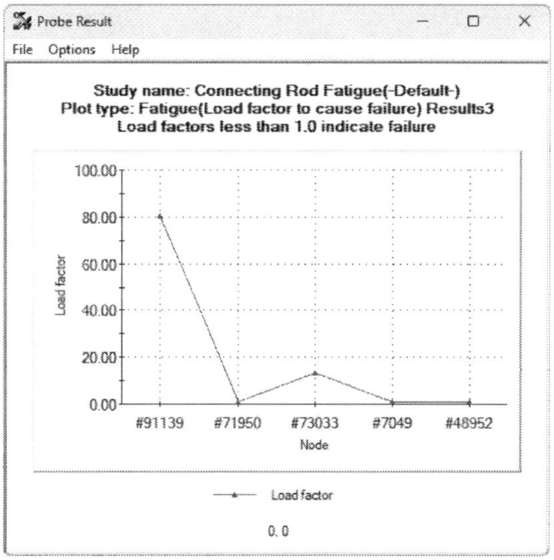

Figure 5-63 *Load factor vs. node graph*

Although, the static study shows that the part would function successfully for a single loading case. But, the fatigue analysis actually shows the performance of the part over a period of time. According to the fatigue study, the part will fail before the desired 1,000,000 cycles. So, you need to reduce the number of cycles.

It is precisely clear from the above mentioned steps that fatigue analysis of a part is extremely crucial to predict the success or failure of a part over an extended period.

Saving the File

After performing the study on the model, you need to save the results and files in the specified folder and then exit the SOLIDWORKS Simulation.

1. Choose the **Save** button in the menu bar and save all the studies.

2. Exit the SOLIDWORKS Simulation.

Tutorial 4

In this tutorial, you will open *c05_tut04* file that was downloaded earlier and perform buckling analysis on an I-beam. Also, you will measure the Critical Load on a column using SOLIDWORKS. The pressure-loaded I-beam column will be subject to three different environments with different end conditions: Fixed-Free, Fixed-Pinned along X-axis, and Fixed-Pinned along X and Y axes. You will perform these three buckling analyses on the I-beam component. The boundary conditions are given in Figure 5-64 and the model is made up of AISI 304.

(Expected time: 1hr)

Figure 5-64 *I-beam*

The following steps are required to complete this tutorial:

a. Start a new part file and import the part.
b. Start a new buckling study.
c. Apply material, fixture, and load.
d. Create mesh.
e. Run the study.
f. Repeat the process twice with different fixtures.
g. Compare the result.

Starting SOLIDWORKS and Importing the Model for Study

You need to start SOLIDWORKS and import the I-beam model (c05-tut04) for analysis.

1. Start SOLIDWORKS Simulation and choose **Open** from the **File** menu in the SOLIDWORKS menus; the **Open** dialog box is displayed.

2. Select the model (c05-tut04) and choose the **Open** button in the dialog box; the selected model is opened in SOLIDWORKS and is ready to start simulation study.

Starting Simulation Study

In this section, you need to start the study using default configuration of the model.

1. Select the **New Study** option from the **New Study** flyout of the **Simulation CommandManager;** the **Study PropertyManager** is displayed.

   Alternatively, you can invoke the **Study PropertyManager** by selecting the **Study** option from the **Simulation** menu.

2. Select the **Buckling** option from the **Study PropertyManager** and choose **OK**; the **Simulation CommandManager** and Simulation Tree will be displayed.

Applying Material

Now, you need to apply AISI 304 material to the model.

1. Right-click on the **c05_tut04** part node in the Simulation Tree and choose the **Apply/Edit Material** option from the shortcut menu; the **Material** dialog box is displayed.

2. Expand the **SOLIDWORKS Materials** node in the **Material** dialog box. Next, expand the **Steel** sub-node under it and choose the **AISI 304** material; all the properties of the selected material are displayed in the right side of the dialog box.

3. Choose the **Apply** button from the dialog box; the AISI 304 material is applied to the model. Next, choose the **Close** button to close the **Material** dialog box.

Applying Boundary Conditions

1. Right-click on the Fixtures in the Simulation Tree and choose the **Fixed Geometry** option from the shortcut menu; the **Fixture PropertyManager** is displayed.

2. Make sure the **Fixed Geometry** button is chosen in the **Standard (Fixed Geometry)** rollout. Next, select the side face of the I-beam, refer to Figure 5-65. Choose the **OK** button.

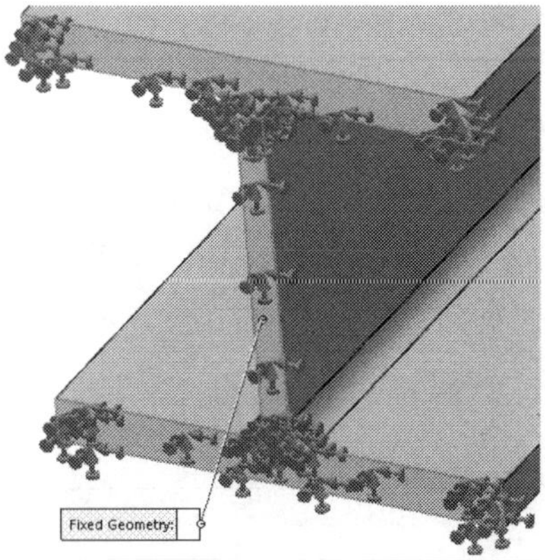

Figure 5-65 *Applying fixtures to bottom face*

3. Now, right-click on the **External Loads** node in the Simulation Tree and choose the **Pressure** option from the shortcut menu displayed; the **Pressure PropertyManager** is displayed.

4. Select the other side face of the beam, refer to Figure 5-66. Set the units to **N/mm ^ 2 (MPa)** in the **Unit** drop-down list in the **Pressure Value** rollout.

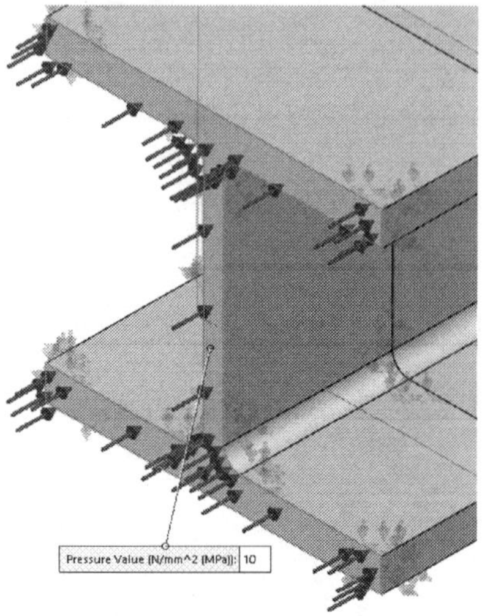

Figure 5-66 *Applying pressure load to top face*

5. Enter **10** in the **Pressure Value** edit box. If the pressure is not pointing downward, reverse the direction by selecting the **Reverse Direction** check box.

6. Choose the **OK** button to exit the **Pressure PropertyManager.**

These boundary conditions simulate a fixed-free condition, where one end is fixed by a rigid support and the other end is completely free.

Creating the Mesh and Running the Study

Since the model is simple, the default mesh will provide good results and consume less computing time as well.

1. Right-click on the **Mesh** folder in the Simulation Tree and choose the **Create Mesh** option from the shortcut menu; the **Mesh PropertyManager** is displayed.

2. Choose the **OK** button from the PropertyManager; the **Mesh Progress** window is displayed and the meshed model is created, refer to Figure 5-67.

3. Right-click on the **Buckling 1** node in the Simulation Tree and choose the **Run** option from the shortcut menu displayed. When the study is completed, a folder with the name **Results** is added to the Simulation Tree.

The **Amplitude1** plot is displayed by default and shows the deformation of the beam under the applied constraints, refer to Figure 5-68 for the plot.

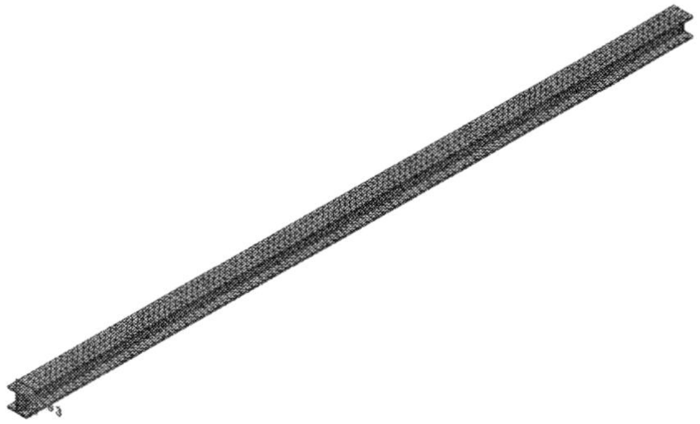

Figure 5-67 The meshed model

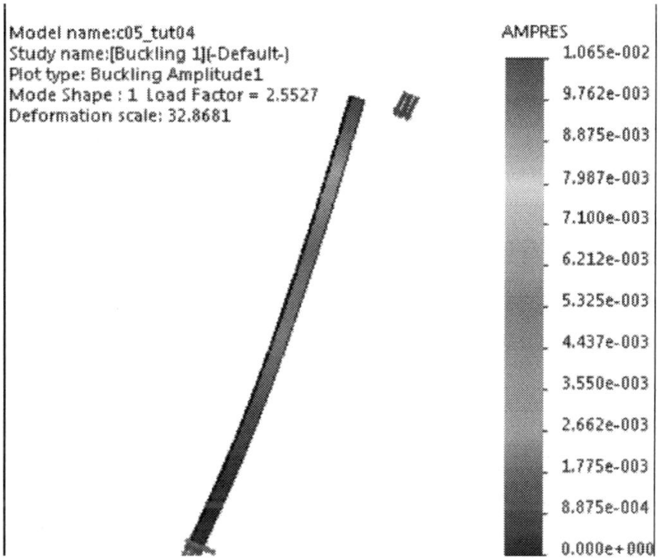

Figure 5-68 AMPRES plot of the I-beam for case 1

4. Right-click on **Amplitude1** node under the **Results** folder and choose the **Animate** option from the shortcut menu displayed; the **Animation PropertyManager** is displayed. You can change the speed of the animation by dragging the **Speed** slider.

Finding Buckling Load Factor (BLF)

The buckling load factor (BLF) is the factor of safety against buckling or the ratio of the buckling loads to the applied loads. It varies for different fixtures despite having the same load. You are going to use SOLIDWORKS to find the BLF for each case.

1. Right-click on the **Results** node and choose the **List Buckling Factor of Safety** option from the displayed shortcut menu; the **List Modes** dialog box is displayed, as shown in Figure 5-69. Choose the **Close** button from the dialog box.

It is evident from Figure 5-69 that the value given under the **Buckling Factor of Safety** column which is 2.5527 for the first case.

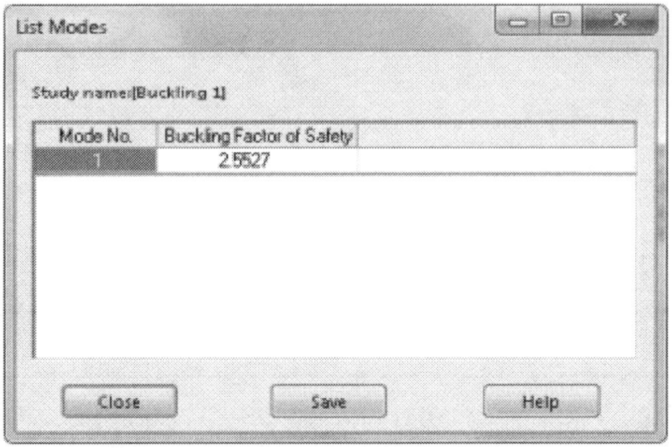

***Figure 5-69** The **List Modes** dialog box*

Creating a New Pinned Fixture

In this section, you will simulate a fixed-pin support case. To do so, first need to create a pinned connection.

1. Choose the **Features** tab in the **SOLIDWORKS CommandManager** of the Model drawing area.

Note

*To return the beam to default position, right-click on **Amplitude** and choose **Hide**, and then click anywhere in the drawing area.*

2. Choose the **Reference Geometry** option from the **CommandManager** and choose the **Plane** option from the drop-down menu; the **Plane PropertyManager** is displayed.

3. Select the load-bearing top face of the I-beam for the first reference. Enter **50** in the **Offset Distance** edit box and select the **Flip offset** check box. Next, choose the **OK** button; a plane is created in the model, refer to Figure 5-70.

4. Next, select **Curves** from the **Command Manager** by keeping the new plane selected and then select the **Split Line** option from the drop-down menu; the **Split Line PropertyManager** is displayed.

5. Now, select the **Intersection** radio button under the **Type of Split** rollout and select all lateral faces of the I-beam including the fillets in the **Faces/Bodies to Split** selection box in the **Selections** rollout. Now, choose the **OK** button; a split is created in the model, refer to Figure 5-71.

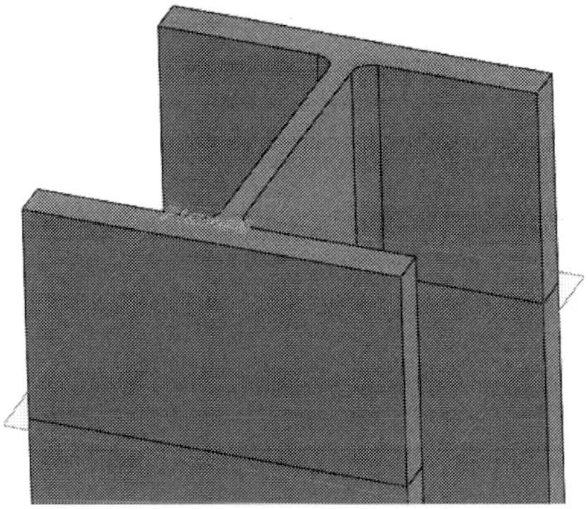

Figure 5-70 Newly created plane

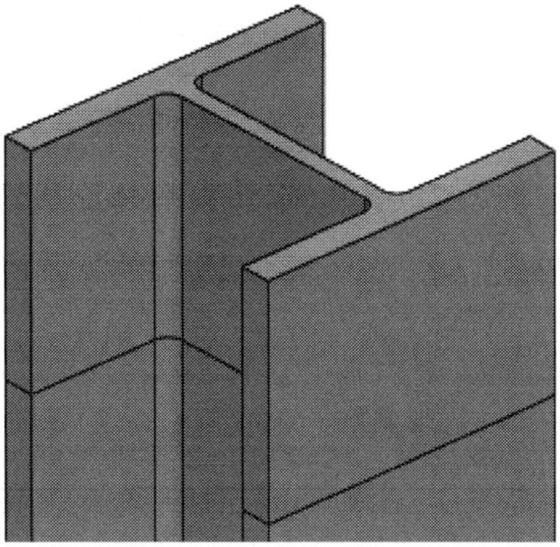

Figure 5-71 Split Faces

Note

If split is created successfully, you will see the "I" shape, highlighted in yellow, at the split section.

6. Choose the **Buckling 1** Study tab to enter the Simulation environment.

7. Right-click on **Fixtures** in the Simulation Tree; a shortcut menu is displayed. Next, choose the **Advanced Fixtures** option from the shortcut menu.

8. In the **Advanced (Use Reference Geometry)** rollout, select all the 16 newly created faces in the **Faces, Edges, Vertices for Fixtures** selection box.

9. Now, select any edge of the model along the X-axis to restrict the motion along this axis in the **Face, Edge, Plane, Axis for Direction** selection box. Choose the **OK** button, refer to Figure 5-72.

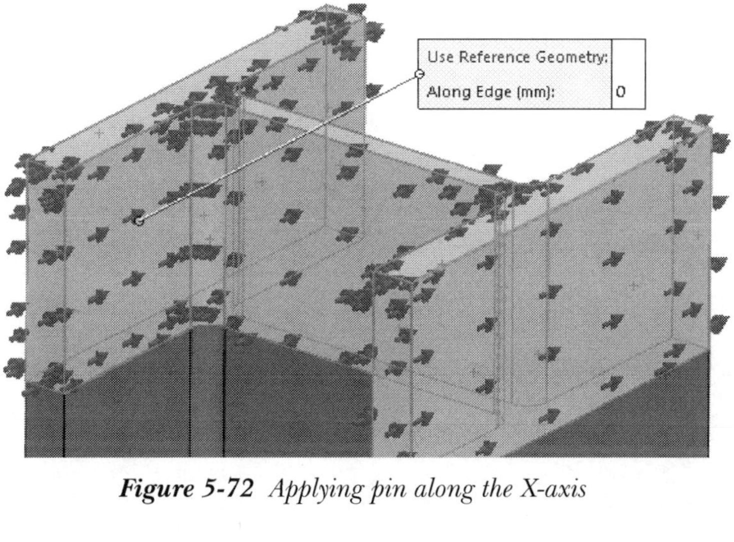

Figure 5-72 Applying pin along the X-axis

This combination simulates a pinned support where translation is only allowed along the vertical axis. In this case, the y-axis and the z-axis are free.

Creating the Mesh and Performing Study

You need to run the study for the newly applied restraints.

1. Now, choose the **Run this Study** button from the **Simulation CommandManager;** the **Amplitude1** plot is obtained after solving the study, as shown in Figure 5-73.

Finding BLF

The beam bends with the new fixture because of different effective loads applied on the beam. So, you need to check BLF in this section.

1. Right-click on the **Results** node in the Simulation Tree; a shortcut menu is displayed.

2. Choose the **List Buckling Factor of Safety** option from the shortcut menu; the **List Modes** dialog box is displayed.

   Notice, the value under the **Buckling Factor of Safety** column in the dialog box, which is 6.6403 in this case.

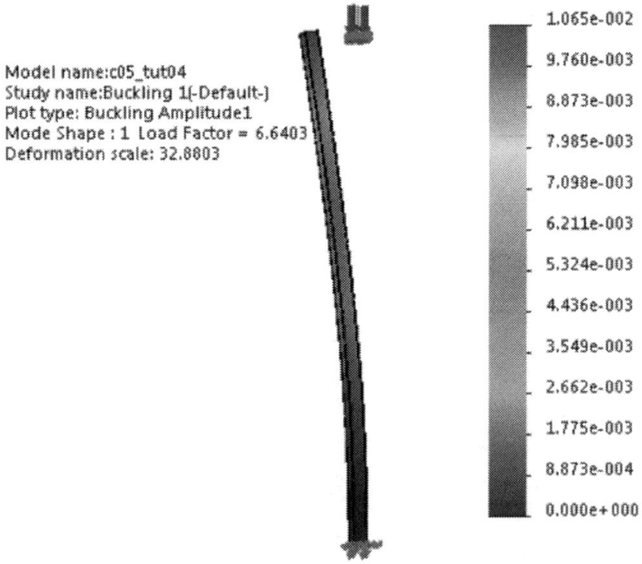

Figure 5-73 *AMPRES plot of the I-beam for case 2*

Applying the Third Set of Boundary Conditions

You will finally create a fixture to restrict motion along the Y-axis.

1. Right-click on **Fixtures** in the Simulation Tree; a shortcut menu is displayed. Select the **Advanced Fixtures** option from the shortcut menu.

2. In the **Advanced(Use Reference Geometry)** rollout, again select all the 16 newly created faces in the **Faces, Edges, Vertices for Fixtures** selection box.

3. Now, select the **Face, Edge, Plane, Axis for Direction** selection box. Select any edge from the model along the Y-axis to restrict any motion along the Y-axis. Choose the **OK** button.

   This combination simulates a pinned support where translation is only allowed along the vertical axis and restricts all motion along the horizontal directions.

Creating the Mesh, Running the Study and Finding BLF

In this section, you will run the study for the newly applied restraints.

1. Choose the **Run this Study** button from the **Simulation CommandManager**; the AMPRES plot is obtained, as shown in Figure 5-74.

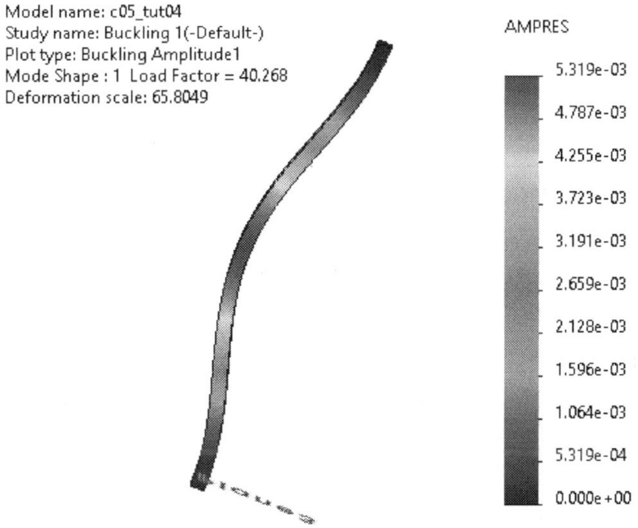

Figure 5-74 *AMPRES plot of the I-beam for case 3*

2. Right-click on the **Results** folder in the Simulation Tree and then choose the **List Buckling Factor of Safety** option from the shortcut menu; the **List Modes** dialog box is displayed.

   Notice, the value under the Buckling Factor of Safety column, which is 40.271 in this case.

Analyzing the Results

Using the BLF values obtained, you will find the critical load at which the I-beam will buckle for each of the three cases.
You will use the formula to obtain the following table:

Critical Load = Buckling Load Factor x Applied Load

Type	Buckling Load Factor	Applied Load	Critical Load
Fixed-Free	2.5527	10 MPa	25.527 MPa
Fixed-Pinned (X-Axis)	6.6403	10 MPa	66.403 MPa
Fixed-Pinned (X, Y-Axis)	40.271	10 MPa	402.71 MPa

As expected, in the third case where more direction constraints are applied, the critical load is highest. Therefore, the beam will initially bear the load but will fail at the last.

Saving the Study

After performing the study on the model, you need to save the results and files in the specified folder and then exit SOLIDWORKS Simulation.

1. Choose the **Save** button in the menu bar and save all the studies.

2. Exit SOLIDWORKS Simulation.

Self-Evaluation Test

Answer the following questions and then compare them to those given at the end of this chapter:

1. The number of buckling modes can be edited using the ______________ option from the shortcut menu displayed on right-click on the **Buckling Study** node in the Simulation Tree.

2. For a ______________ type loading event, the minimum stress value is a factor of the maximum stress.

3. The **Symmetry Fixtures** options are available in the ______________ rollout in the **Fixtures Property Manager.**

4. Besides the ______________ tool in the **Fixtures PropertyManager**, the **Symmetry Check** tool or the **Surface Cut** tool can also be used to split symmetric parts.

5. A ______________ type connector connects cylindrical faces of two components.

6. You can use the ______________ list box to list the factors of safety against buckling.

7. Interference during simulation can occur between two load-bearing faces which have a bonded contact. (T/F)

8. SOLIDWORKS Simulation allows you to create a fatigue study without creating any previous studies. (T/F)

9. While using the **Cyclic Symmetry** fixture option, you need to select the faces about which the model is symmetric. (T/F)

10. If the alternating stress at a node lies below the S-N curve for that particular material then fatigue failure will be predicted at that point. (T/F)

Review Questions

Answer the following questions:

1. Which of the following is the K value in Euler's formula for the fixed-pinned support?

   (a) 1 (b) 0.7071
   (c) 0.5 (d) 2.46

2. Which of the following connectors cannot be utilized while performing a buckling study?

   (a) Bolt (b) Pin
   (c) Elastic Support (d) Spring

3. Which of the following studies should not be setup using symmetry approach of FEA?

   (a) Dynamic studies (b) Thermal studies
   (c) Static studies (d) Buckling studies.

4. The two types of S-N Curves that can be defined from a material's Elastic Modulus in SOLIDWORKS are Austenitic Steel Curves and ______________.

5. Custom S-N curves for a material can be defined using the ______________ tool under the Study node in the Simulation Tree.

6. A value of 100% ______________ predicts that the particular location will be completely consumed by the loading events in a damage.

7. When applying Cyclic Symmetry, the selected ______________ should lie at the intersection of the two selected planes/faces.

8. Before specifying a bearing connector, ______________ must be created on the shaft to avoid unwanted added stiffness in motion.

9. You can modify buckling results to display only the amplitude along a particular axis. (T/F)

10. When applying contacts between parts, local contact conditions override global and component contact conditions. (T/F)

EXERCISES

To perform the exercises of this chapter, you need to download the input file from *www.cadcim.com*. The complete path for downloading the file is:

Textbooks > CAE Simulation > SOLIDWORKS Simulation > SOLIDWORKS Simulation 2022: A Tutorial Approach > Input Files > c05_simulation_2022_input.zip

Extract the downloaded zipped file at the location *C:\SOLIDWORKS Simulation\c05.*

Exercise 1

In this exercise, you will use the *c05_exr01* file and create the cyclic symmetry and then perform the static structural study. The cooling fan model is made up of PP Homopolymer material, as shown in Figure 5-75. The boundary conditions will be applied such as 2 N-m torque on the lateral face and bearing fixture on the inner cylindrical face of the fan, as shown in Figure 5-76 and Figure 5-77. Also, to understand the importance of geometry optimization, you will optimize the model, and then evaluate the results. **(Expected time: 45 min)**

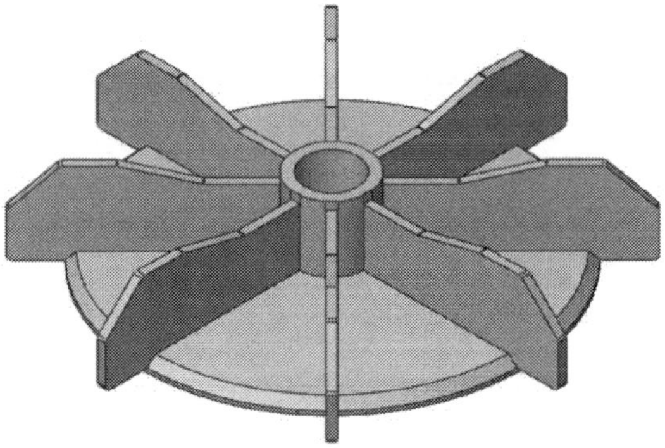

Figure 5-75 *The full model of cooling fan*

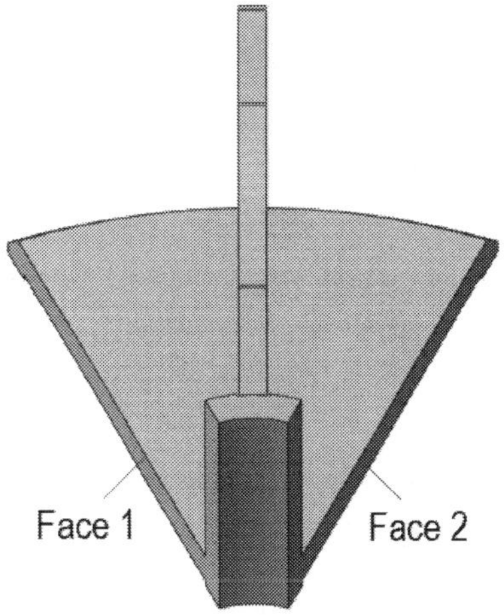

Figure 5-76 The 1/4 model for Exercise 1

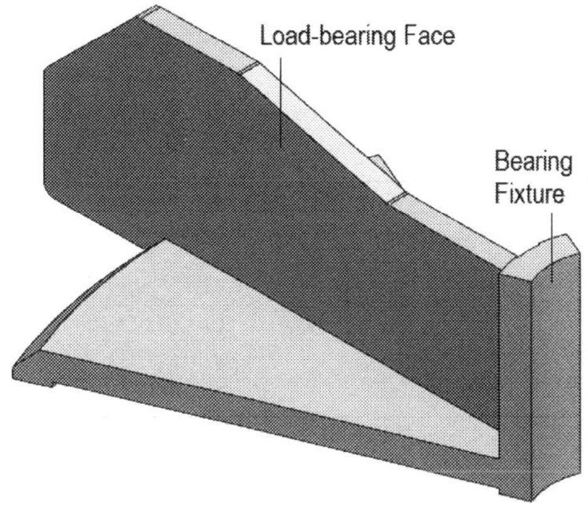

Figure 5-77 The boundary conditions for Exercise 1

Exercise 2

In this exercise, you will use the *c05_exr02* file and create the contacts and perform the static structural study. The model and its boundary conditions are given in Figure 5-78. The load is applied along the annotated edge. The model is made by the Alloy Steel (SS) material. You will run a static structural study, optimize the model, and then evaluate the results.

(Expected time: 1 hr)

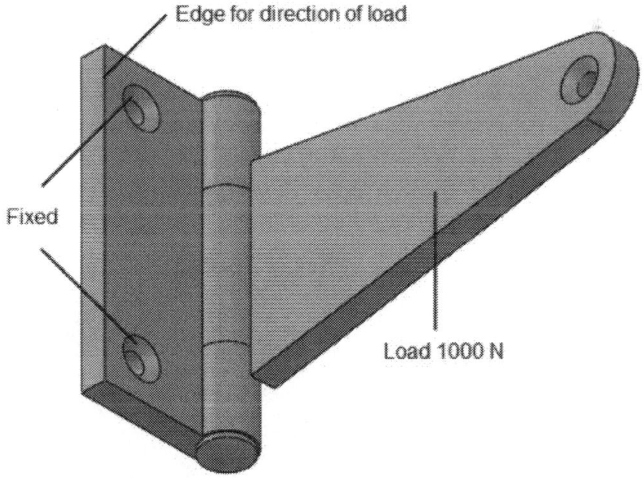

Figure 5-78 The model for Exercise 2

Exercise 3

In this exercise, you will use *c05_exr03* file to create supports and connections and then perform buckling analysis on the Table model. Also, you will measure the critical load on a column. 500 N load will be applied on top of the model and which is fixed at the bottom. The boundary conditions are given in Figure 5-79 and the model is made of AISI 304.

(Expected time: 1 hr)

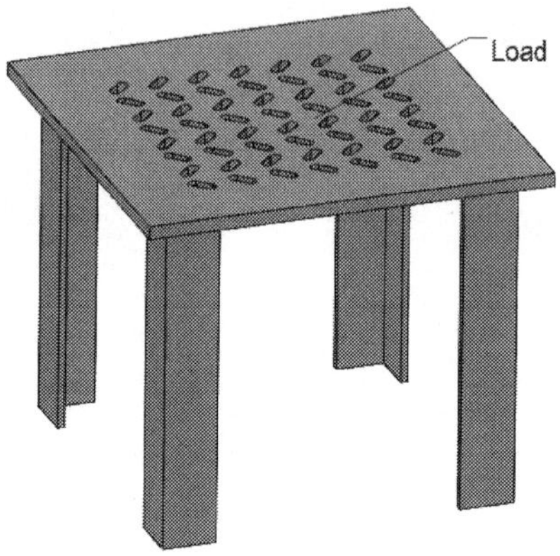

Figure 5-79 *The model for Exercise 3*

Exercise 4

In this exercise, you will use the *c05_exr04* file to create supports, connections, and perform fatigue analysis on an assembly model. The pressure of 5 N/mm2 is applied on the piston and another end is kept fixed. The boundary conditions are given in Figure 5-80 and the model is made of AISI 304 stainless steel material.

(Expected time: 1 hr)

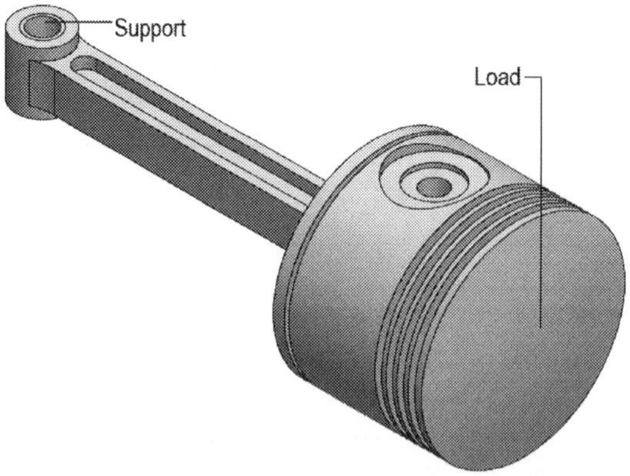

Figure 5-80 *The model for Exercise 4*

Answers to Self-Evaluation Test

1. **Properties**, **2**. **Loading Ratio**, **3**. **Advanced Fixtures**, **4**. **Split**, **5**. **Pin**, **6**. **List Modes**, **7**. F, **8**. F, **9**. T, **10**. F

Chapter 6

Frequency Analysis

Learning Objectives

After completing this chapter, you will be able to:

- *Perform frequency analysis*
- *Apply boundary conditions*
- *Apply different types of constraints*
- *Generate the result of frequency analysis*

FREQUENCY ANALYSIS

Frequency analysis is a modal analysis and it provides information regarding problematic vibrational response a structure is likely to experience at particular frequencies. It provides two results, the resonant or natural frequencies of your model and the mode shapes that will occur at those resonant frequencies.

From the design perspective, frequency analysis is most often used to confirm whether your structural design has any resonant frequencies or it is near the operating range of externally applied vibrational loading for your model. For example, if you are designing a support for a rotatory drive which runs at 5000 rpm, and the support has a natural frequency at 115 Hz, you are likely to experience performance problems. You can use the mode shape provided by the analysis to review the vibration and improve the design. Frequency analysis does not include any dynamic loading applied and also does not provide any information about the vibratory response of the dynamic loading. To check this type of information, you need to perform dynamic analysis.

Assumptions of Frequency Analysis

There are many assumptions related to loads and material while performing frequency analysis. These are listed below:

- Distributional assumptions about the residuals for examples include independence, normality, and constant variance in a linear model.
- Relation between loads and induced responses is linear.
- Hooke's law is applicable.
- All material properties such as Young's modulus and Poisson's ratio remain constant throughout the study.

Starting Frequency Analysis in SOLIDWORKS Simulation

To perform frequency analysis in SOLIDWORKS Simulation, you need to start a new study. To start the study, choose the **New Study** tool from the **New Study** flyout in the **Simulation CommandManager,** refer to Figure 6-1 ; the **Study PropertyManager** is displayed. Choose the **Frequency** button from the **General Simulation** rollout to start the frequency study and then choose **OK**.

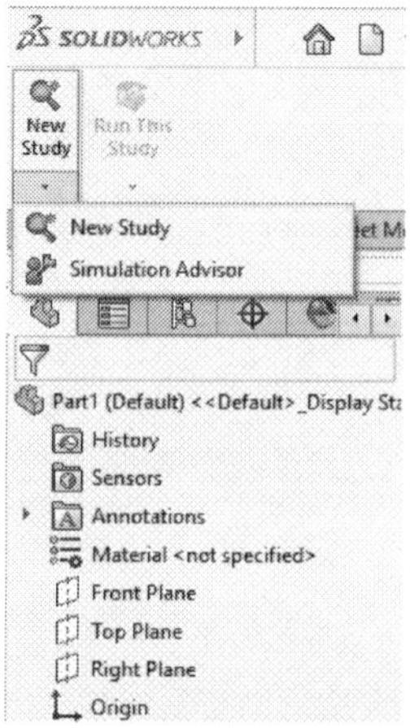

***Figure 6-1** The **New Study** flyout*

Specifying Frequency Analysis Settings in SOLIDWORKS Simulation

After generating mesh for the model, it is required to specify the settings needed to run frequency analysis. To do so, right-click on the **Frequency** node in the **Simulation Tree**; a shortcut menu will be displayed. Choose the **Properties** option from the shortcut menu; the **Frequency** dialog box will be displayed, as shown in Figure 6-2. In this dialog box, specify the number of frequencies and upper limit of the frequency. You can also specify the flow and thermal effect in the analysis by using the options available in this dialog box.

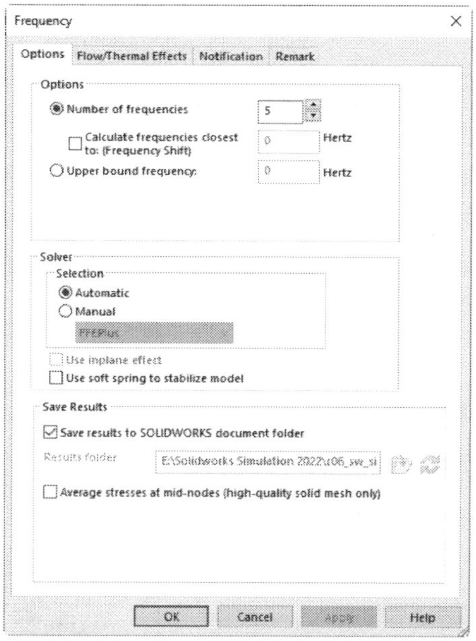

*Figure 6-2 The **Frequency** dialog box*

Plotting the Deformed Shape (Mode Shape)

You can plot a deformed shape at each of its mode. However, before plotting the deformed shape, you need to check the graph in between the number of modes and frequency. To create the graph, right-click on the **Results** node in the **Simulation Tree**; a shortcut menu will be displayed, refer to Figure 6-3. Choose the **Define Frequency Response Graph** option from the shortcut menu; the **Frequency Analysis Graph PropertyManager** will be displayed. Choose the **OK** button from the PropertyManager; the **Frequency vs** window will be displayed which shows the graph between frequencies and number of modes, refer to Figure 6-4. In this window, you can check the frequency at every mode. You can also plot the mode shape of a structural component at a particular frequency. To do so, right-click on the **Results** node in the **Simulation Tree**; a shortcut menu will be displayed, refer to Figure 6-5. Choose the **Define Mode Shape/Amplitude Plot** option from the shortcut menu; the **Mode Shape/Amplitude Plot PropertyManager** will be displayed. You can specify the required options in this PropertyManager and then choose the **OK** button to obtain the plot, refer to Figure 6-6.

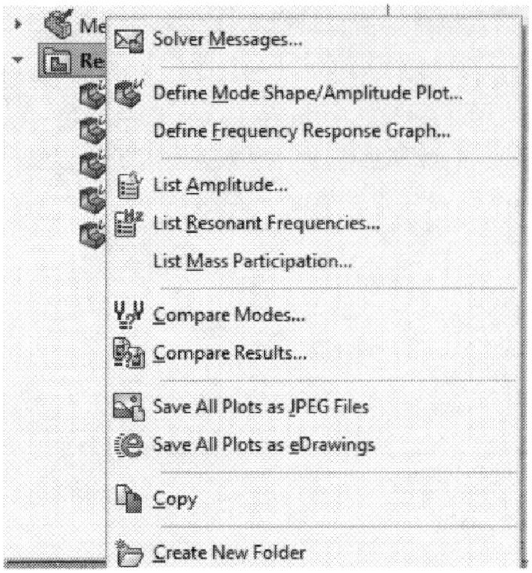

Figure 6-3 *The shortcut menu*

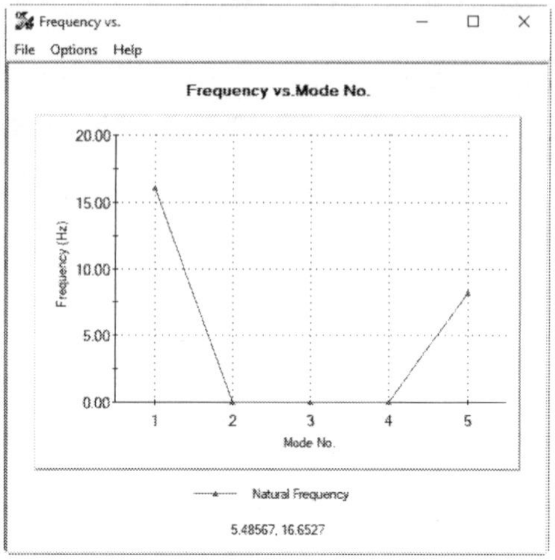

Figure 6-4 *The frequency vs Mode no.plot*

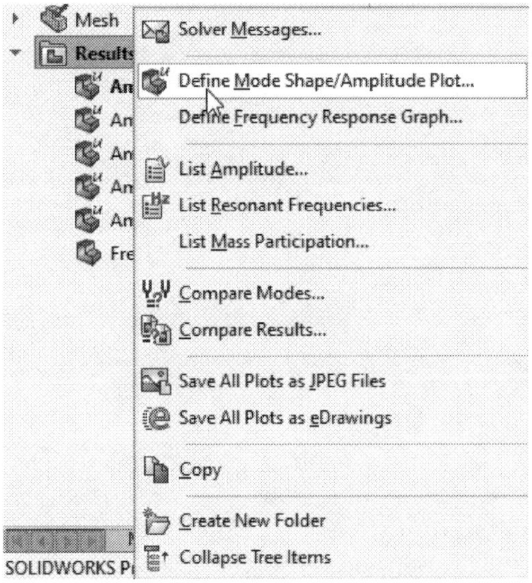

Figure 6-5 *The shortcut menu*

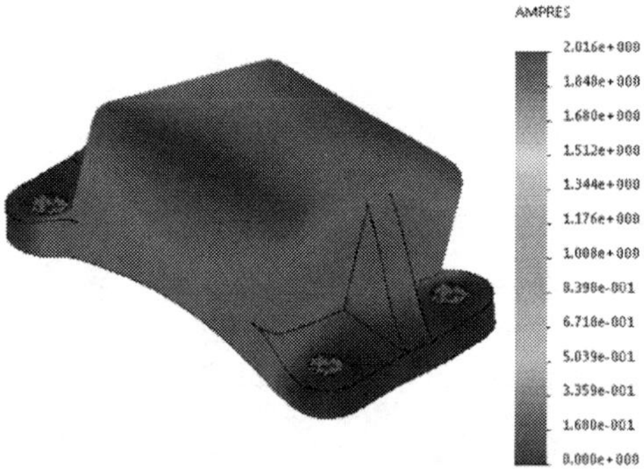

Figure 6-6 *The Amplitude plot*

TUTORIALS

To perform the tutorials, you need to download the input files from www.cadcim.com. The complete path for downloading the files is as follows:

Textbooks > CAE Simulation > SOLIDWORKS Simulation > SOLIDWORKS Simulation 2022: A Tutorial Approach > Input Files > c06_simulation_2022_input.zip

Extract the downloaded zipped file and save it.

Tutorial 1

In this tutorial, you will open the *c06_tut01* file downloaded earlier and perform the Frequency Study on the model given in Figure 6-7. You will generate mesh with default element size and then find 5 natural frequencies and their respective mode shapes. You will apply Chrome Stainless Steel material to the model and fixed restraint on the surface of the spherical wall, as shown in Figure 6-7. **(Expected time: 45 min)**

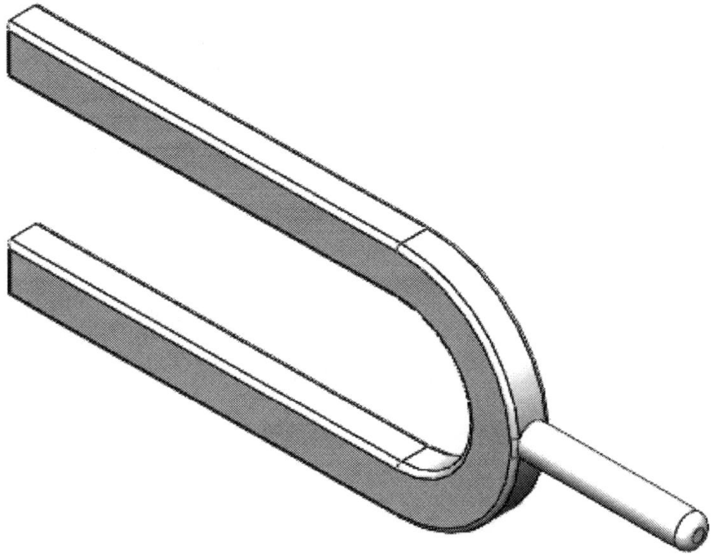

Figure 6-7 *Model for frequency analysis*

The following steps are required to complete this tutorial:

a. Start SOLIDWORKS Simulation and import the model.
b. Start a new frequency study.
c. Apply material and fixture.
d. Create mesh.
e. Run the study.
f. View the results.
g. Generate the report.
h. Save the model.

Starting SOLIDWORKS Simulation and Importing the Model

Now, you need to import the model *(c06_tut01)* for analysis.

1. Choose **Open** from the **File** menu in the **SOLIDWORKS** menus; the **Open** dialog box is displayed.

2. Select the model *(c06_tut01)* and choose the **Open** button from the dialog box; selected model is opened in SOLIDWORKS.

Starting a New Study and Defining the Frequency Study Properties

Since you want to determine 5 natural frequencies and their respective mode shapes, you need to start a frequency analysis.

1. Start SOLIDWORKS Simulation and choose the **New Study** tool from the **Study** flyout in the **Simulation CommandManager**; the **Study PropertyManager** is displayed.

2. Choose the **Frequency** button from the **General Simulation** rollout and then enter the name of the study as **Tuning Fork** in the **Study name** edit box of the PropertyManager.

3. Next, choose the **OK** button; a new node named **Tuning Fork** is added to the Simulation Tree.

4. Right-click on the **Tuning Fork** node in the Simulation Tree; a shortcut menu is displayed.

5. Choose the **Properties** option from this shortcut menu; the **Frequency** dialog box is displayed, as shown in Figure 6-8.

6. Enter **5** in the **Number of Frequencies** edit box and accept default settings for rest of the options in the dialog box and then choose the **OK** button, refer to Figure 6-8.

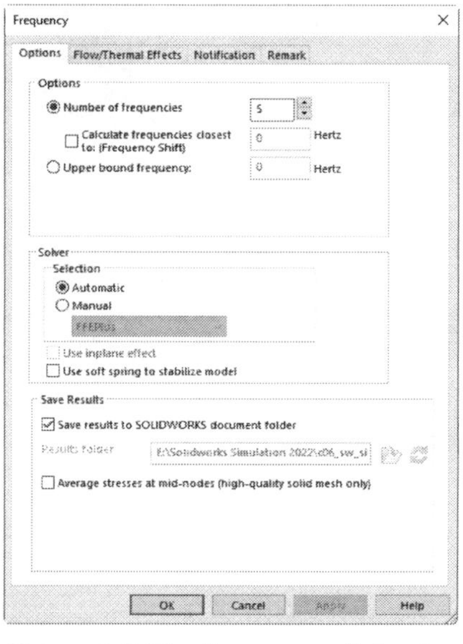

***Figure 6-8** The **Frequency** dialog box*

Applying the Material

Now, you need to apply the material to the model.

1. Choose the **Apply Material** button from the **Simulation CommandManager**; the **Material** dialog box is displayed.

2. Expand the **SOLIDWORKS Materials** node in the **Material** dialog box. Next, expand the **Steel** sub-node and choose **Chrome Stainless Steel**; all the properties of the selected material are displayed on the right in the dialog box.

3. Choose the **Apply** button from the dialog box; the Chrome Stainless Steel material is applied to the model. Next, close the **Material** dialog box.

Applying Constraints

After applying the material to the model, you need to apply constraints to fix the cylindrical filleted surface of the model.

1. Choose the **Fixed Geometry** tool from the **Fixtures Advisor** flyout of the **Simulation CommandManager**; the **Fixture PropertyManager** is displayed on the left in the drawing area.

2. Make sure that the **Fixed Geometry** button is chosen in the **Standard (Fixed Geometry)** rollout of the PropertyManager.

3. Select the cylindrical filleted surface of the model; the selected face is fixed and the fixed geometry callout is attached to it, as shown in Figure 6-9.

4. Choose the **OK** button from the PropertyManager.

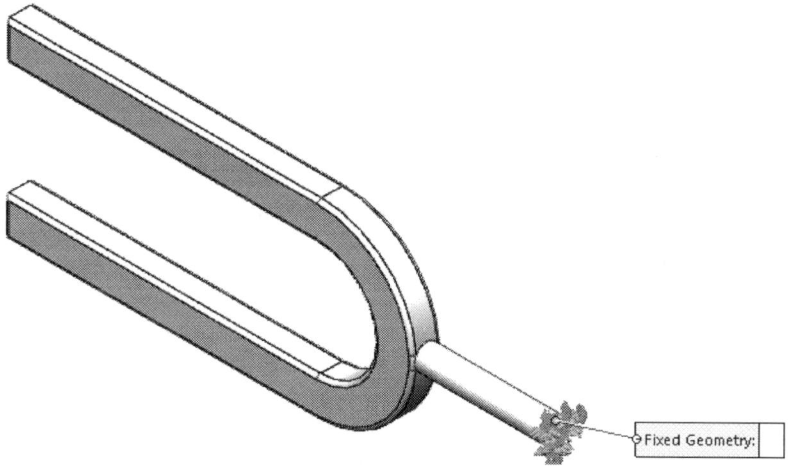

Figure 6-9 *The face of the model is fixed*

Creating Mesh

After applying the boundary condition to the model, you need to create the mesh.

1. Choose the **Create Mesh** tool from the **Run This Study** flyout in the **Simulation CommandManager**; the **Mesh PropertyManager** is displayed.

2. Accept all the default settings and choose the **OK** button from the PropertyManager; the **Mesh Progress** window is displayed. Once the process of meshing is done, the meshed model is displayed in the drawing area, as shown in Figure 6-10.

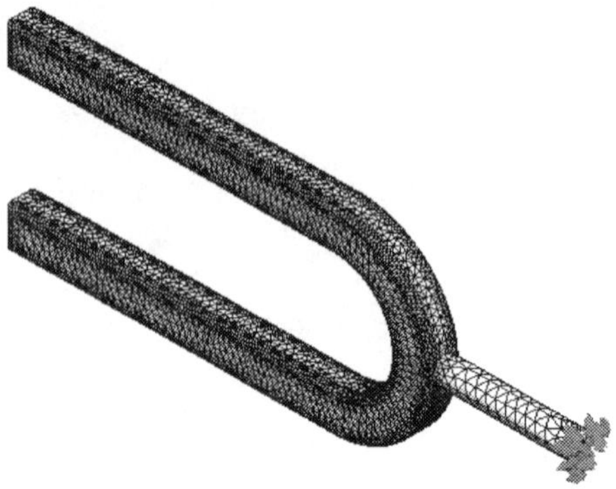

Figure 6-10 *The model after meshing*

Running the Study

Now, you need to run the analysis on the model.

1. Choose the **Run This Study** tool from the **Simulation CommandManager**; the **Tuning Fork** window is displayed and the process of solving the FE model is started. Once the analysis process is completed, a new node named **Results** is added to the **Simulation Tree**. Also, the resultant model is displayed in the drawing area with the mode shape 1 contour, as shown in Figure 6-11.

2. Right-click on the **Amplitude 1 (-Res Amp-)** under the **Results** node; a shortcut menu is displayed. Choose the **Animate** option; the animation starts with default settings. Also, the **Animation PropertyManager** is displayed on the left of the screen enabling you to change the settings according to your requirement.

   Similarly, you can check or plot the result for the other mode shapes such as mode 1, mode 2, mode 3, mode 4, and mode 5. Note that to activate the result plot, you need to double click on the result node in the Simulation Tree.

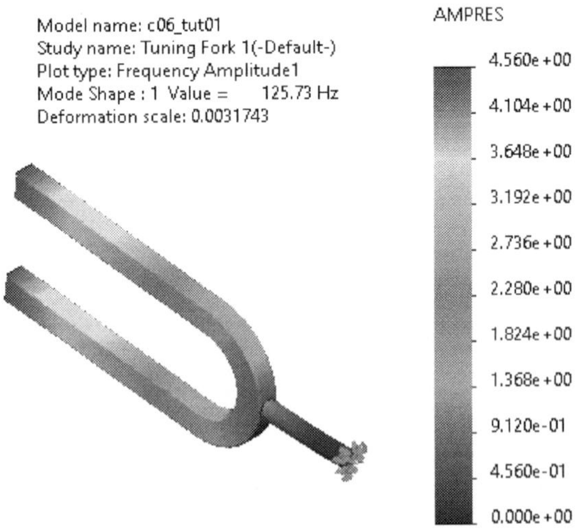

Figure 6-11 *The resultant model with mode shape 1 contour*

Generating the Report

Now, you need to generate the report.

1. Choose the **Report** tool from the **Simulation CommandManager**; the **Report Options** dialog box is displayed.

2. Select all the sections that you want to add to the report by selecting the respective check boxes available in the **Report sections** area of the dialog box.

   As discussed earlier, you can also add information such as designer name, company, URL, logo, and so on to the header of the report. To do so, you need to enter information in the corresponding edit boxes enabled after selecting the corresponding check boxes.

3. Choose the **Publish** button to generate the report in Word format. As soon as the analysis report is generated and displayed in Word document, the **Report** node is added to the Simulation Tree. Close the report after viewing.

Saving the File

After performing all the actions and achieving the desired results, you need to save the model.

1. Choose **File > Save** from the menu bar to save the simulation files.

Tutorial 2

In this tutorial, you will open *c06_tut02* file downloaded earlier and perform the static and modal analysis on the model of support bracket given in Figure 6-12. You will generate the mesh by using shell elements and find 5 natural frequencies and their respective mode shapes. The material used to perform study is Galvanized Steel. Boundary conditions for the analysis are shown in Figure 6-12 and fixed supports are to be applied on all holes of the model.

(Expected time: 45 min)

The following steps are required to complete this tutorial:

a. Start SOLIDWORKS Simulation and import the model.
b. Start a new static study.
c. Apply material and fixture.
d. Create mesh.
e. Run the study.
f. View the results.
g. Start a new frequency study.
h. Copy the results from the static study.
i. Generating the report.
j. Save the model.

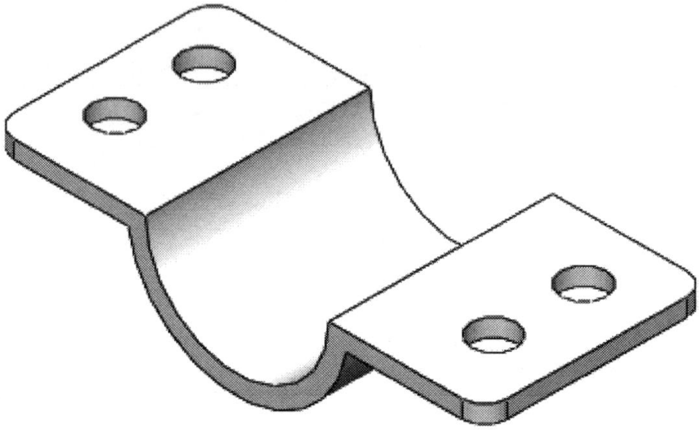

Figure 6-12 *Model for study*

Starting SOLIDWORKS Simulation and Importing the Model

Now, you need to import the model *(c06_tut02)* for analysis.

1. Choose **Open** from the **File** menu in the **SOLIDWORKS** menus; the **Open** dialog box is displayed.

2. Select the model *(c06_tut02)* and choose the **Open** button from the dialog box; selected model is opened in SOLIDWORKS.

Starting a New Static Study

Since you want to find stress and five modes of vibration, you need to run both static and frequency study on the model.

1. Start SOLIDWORKS Simulation and choose the **New Study** tool from the **New Study** flyout in the **Simulation CommandManager**; the **Study PropertyManager** is displayed.

2. Choose the **Static** button from the **General Simulation** rollout and then enter the name of the study as **Static Analysis** in the **Study name** edit box of the PropertyManager. Next, choose **OK**; a new node named **Static Analysis** is added to the Simulation Tree.

Applying the Material

Now, you need to apply the Galvanized Steel material to the model.

1. Choose the **Apply Material** button from the **Simulation CommandManager**; the **Material** dialog box is displayed.

2. Expand the **SOLIDWORKS Materials** node of the **Material** dialog box. Next, expand the **Steel** sub-node and choose the **Galvanized Steel** material; all the properties of the selected material are displayed on the right of the dialog box.

3. Choose the **Apply** button from the dialog box; the Galvanized Steel material is applied to the model. Next, close the **Material** dialog box.

Applying Constraints

After applying the material to the model, you need to apply constraints to fix the holes of the model.

1. Choose the **Fixed Geometry** tool from the **Fixtures Advisor** flyout of the **Simulation CommandManager**; the **Fixture PropertyManager** is displayed at the left of the drawing area.

2. Next, choose the **Fixed Geometry** button in the **Standard (Fixed Geometry)** rollout of the PropertyManager.

3. Select the upper cylindrical edge of all the holes in the model; the selected edges are displayed in the **Faces, Edges, Vertices for Fixture** selection box and the fixed geometry fixture symbol is attached to the edges, as shown in Figure 6-13.

4. Choose the **OK** button from the PropertyManager.

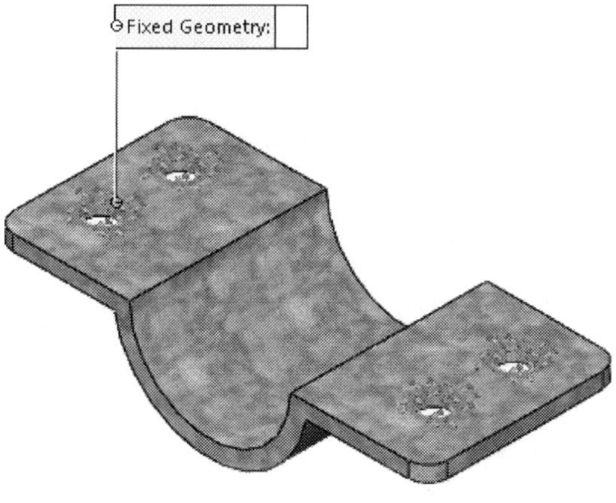

Figure 6-13 *The holes of the model are fixed*

Applying Loads

After applying the material and constraints to the model, you need to apply load to it.

1. Choose the **Bearing Load** button from the **External Loads Advisor** flyout of the **Simulation CommandManager**; the **Bearing Load PropertyManager** is displayed on the left in the drawing area.

2. Select the top semi-circular face of the model to apply the bearing load, refer to Figure 6-14. Now, you need to select coordinate system to apply the bearing load to the model.

3. Expand the design tree and then select coordinate system in the **Select Coordinate System** selection box of the **Selected Entities** rollout of the **Bearing Load PropertyManager.**

You can also create the coordinate system at the center of the cylindrical face. Now, you need to specify the direction of force.

4. Choose the **Y-Direction** button from the **Bearing Load** rollout of the PropertyManager and enter **150** in the edit box to specify the bearing load in the Y-direction.

   If the direction of the applied force is vertically upward, you need to reverse it.

5. Select the **Reverse direction** check box from the **Bearing Load** rollout in the PropertyManager.

6. Select **SI** from the **Unit** drop-down list of the **Units** rollout in the PropertyManager.

7. Accept the other default settings and choose **OK** to apply the load. Figure 6-14 shows the model after applying the bearing load.

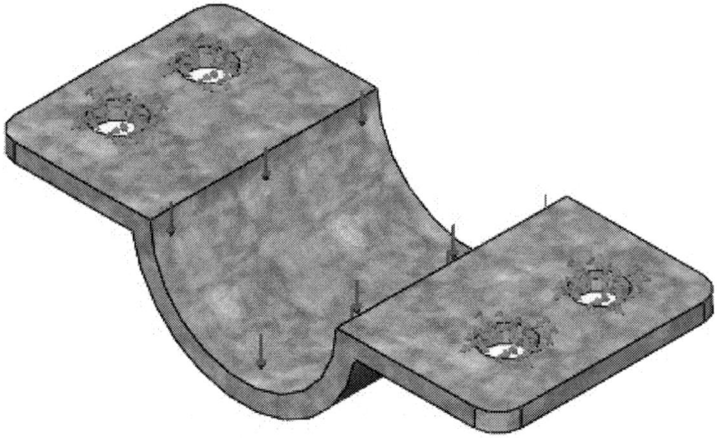

Figure 6-14 *Model after applying the bearing load*

Creating the Mesh

After applying the boundary conditions to the model, you need to create the mesh.

1. Choose the **Create Mesh** tool from the **Run This Study** flyout of the **Simulation CommandManager**; the **Mesh PropertyManager** is displayed.

2. Accept all the default settings and choose the **OK** button from the PropertyManager; the **Mesh Progress** window is displayed. Once the process of meshing is done, the meshed model is displayed in the drawing area. This is the finite element model created using the solid elements. Now, you need to convert solid elements into the shell elements.

3. Select the **c06_tut02 (-Galvanized Steel-)** node from the Simulation Tree. Next, right-click on it; a shortcut menu is displayed.

4. Choose the **Define Shell By Selected Faces** option from the shortcut menu; the **Shell Definition PropertyManager** is displayed.

5. Select all top faces of the model and enter **2.5** in the **Shell thickness** edit box. Make sure **mm** is selected in the Unit drop-down. Choose the **OK** button from the PropertyManager.

   Now, you need to re-mesh the model because you have changed the element type in the finite element model.

6. Select the **Mesh** node from the Simulation Tree and right-click; a shortcut menu is displayed.

7. Choose the **Create Mesh** option from the shortcut menu; the **Mesh PropertyManager** is displayed. Accept the default settings and choose the **OK** button from the PropertyManager.

   You will notice that a shell mesh node is created in the Simulation Tree and all the created elements of the mesh are shell elements. The finite element model with shell elements is shown in Figure 6-15.

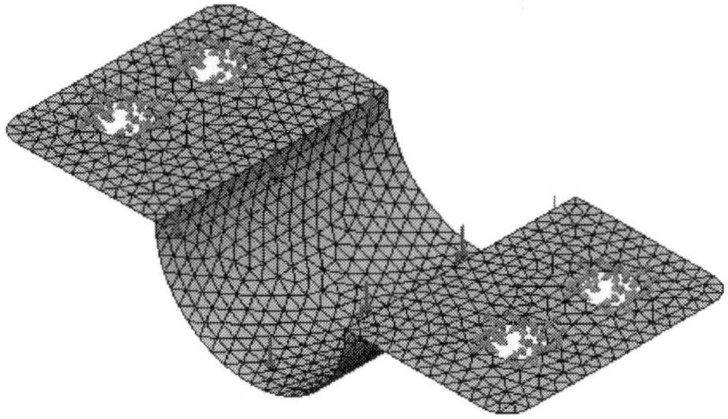

Figure 6-15 *The FE model with shell elements*

Running the Study

Now, you need to run study on the model to check the response of the static study on the model.

1. Choose the **Run This Study** tool from the **Simulation CommandManager**; the **Static Analysis** window is displayed and the process of solving the finite element model gets started. Once the analysis process is completed, a new node named **Results** is added to the Simulation Tree. Also, the resultant model is displayed in the drawing area with the von Mises stress contour, as shown in Figure 6-16.

2. Right-click on the **Stress 1 (-vonMises-)** under the **Results** node; a shortcut menu is displayed. Select the **Animate** option; the animation starts with the default settings. Also, the **Animation PropertyManager** is displayed on the left of the drawing area enabling you to change the settings according to your requirement.

   Similarly, you can check or plot the other results for the analysis.

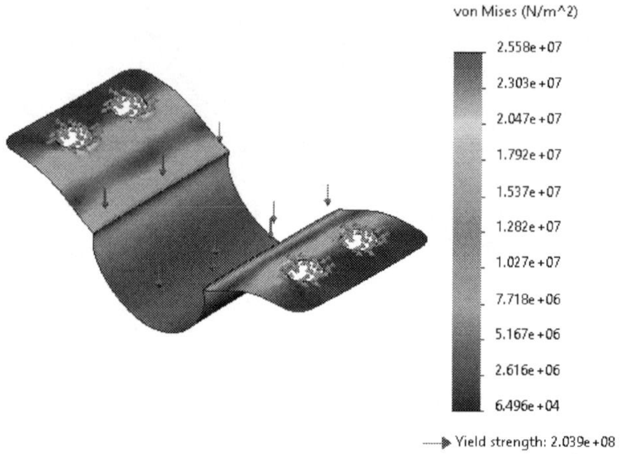

Figure 6-16 *The resultant model with von Mises stress contour*

Starting a New Frequency Study

Now, start a new Frequency Study.

1. Choose the **New Study** tool from the **New Study** flyout in the **Simulation CommandManager;** the **Study PropertyManager** is displayed.

2. Choose the **Frequency** button from the **General Simulation** rollout and then enter the name of the study as **Frequency Analysis** in the **Study name** edit box of the PropertyManager. Next, choose **OK**; a new node named **Frequency Analysis** is added to the Simulation Tree.

Copying the Result from Existing Study

Now, you need to setup the material properties, boundary conditions, and meshing results to the newly added study. In SOLIDWORKS Simulation, you can do these operations separately or you can also copy the results from the existing study.

1. Select the **Material** node in the static analysis study and then drop on the recently added frequency study at the bottom of the SOLIDWORKS window; the same material gets applied to the frequency study.

   Similarly, you need to transfer the Fixture and Mesh from static study to the frequency study. After transferring the material, fixture, and mesh results to the newly added frequency study, you will solve this study.

Running the Study

Now, you need to run the study with the newly added conditions.

1. Choose the **Run This Study** tool from the **Simulation CommandManager**; the **Frequency Analysis** window is displayed and the process of solving the FE model gets started. Once the analysis process is completed, a new node named **Results** is added to the Simulation Tree.

Also, the result of Amplitude1 is displayed in the drawing area, as shown in Figure 6-17.

2. Right-click on the **Amplitude 1 (-Res Amp-)** under the **Results** node; a shortcut menu is displayed. Choose the **Animate** option; the animation starts with the default settings. Also, the **Animation PropertyManager** is displayed on the left of the screen enabling you to change the settings according to your requirement.

Similarly, you can check or plot the result for the other mode shapes also such as mode 1, mode 2, mode 3, mode 4, and mode 5.

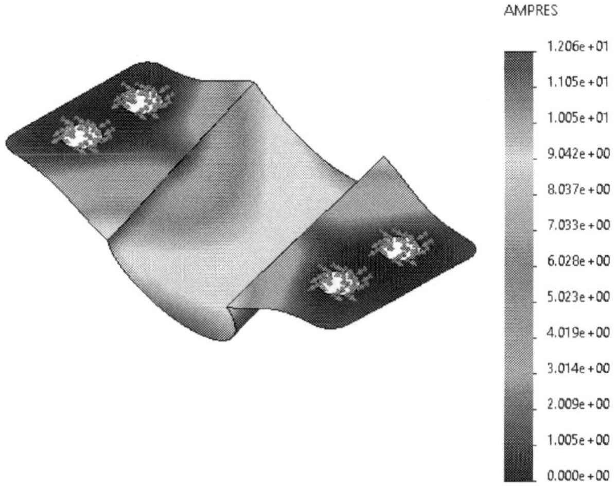

Figure 6-17 *Amplitude1 contour plot*

Defining Frequency Analysis Graph

Now, you need to check the graph of frequency analysis against mode. This graph will help you to check the frequency at a particular mode number.

1. Right-click on the **Results** node in the Simulation Tree; a shortcut menu is displayed.

2. Choose the **Define Frequency Response Graph** option from the shortcut menu; the **Frequency Analysis Graph PropertyManager** is displayed.

3. Select the **Mode Number** radio button from the **Frequency vs** rollout and then choose the **OK** button from the PropertyManager; the **Frequency vs** window is displayed and shows frequency vs mode number graph, refer to Figure 6-18.

By using this graph, you can check the frequency at any mode number included in the study.

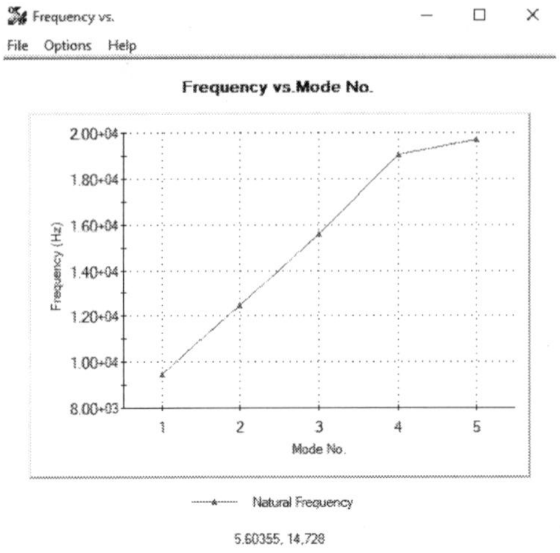

Figure 6-18 Frequency vs Mode No. graph

Listing Amplitudes and Resonant Frequencies

In SOLIDWORKS Simulation, you can also create a list of amplitudes and resonant frequencies to know the structure response at any frequency.

1. Right-click on the **Results** node in the Simulation Tree; a shortcut menu is displayed.

2. Choose the **List Amplitudes** option from the shortcut menu; the **List Results PropertyManager** is displayed.

3. Select the **AMPRES: Resultant Amplitude** option in the **Component** drop-down list. You can also set the mode number of the frequency study by using the **Mode shape** edit box in the **Component** rollout.

4. Choose the **OK** button from the PropertyManager; the **List Results** window is displayed and you can see the results for the selected mode number of the study, refer to Figure 6-19.

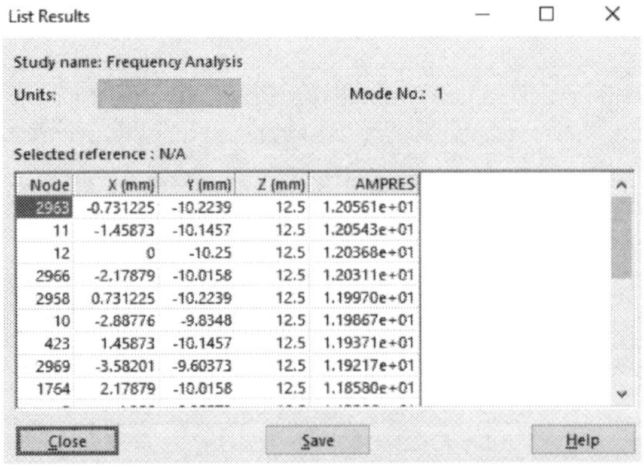

*Figure 6-19 The **List Results** window*

Similarly, you can check the resonant frequencies of the study.

5. Right-click on the **Results** folder in the Simulation Tree; a shortcut menu is displayed.

6. Choose the **List Resonant Frequencies** options from the shortcut menu; the **List Modes** window is displayed. In this window, you can see the resonant frequencies for all the modes of the study, refer to Figure 6-20.

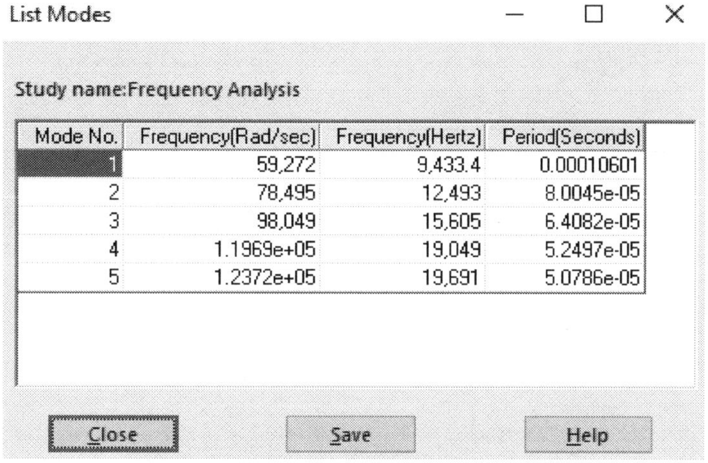

Mode No.	Frequency(Rad/sec)	Frequency(Hertz)	Period(Seconds)
1	59,272	9,433.4	0.00010601
2	78,495	12,493	8.0045e-05
3	98,049	15,605	6.4082e-05
4	1.1969e+05	19,049	5.2497e-05
5	1.2372e+05	19,691	5.0786e-05

Figure 6-20 *The* ***List Modes*** *window*

Generating Report

Now, you need to generate report.

1. Choose the **Report** button from the **Simulation CommandManager**; the **Report Options** dialog box is displayed.

2. Select all the sections that you want to add to the report by selecting their respective check boxes available in the **Report sections** area of the dialog box.

   You can also add information such as designer name, company name, URL, logo, and so on to the header of the report. To do so, you need to enter information in the respective edit boxes enabled after selecting the corresponding check boxes.

3. Choose the **Publish** button to generate the report in Word format. As soon as the analysis report is generated and displayed in Word document, the **Report node** is added to the Simulation Tree. Close the report after viewing.

Saving the File

After all the actions are performed and the desired solution is achieved, you need to save the model.

1. Choose **File > Save** from the menu bar to save the simulation files.

Tutorial 3

In this tutorial, you will open the *c06_tut03* file downloaded earlier and perform the static study on the model given in Figure 6-21. You will generate the mesh with default element size. After performing the static study, perform the frequency study and calculate 5 natural frequencies and their respective mode shapes. The material to be used is Chrome Stainless Steel. Fixed restraint and loads of 5000 N will be applied on the surface, as shown in Figure 6-21.

(Expected time: 45 min)

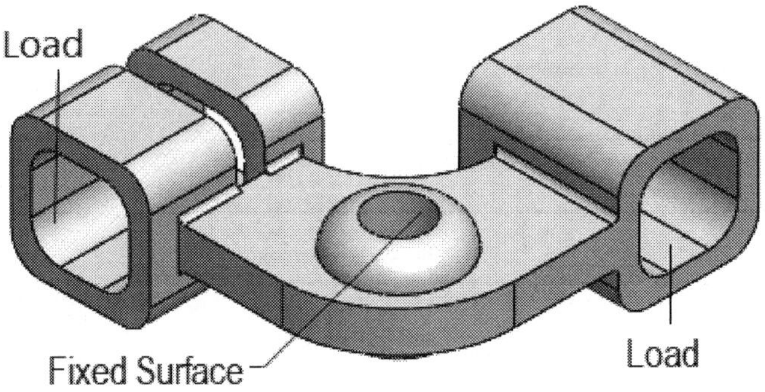

Figure 6-21 *Model for frequency analysis*

The following steps are required to complete this tutorial:

a. Start SOLIDWORKS Simulation and import the model.
b. Start a new static study.
c. Apply material and fixture.
d. Create mesh.
e. Run the study.
f. View the results.
g. Start a new frequency study.
h. Copy the results from the static study.
i. Generate the report.
j. Save the model.

Starting SOLIDWORKS Simulation and Importing the Model

Now, you need to import the model *(c06_tut03)* for analysis.

1. Choose **Open** from the **File** menu in the **SOLIDWORKS** menus; the **Open** dialog box is displayed.

2. Select the model *(c06_tut03)* and choose the **Open** button from the dialog box; selected model opens in SOLIDWORKS.

Starting a New Static Study

To find stress and five modes of vibration, you will require to run both static and frequency analysis on the model.

1. Start SOLIDWORKS Simulation and choose the **New Study** tool from the **New Study** flyout in the **Simulation CommandManager**; the **Study PropertyManager** is displayed.

2. Choose the **Static** button if not already chosen from the **General Simulation** rollout and then enter the name of the study as **Static Analysis** in the **Study name** edit box of the PropertyManager. Next, choose **OK**; a new node named **Static Analysis** is added to the Simulation Tree.

Applying the Material

Now, you need to apply the Chrome Stainless Steel material to the model.

1. Choose the **Apply Material** button from the **Simulation CommandManager**; the **Material** dialog box is displayed.

2. Expand the **SOLIDWORKS Materials** node in the **Material** dialog box. Next, expand the **Steel** sub-node and choose the **Chrome Stainless Steel** material; all properties of the selected material are displayed on the right in the dialog box.

3. Choose the **Apply** button from the dialog box; the Chrome Stainless Steel material is applied to the model. Next, close the **Material** dialog box.

Applying Constraints

After applying the material to the model, you need to apply constraints to fix the mentioned surface of the model.

1. Choose the **Fixed Geometry** tool from the **Fixtures Advisor** flyout of the **Simulation CommandManager**; the **Fixture PropertyManager** is displayed at the left of the drawing area.

2. Make sure that the **Fixed Geometry** button is chosen in the **Standard (Fixed Geometry)** rollout of the PropertyManager.

3. Select the inner cylindrical surface of the hole shown in Figure 6-21; the selected surface gets fixed and the fixed geometry fixture is attached to the selected surface, as shown in the Figure 6-22.

4. Choose the **OK** button from the PropertyManager.

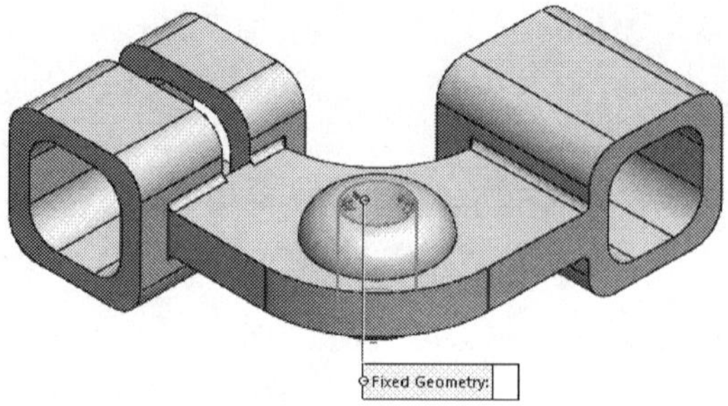

***Figure 6-22** The fixed geometry fixture attached to cylindrical surface of the model*

Applying Loads

After applying the material and constraints to the model, you need to apply loads to it.

1. Choose the **Force** tool from the **External Loads Advisor** flyout of the **Simulation CommandManager**; the **Force/Torqe PropertyManager** is displayed on the left in the drawing area.

2. Select the faces of the model to apply the force load, refer to Figure 6-23, in the **Faces, and Shell Edges for Normal Force** selection box. Also, the force callout is attached to the selected face. Now, you need to specify the direction of force.

3. Select the **Selected direction** radio button and then select an edge in Y direction from the model to apply the load in the Y direction. If the direction of the applied force is vertically upward, you need to reverse it using the **Reverse direction** check box from the **Force** rollout.

4. Select the **SI** option from the **Unit** drop-down list of the **Units** rollout and enter **5000** in the **Along Edge** edit box of the PropertyManager. Accept other default settings and choose the **OK** button to apply the load. Figure 6-23 shows the view of the model.

Creating the Mesh

After applying the boundary condition to the model, you need to create mesh.

1. Choose the **Create Mesh** tool from the **Run This Study** flyout of the **Simulation CommandManager**; the **Mesh PropertyManager** is displayed.

2. Accept all the default settings and choose the **OK** button from the PropertyManager; the **Mesh Progress** window is displayed.

   Once the process of meshing is completed, the meshed model is displayed in the drawing area, as shown in Figure 6-24.

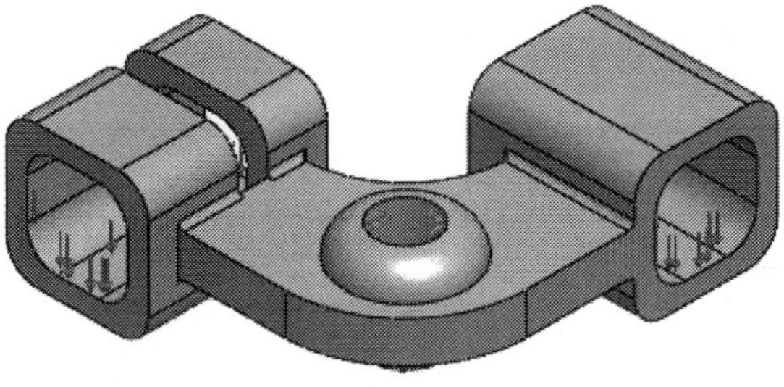

Figure 6-23 Model after applying the force load

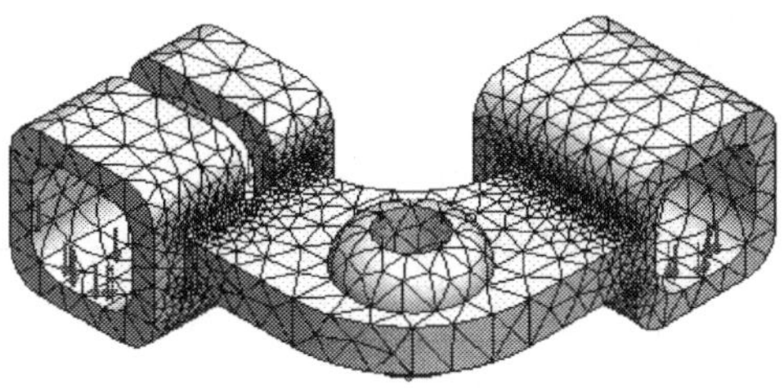

Figure 6-24 The model after meshing

Running the Study

Now, you need to run the analysis on the model.

1. Choose the **Run This Study** tool from the **Simulation CommandManager**; the **Static Analysis** window is displayed and the process of solving the finite element model gets started. Once the analysis process is completed, a new node named Results is added to the Simulation Tree. Also, the resultant model is displayed in the drawing area with the stress result contour, as shown in Figure 6-25.

2. Right-click on **Stress 1 (-vonMises-)** under the **Results** node; a shortcut menu is displayed. Choose the **Animate** option; the animation starts with the default settings. Also, the **Animation PropertyManager** is displayed on the left in the screen enabling you to change the settings according to your requirement.

   Similarly, you can check or plot the result for other results such as displacement and strain.

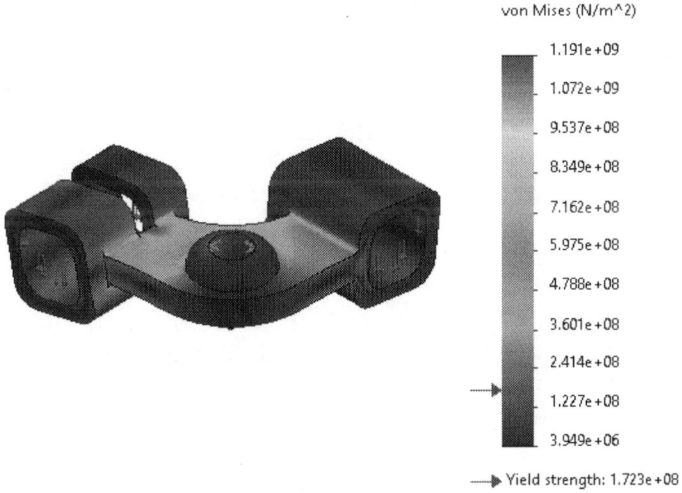

Figure 6-25 The resultant model with von Mises stress

Starting a New Frequency Study

To find the stress and five modes of the vibration, you need to run both static and frequency analysis on the model.

1. Choose the **Simulation** tab from the **CommandManager**; the **Simulation CommandManager** is displayed.

2. Choose the **New Study** tool from the **New Study** flyout in the **Simulation CommandManager**; the **Study PropertyManager** is displayed.

3. Choose the **Frequency** button, if not chosen, and then enter **Frequency Analysis** as the name of the study in the **Study name** edit box of the PropertyManager. Next, choose the **OK** button; a new tree named **Frequency Analysis** is added to the Simulation Tree.

Copying the Result from Existing Study

Now, you need to set the material properties, boundary conditions, and mesh to perform the study. In SOLIDWORKS Simulation, you can either separately set the above mentioned or you can also copy the results from the existing study.

1. Select the material sub node under the **Static Analysis** node in the Simulation Tree and then drop it on the recently added frequency study; the picked material gets applied to the frequency study.

   Similarly, you need to transfer, fixture, and mesh from the static study to the frequency study. After transferring the material, fixture, and mesh to the newly added frequency study, you need to solve this study.

Running the Study

Now, you need to run the study with the newly added conditions from the static study.

1. Choose the **Run This Study** button from the **Simulation CommandManager**; the **Frequency Analysis** window is displayed and the process of solving the FE model gets started. Once the analysis process is completed, a new node named **Results** is added to the Simulation Tree. Also, the Amplitude1 contour plot is displayed in the drawing area, as shown in Figure 6-26. Figure 6-26 through Figure 6-30 show the contours of the mode shapes of the model.

2. Right-click on **Amplitude 1 (-Res Amp-)** under the **Results** node; a shortcut menu will be displayed. Choose the **Animate** option; the animation starts with the default settings. Also, the **Animation PropertyManager** is displayed on the left of the screen enabling you to change the settings according to your requirement.

   Similarly, you can check or plot the result for the other mode shapes such as mode 1, mode 2, mode 3, mode 4, and mode 5. Note that if you get any error related to solver while solving the study, right-click on the **Frequency Analysis** node in the Simulation Tree to rectify it; a shortcut menu is displayed. Choose the **Properties** option from this menu; the **Frequency** dialog box is displayed. Select the **Automatic Solver Selection** check box in the **Solver** area of this dialog box and then choose the **OK** button to close the dialog box.

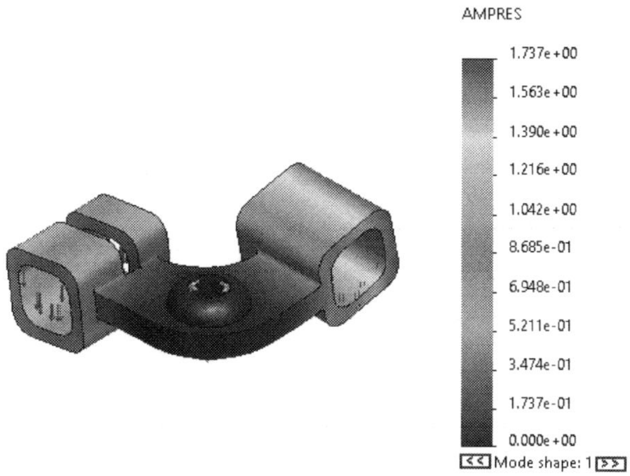

***Figure 6-26** The resultant model with mode shape 1 contour*

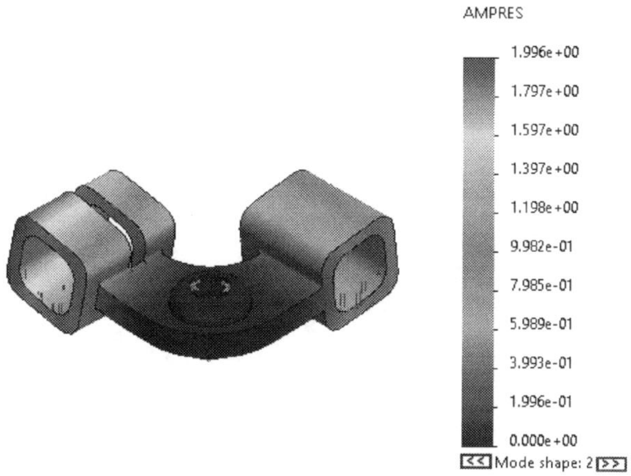

***Figure 6-27** The resultant model with mode shape 2 contour*

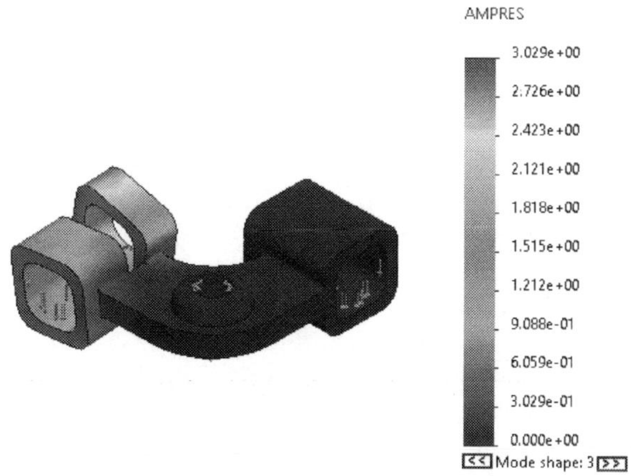

***Figure 6-28** The resultant model with mode shape 3 contour*

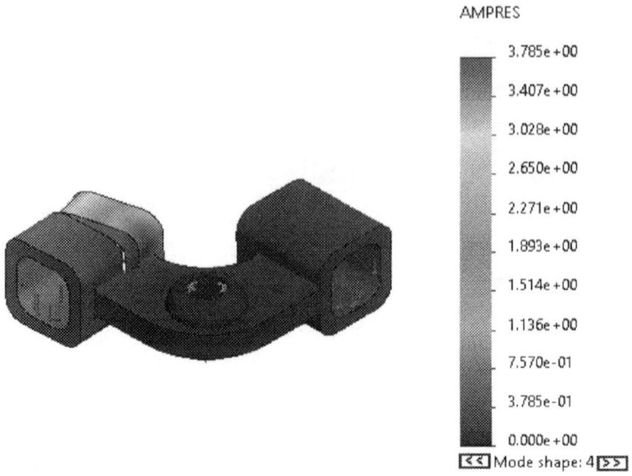

Figure 6-29 *The resultant model with mode shape 4 contour*

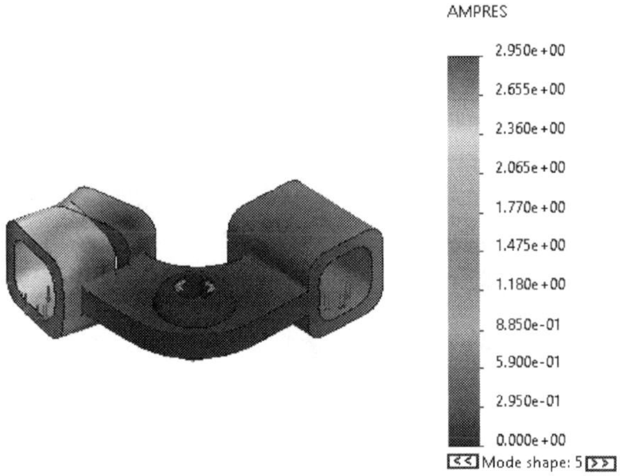

Figure 6-30 *The resultant model with mode shape 5 contour*

Defining Frequency Analysis Graph

Now, you need to check the Frequency vs Mode No. graph. This graph will help you to check the frequency at a particular mode number.

1. Right-click on the **Results** node in the Simulation Tree; a shortcut menu is displayed.

2. Choose the **Define Frequency Response Graph** option from the shortcut menu; the **Frequency Analysis Graph PropertyManager** is displayed.

3. Select the **Mode No.** radio button from the **Frequency Vs** rollout and then choose the **OK** button from the PropertyManager; the **Frequency Vs** window is displayed and shows the Frequency vs Mode No. graph, refer to Figure 6-31.

   By using this graph, you can check the frequency at any mode number included in the study.

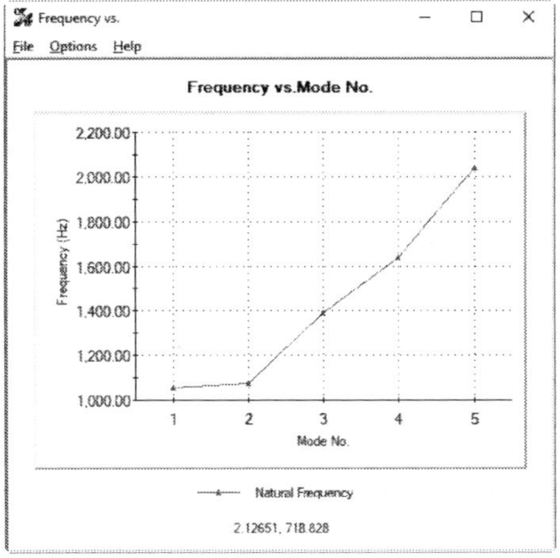

Figure 6-31 *Frequency vs Mode No. graph*

Listing Amplitudes and Resonant Frequencies

In SOLIDWORKS Simulation, you can also create a list of amplitudes and resonant frequencies to know the structure response at any frequency.

1. Right-click on the **Results** node in the Simulation Tree; a shortcut menu is displayed.
2. Choose the **List Amplitude** option from the shortcut menu; the **List Results PropertyManager** is displayed.
3. Select the **AMPRES: Resultant Amplitude** option in the **Component** drop-down list. You can also set the mode number of the frequency study by using the **Mode Shape** edit box in the **Component** rollout.
4. Choose the **OK** button from the PropertyManager; the **List Results** window is displayed and you can see the results for the selected mode number of the study, refer to Figure 6-32.

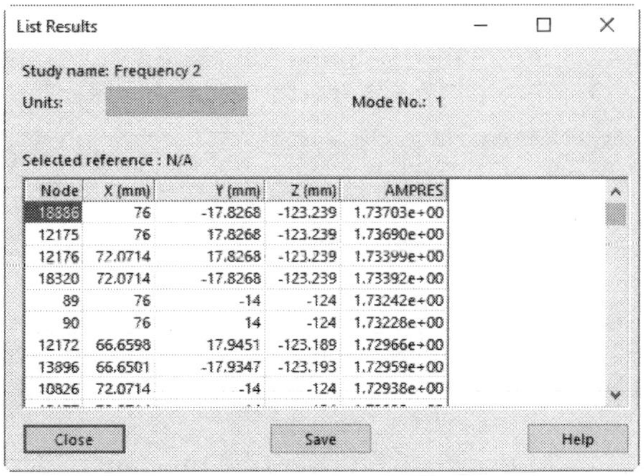

Node	X (mm)	Y (mm)	Z (mm)	AMPRES
18336	76	-17.8268	-123.239	1.73703e+00
12175	76	17.8268	-123.239	1.73690e+00
12176	72.0714	17.8268	-123.239	1.73399e+00
18320	72.0714	-17.8268	-123.239	1.73392e+00
89	76	-14	-124	1.73242e+00
90	76	14	-124	1.73228e+00
12172	66.6598	17.9451	-123.189	1.72966e+00
13896	66.6501	-17.9347	-123.193	1.72959e+00
10826	72.0714	-14	-124	1.72938e+00

Figure 6-32 List Results *window*

Similarly, you can check the resonant frequencies of the study.

5. Right-click on the **Results** node in the Simulation Tree; a shortcut menu is displayed.

6. Choose the **List Resonant Frequencies** option from the shortcut menu; the **List Modes** window is displayed. In this window, you can see the resonant frequencies for all the modes of the study, refer to Figure 6-33.

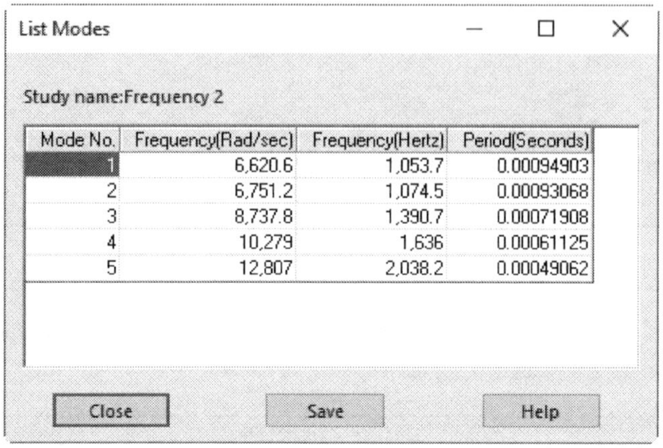

Figure 6-33 *The* ***List Modes*** *window*

Generating the Report

Now, you need to generate the report.

1. Choose the **Report** button from the **Simulation CommandManager**; the **Report Options** dialog box is displayed.

2. Select all the sections that you want to add to the report by selecting their respective check boxes available in the **Report sections** area of the dialog box.

   As discussed earlier, you can also add information such as designer name, company name, URL, logo, and so on to the header of the report. To do so, you need to add information in the edit boxes that will be enabled after selecting their corresponding check boxes.

3. Choose the **Publish** button to generate the report in the Word format. As soon as the analysis report is generated and displayed in a Word document, the **Report** node is added to the Simulation Tree. Close the report after viewing.

Saving the File

After all the actions are performed and the desired solution is achieved, you need to save the model.

1. Choose **File > Save** from the menu bar to save the simulation files.

Self-Evaluation Test

Answer the following questions and then compare them to those given at the end of this chapter:

1. The material looses its _________ property when the intensity of the stress exceeds the elastic limit.

2. When you choose the _________ button in the **Report option** dialog box, the report will be generated in the Word format.

3. Frequency analysis determines the response of the structure to the time dependent vibrations. (T/F)

4. To perform the frequency analysis, you need to perform static analysis as well. (T/F)

5. In SOLIDWORKS Simulation, you can add load to the frequency analysis. (T/F)

6. The frequencies and loads at which body tends to vibrate are natural frequencies. (T/F)

7. Frequency analysis is a type of analysis that can help you avoid resonance. (T/F)

8. Analysis is a process used to simulate design performance of the field. (T/F)

9. FEA can help you design faster, better, safer, and economically by reducing traditional, expensive design cycles. (T/F)

10. You can edit, add, or delete features, and define a study by dragging and dropping the icon of an existing study onto the top icon of the SOLIDWORKS Simulation Tree. (T/F)

Review Questions

Answer the following questions:

1. At every _________, there is a corresponding mode shape displayed in the results.

2. Frequency analysis is used to determine the vibration characteristics of a structural component. (T/F)

3. In frequency analysis, you cannot find the stress results induced in the model without adding the load. (T/F)

4. In SOLIDWORKS Simulation, you can list the resonant frequency by right-clicking on the **Results** node in the **FeatureManager Design Tree.** (T/F)

5. In frequency analysis, you cannot find the results of stress induced in the model. (T/F)

6. In Frequency study, the field conditions vary with time. (T/F)

7. The frequencies at which body tends to vibrate are natural frequencies. (T/F)

8. Frequency is the number of times an event occurs. (T/F)

EXERCISES

To perform the exercises of this chapter, download the input file from www.cadcim.com. The complete path for downloading the file is:

Textbooks > CAE Simulation > SOLIDWORKS Simulation > SOLIDWORKS Simulation 2022: A Tutorial Approach > Input Files > c06_simulation_2022_input.zip

Extract the downloaded **c06_simulation_2022_input** zipped file at the location *C:\SOLIDWORKS Simulation\c06.*

Exercise 1

In this exercise, you will use *c06_exr01* file to perform the frequency analysis and calculate the resonant frequencies at first 5 modes. To perform the analysis, import the model. Figure 6-34 shows the model for the exercise. Next, start a Frequency Study and assign AISI 1045 Steel, cold drawn material from the SOLIDWORKS material library to the model. You need to consider all the holes as fixed support to perform the analysis. **(Expected time: 45 min)**

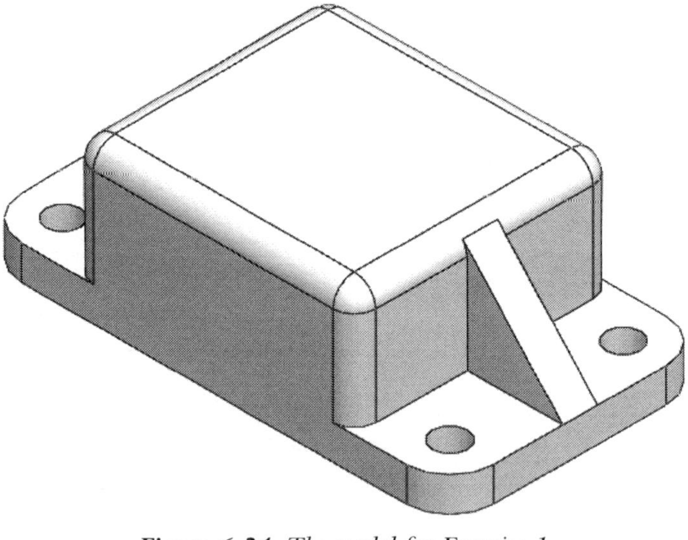

Figure 6-34 *The model for Exercise 1*

Exercise 2

In this exercise, you will use *c06_exr02* file to perform the frequency analysis of the model shown in Figure 6-35. The boundary conditions applied on the model are also shown in Figure 6-35. The model is made up of alloy steel material. You need to consider mentioned holes as Fixed support. After performing the Frequency Study, you need to check the mode shapes and frequencies of the model at 5 number of modes. **(Expected time: 45 min)**

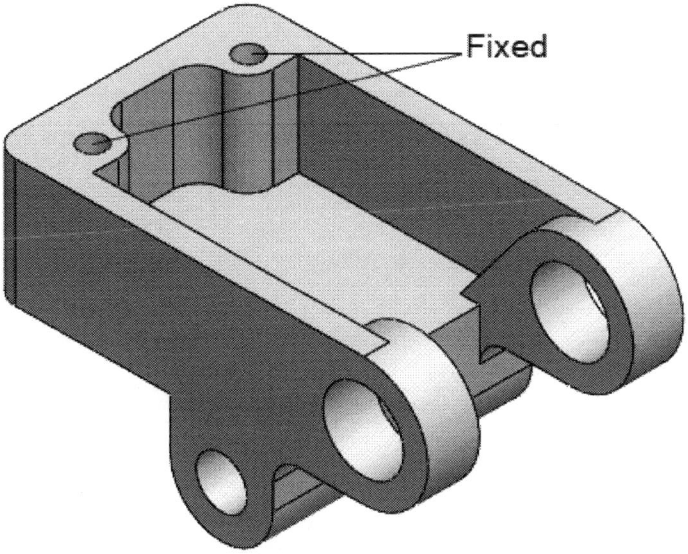

Figure 6-35 *The model for Exercise 2*

Answers to Self-Evaluation Test

1. **Elastic**, **2**. **Publish**, **3**. T, **4**. F, **5**. T, **6**. T, **7**. T, **8**. T, **9**. T, **10**. T

Chapter 7

Thermal Analysis

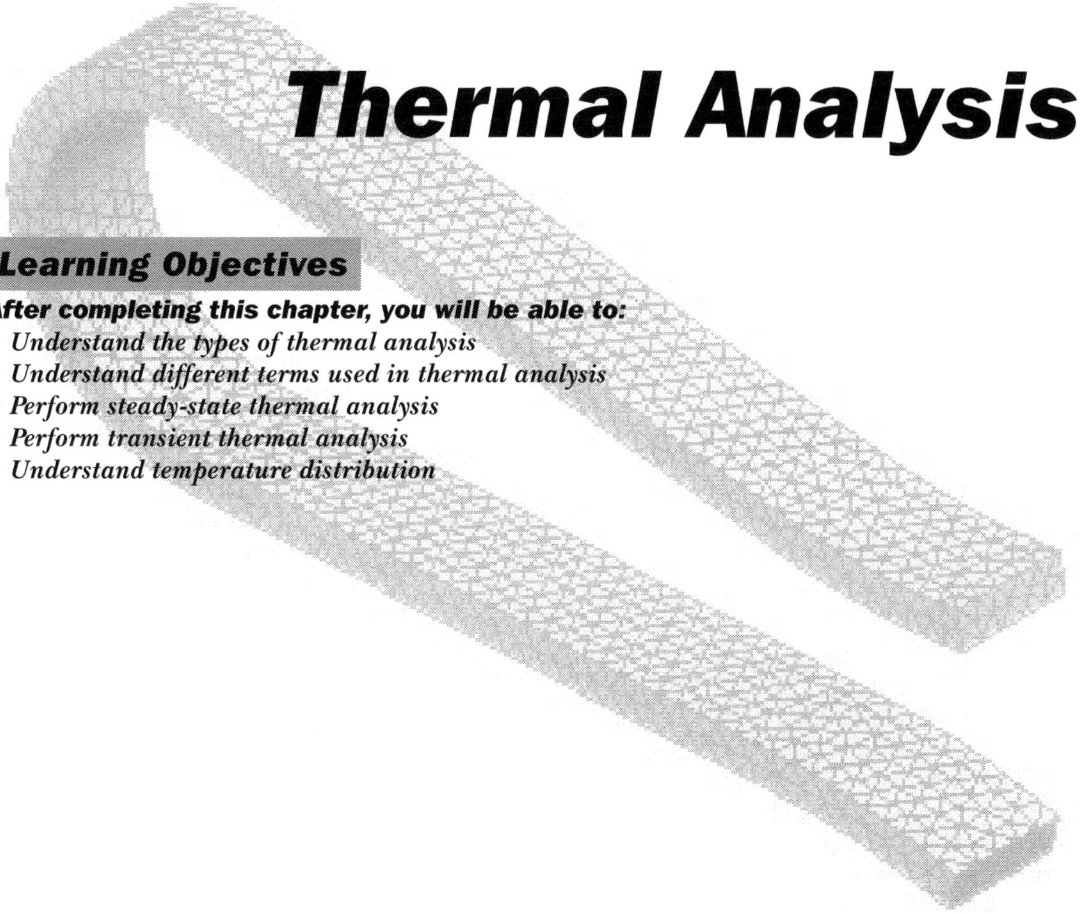

Learning Objectives

After completing this chapter, you will be able to:

- *Understand the types of thermal analysis*
- *Understand different terms used in thermal analysis*
- *Perform steady-state thermal analysis*
- *Perform transient thermal analysis*
- *Understand temperature distribution*

THERMAL ANALYSIS

A model passes through several stages in product development before production. Thermal analysis is one of them and plays an important role in product development. Various products such as engines, refrigerators, heat exchangers, and so on are designed based on the results of this analysis.

Thermal analysis is used to study the temperature distribution and related thermal quantities in the model. In this analysis, all heat transfer modes, namely Conduction, Convection, and Radiation are analyzed. Thermal analysis analyzes the following:

1. Temperature distribution
2. Amount of heat loss or gain
3. Thermal gradients
4. Thermal fluxes

This analysis is used in many engineering industries such as automobile, piping, electronic, power generation, and so on. In SOLIDWORKS Simulation, two types of thermal analysis can be carried out, namely Steady-State and Transient Thermal analysis.

The following are the basic steps required to perform the thermal analysis:

1. Set the analysis preference.
2. Create or import solid model.
3. Define element attributes (element types, real constants, and material properties).
4. Mesh the solid model.
5. Specify the analysis type, analysis options, and the loads to be applied.
6. Solve the analysis problem.
7. Post-process results.

IMPORTANT TERMS USED IN THERMAL ANALYSIS

Before conducting thermal analysis, you should be familiar with the basic concepts and terminologies of thermal analysis. Following are some of the important terms used in thermal analysis:

Heat Transfer Modes

Whenever two bodies having different temperatures comes in contact with each other, heat transfer takes place from the body of the higher temperature to the body of the lower temperature. There are three modes of heat transfer: Conduction, Convection, and Radiation.

Conduction

Conduction is the process of heat transfer between bodies in contact. For example, hold an iron rod on a flame at one end and then wait for sometime, the iron rod will be heated on the other end and you will also feel the heat on your palm.

Convection
Convection is the process of heat transfer in which the medium of heat transfer is a fluid or gas. For example, heating up water using an electric water heater is a good example of heat convection. In this case, water takes heat from the heater.

Radiation
Radiation is mode of heat transfer in which the heat is transferred through space. Heat from the sun is a perfect example of this mode of heat transfer.

Thermal Gradient
The thermal gradient is the rate of change in temperature per unit length. It is a physical quantity

Thermal Flux
The thermal flux is defined as the rate of heat transfer per unit cross-sectional area. It is denoted by $¢_q$.

Bulk Temperature
It is the temperature of a fluid flowing outside the material. It is denoted by T_b. The Bulk temperature is used in convective heat transfer.

Film Coefficient
It is the measure of the heat transfer through a fluid film.

Emissivity
The emissivity of a material is the ratio of energy radiated by the material to the energy radiated by a black body at the same temperature. Emissivity is the measure of a material's ability to absorb and radiate heat. It is denoted by e. Emissivity is a numerical value without any unit. For a perfect black body, $e = 1$. For any other material, $e < 1$.

Stefan–Boltzmann Constant
The Stefan-Boltzmann constant is a physical constant and defines the power per unit area emitted by a black body as a function of its thermodynamic temperature. It is denoted by sigma.

Thermal Conductivity
The thermal conductivity is the property of a material that indicates its ability to conduct heat. It is denoted by k.

Specific Heat
The specific heat is the amount of heat required per unit mass to raise the temperature of a body by one degree celsius. It is denoted by c.

TYPES OF THERMAL ANALYSIS
In SOLIDWORKS Simulation, two types of thermal analysis can be carried out, namely Steady-State Thermal analysis and Transient Thermal analysis.

Steady-State Thermal Analysis

In Steady-State Thermal analysis, the thermal load does not vary with time and remains constant throughout the period of application. These thermal loads include convection, radiation, heat flow rates, heat fluxes (heat flow per unit area), heat generation rates (heat flowper unit volume), and constant temperature boundaries.

Steady-State Thermal analysis may be either linear or nonlinear with respect to material properties that depend on temperature. The thermal properties of most of the materials do vary with temperature, therefore the analysis usually is nonlinear. Apart from this, the radiation effects or temperature-dependent convection in a model also makes the analysis nonlinear.

The steps to solve a problem related to Thermal analysis are same as that of the Structural analysis except a few steps such as selecting the element type, applying the load, and postprocessing results.

Transient Thermal Analysis

In Transient Thermal analysis, the application of thermal loads is time dependent. Most of the engineering applications, such as engine blocks, pressure vessels, nozzles, and piping systems, need Transient thermal analysis. The process of solving a Transient thermal analysis problem is same as that of a Steady-state thermal analysis with the only difference that in Transient thermal analysis, the load applied on a body is a function of time.

In the Transient thermal analysis, the system is studied under varying thermal loads with respect to time. This analysis is used to find out thermal gradient, temperatures varying with time, and thermal flux.

Transient thermal analysis takes more time compared to other analyses types. It is necessary to understand the basic mechanism of the problem to reduce the time involved in getting its solution. For example, if the problems contain nonlinearity, then you first need to understand how they affect the response of structures by doing Steady-state thermal analysis.

TUTORIALS

Tutorial 1 Steady-State Thermal

In this tutorial, you will create the 3D model of the Car Disk Break Rotor, as shown in Figure 7-1. The dimensions and views of the model are given in Figure 7-2. You can also download this model from www.cadcim.com. The path of the file is as follows:

Textbooks > CAE Simulation > SOLIDWORKS Simulation_2022 > SOLIDWORKS Simulation : A Tutorial Approach > Input Files > c07_simulation__2022_input.zip

Next, you will run a Steady-State thermal analysis on the model, evaluate temperature distribution, and total heat flux for the component on the basis of the boundary conditions given next. Figure 7-3 and Figure 7-4 show the temperature load and convection load for performing thermal study.

Boundary Conditions:
Ambient temperature: 22°C (295K)
Temperature load: 80°C
Convection coefficient: 1.24
Material of the model: Alloy Steel

(Expected time: 45 min)

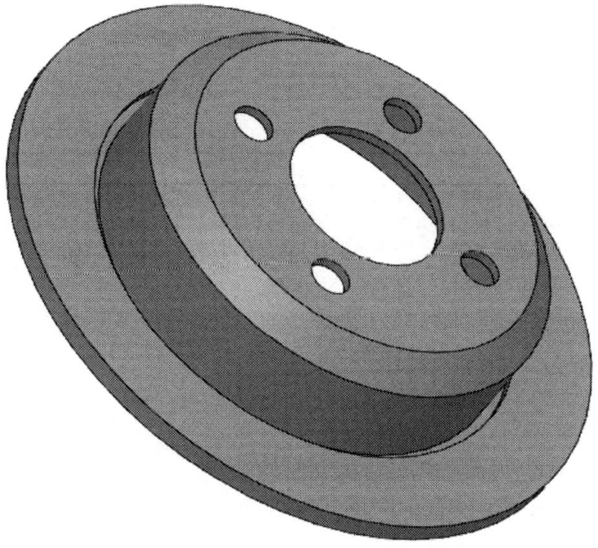

Figure 7-1 *Model for Tutorial 1*

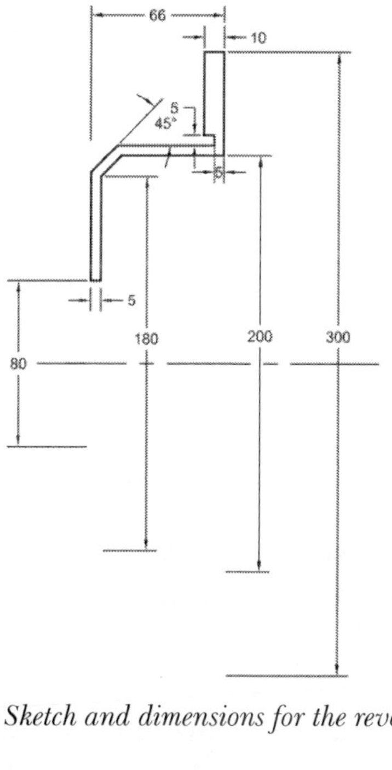

Figure 7-2 *Sketch and dimensions for the revolve feature*

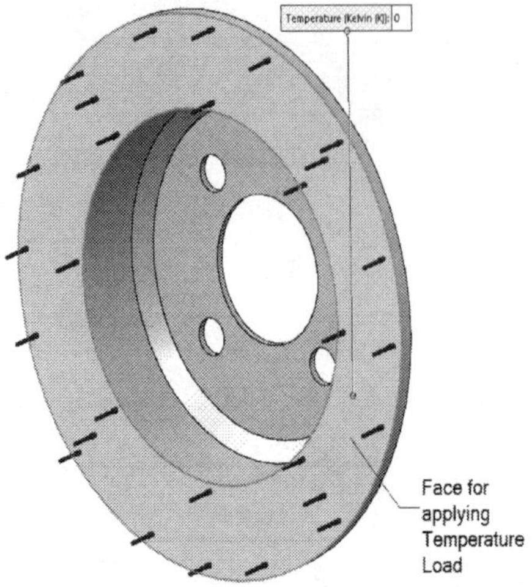

Figure 7-3 *Face for applying the temperature load*

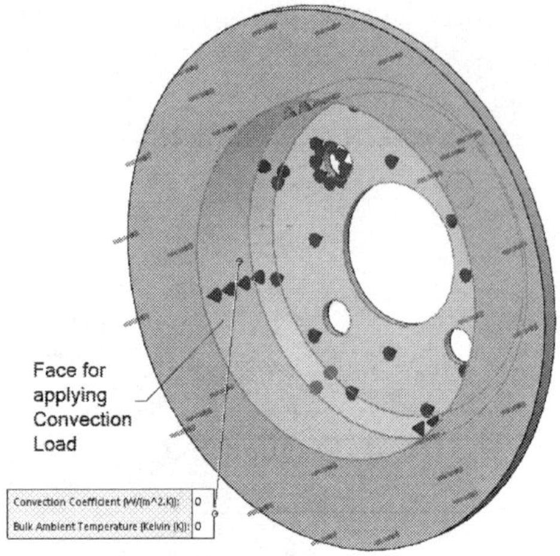

Figure 7-4 *Face for applying the convection load*

Following steps are required to complete this tutorial:

a. Start SOLIDWORKS and create the model.
b. Generate mesh and apply thermal boundary conditions to the model.
c. Analyze the results.
d. Save the model and exit SOLIDWORKS Simulation.

Starting SOLIDWORKS and Creating the Model

First, you need to start SOLIDWORKS and then create the model. For creating the model, set your units to MMGS (millimeter, gram, second).

1. Start a new part file using the **New SOLIDWORKS Document** dialog box and invoke the sketching environment.

2. Choose **Front Plane** as the sketching plane and draw the sketch of the model. For dimensioning, refer to Figure 7-2.

3. Choose the **Revolved Boss/Base** tool from the **Features CommandManager** and make a revolved feature to create the model, refer to Figure 7-1.

4. Save the model with the name **c07_sws_2022_tut01**.

Invoking SOLIDWORKS Simulation and Starting a New Study

After the model is created, you need to add the **SOLIDWORKS Simulation** add-in and start a new study.

1. Start SOLIDWORKS Simulation and choose the **New Study** tool from the **New Study Advisor** flyout in the **Simulation CommandManager**; the **Study PropertyManager** is displayed, refer to Figure 7-5.

2. Choose the **Thermal** button from the **Advanced Simulation** rollout and enter the name of the study as **Car Disk Break Rotor** in the **Study name** edit box in the **Name** rollout. Choose the **OK** button from the **Study PropertyManager.** On doing so, a Simulation Tree is added below the Simulation Tree with **Car Disk Break Rotor** as the name of the study.

3. Right-click on the name of the study node and choose the **Properties** option from the shortcut menu; the **Thermal** dialog box is displayed.

4. Select the **Steady state** radio button from the **Solution type** area in this dialog box, if it is not selected.

5. Click on the **OK** button to close this dialog box.

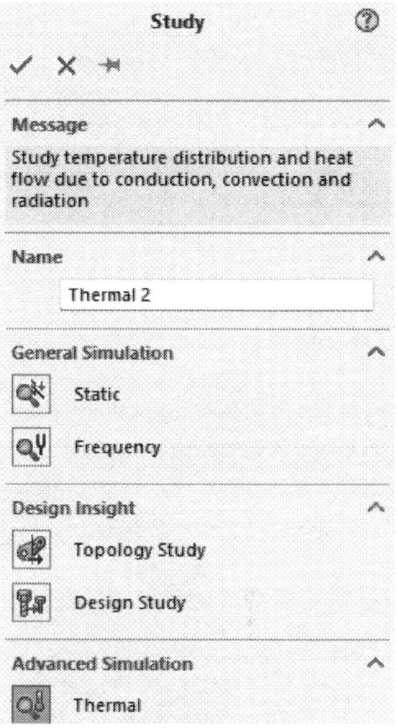

Figure 7-5** Partial view of the **Study PropertyManager

Assigning Material and Generating Mesh

After selecting the type of study, you need to assign material to the model in Simulation Tree, if not assigned in the modeling environment. After that you need to mesh the model.

1. Choose the **Apply Material** tool from the **Simulation CommandManager**; the **Material** dialog box is displayed.

2. Select **Alloy Steel** from the materials displayed. Next, choose the **Apply** button and then the **Close** button to close this dialog box.

   After applying material, you need to mesh the model.

3. Choose the **Create Mesh** tool from the **Run This Study** flyout in the **Simulation CommandManager**; the **Mesh PropertyManager** is displayed, as shown in Figure 7-6.

4. Choose **OK** from the PropertyManager; the **Mesh** progress window is displayed and the mesh with default settings is generated, refer to Figure 7-7.

   Now, you need to know the number of elements created after meshing.

5. Right-click on the **Mesh** node in Simulation Tree and choose **Details** option from the shortcut menu; the **Mesh Details** window is displayed.

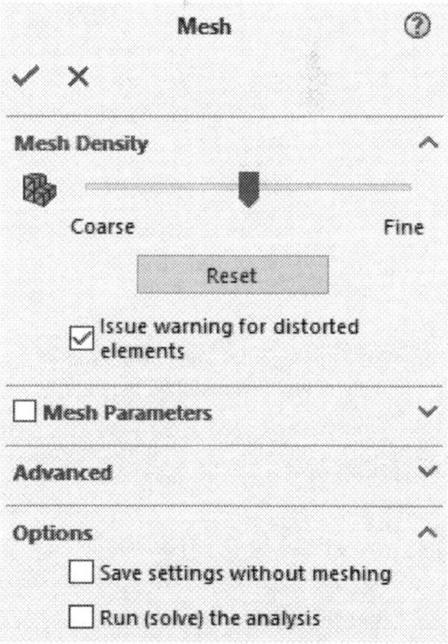

Figure 7-6** The **Mesh PropertyManager

In mesh details window, you get the information about study name, mesh type, mesher used, mesh quality, aspect ratio, total nodes, total elements, and so on. This is evident from the **Mesh Details** window that the total elements are 9738. Now, you need to improve the mesh quality to get more accurate results of thermal study.

6. To improve the mesh quality, choose the **Create Mesh** tool from the **Run This Study** flyout in the **Simulation CommandManager**; the **Mesh PropertyManager** is displayed.

7. Move the Mesh Factor slider in the **Mesh Density** rollout toward Fine; this increase the number of elements and you get increased mesh accuracy.

8. Click **OK** to close this PropertyManager; the **Mesh** progress window is displayed and the mesh is generated with modified settings, refer to Figure 7-8.

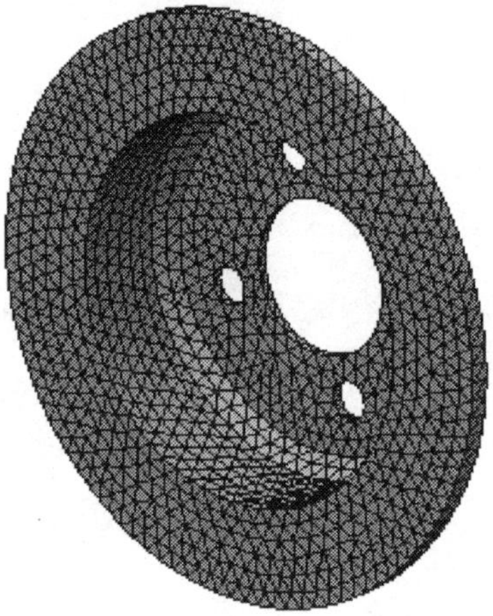

Figure 7-7 *Mesh generated with the default settings*

Figure 7-8 *Mesh generated with the modified settings*

Note that the number of elements in this mesh is 64,389. This number is very large as compared to the element count that was achieved by the mesh generated with the default mesh control settings.

Applying Thermal Boundary Conditions on the Model

After the model is meshed, you need to apply thermal boundary conditions on the model for the analysis. You will apply a temperature load on the front face and convection load on the inner faces.

1. To apply a temperature load, select the **Temperature** tool from the **Thermal Loads** flyout in the **Simulation CommandManager**; the **Temperature PropertyManager** is displayed, as shown in Figure 7-9.

2. Select the front face of the model for applying the temperature load, refer to Figure 7-10; the name of the selected face is displayed in the **Faces, Edges, Vertices, Components for Temperature** display box.

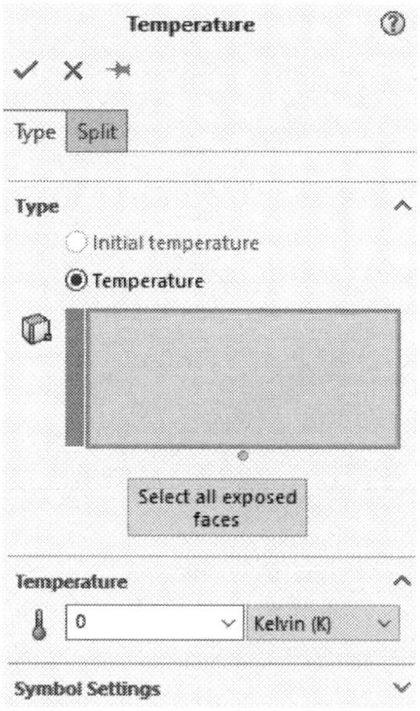

Figure 7-9** The **Temperature PropertyManager

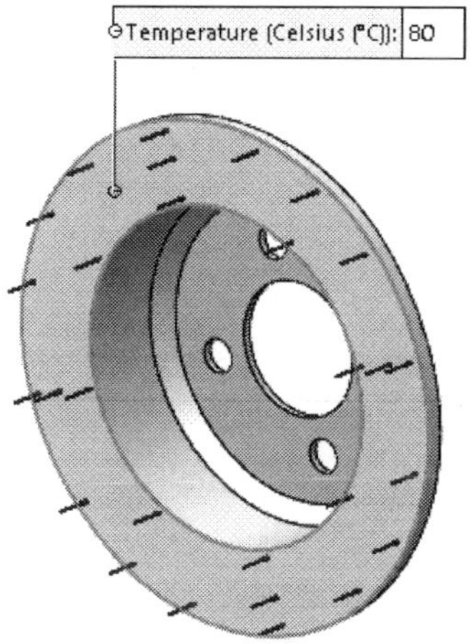

***Figure 7-10** Selecting the front face for applying temperature load*

3. Select the **Celsius(°C)** option from the **Unit** drop-down list and enter **80** in the **Temperature** edit box in the **Temperature** rollout. You can also set the value in **Kelvin(K).**

4. Select the **OK** button to close this PropertyManager.

5. Choose the **Convection** tool from the **Thermal Loads** flyout in the **Simulation CommandManager**; the **Convection PropertyManager** is displayed, refer to Figure 7-11.

6. Select the inner faces of the model to apply the convection load, refer to Figure 7-12; the names of the selected faces get displayed in the **Faces for Convection** selection box in the **Selected Entities** rollout.

7. Enter **1.24** as the value of convection coefficient for air in the **Convection Coefficient** edit box in the **Convection Coefficient** rollout.

8. Enter **22°C or 295.15K** as the ambient temperature in the **Bulk Ambient Temperature** edit box in the **Bulk Ambient Temperature** rollout.

9. Choose the **OK** button to close this PropertyManager.

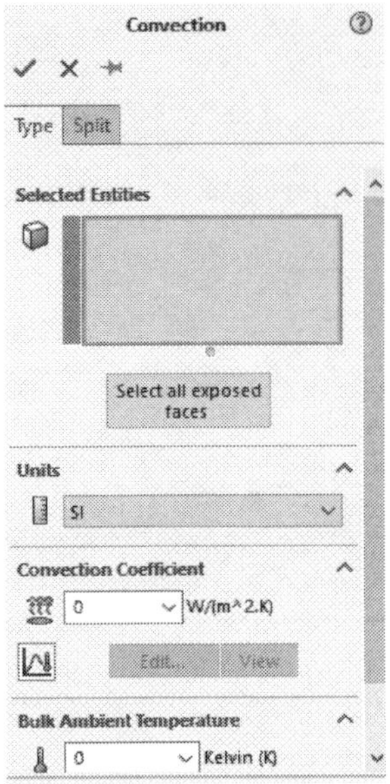

Figure 7-11** The **Convection PropertyManager

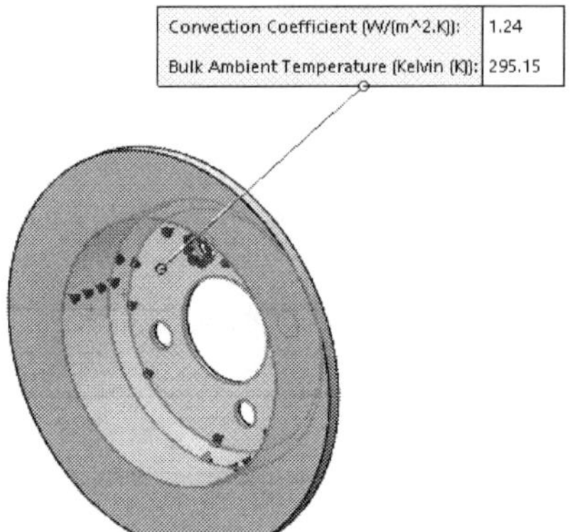

***Figure 7-12** Selecting inner faces for applying the convection load*

Analyzing the Results

After the thermal boundary conditions are applied, it is now important to analyze the behavior of the model with respect to the boundary conditions applied.

1. Click on the **Run This Study** tool from the **Simulation CommandManager** or right-click on the name of the study in Simulation Tree and choose **Run**; the Solver Status message box is displayed, as shown in Figure 7-13 and the **Results** node is added in the Simulation Tree indicating that the solution is done.

Under the **Results** node, the **Thermal1 (-Temperature-)** node is added. Also, the model is displayed in different color contours indicating Figure 7-13 The Solver Status message box the distribution of the temperature in the model. The legend is also displayed in the drawing area indicating values starting from minimum to maximum temperature distribution in the model, refer to Figure 7-14.

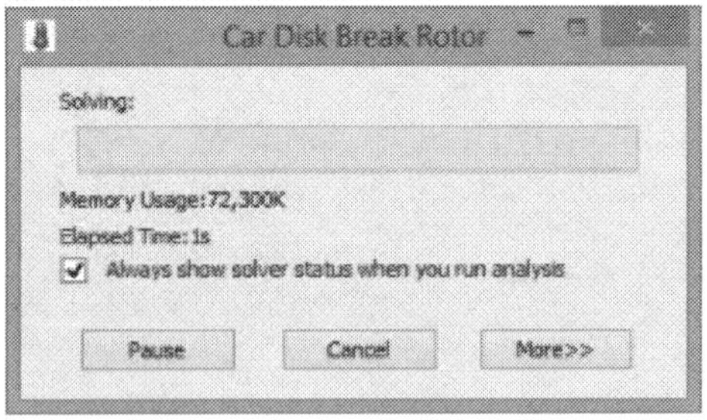

***Figure 7-13** The Solver Status message box*

Note that the temperature in the region where the temperature boundary condition is applied is higher as compared to the regions far from it. The region marked red shows the maximum temperature and the region marked blue shows the minimum temperature in the model. Similarly, other colors display different values of temperature distribution in the model.

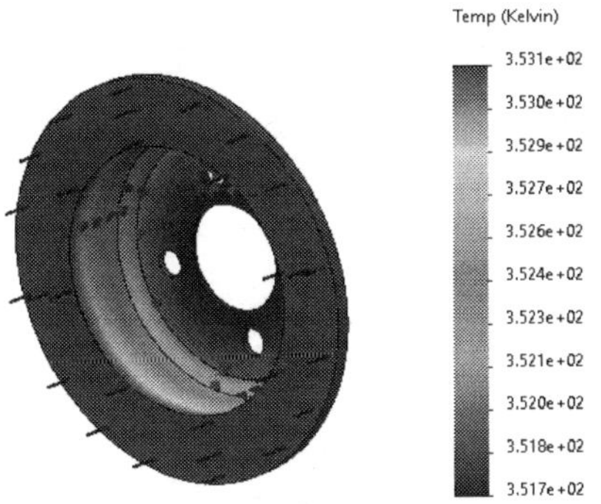

Figure 7-14 *Temperature distribution for the whole body*

2. Right-click on the **Results** node and choose the **Define Thermal Plot** option to invoke the **Thermal Plot PropertyManager,** refer to Figure 7-15.

3. Choose **HFLUXN: Resultant heat flux** from the **Component** drop-down list in the **Display** rollout of the **Definition** tab. Choose the **OK** button to close the PropertyManager; the **Thermal2 (-Res heat flux-)** node is added under the **Results** node and corresponding legend is displayed along with the model, refer to Figure 7-16.

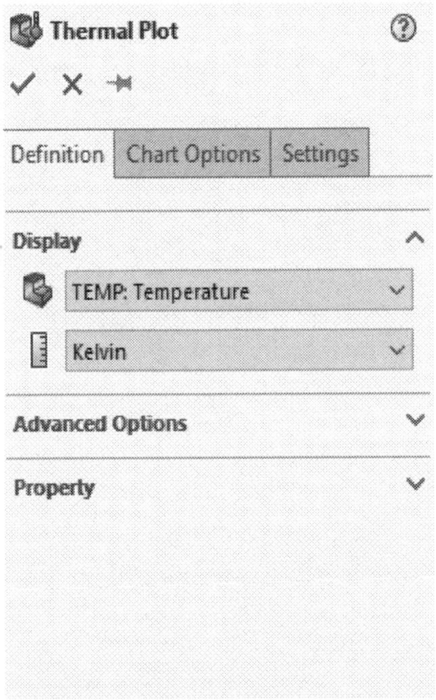

Figure 7-15 *The* ***Thermal Plot PropertyManager***

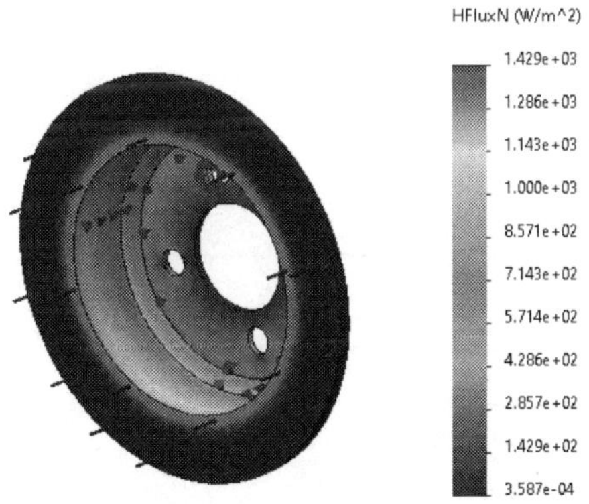

Figure 7-16 *Total heat flux for the whole body*

4. Similarly, choose **HFLUXX: X heat flux** from the **Component** drop-down list in the **Display** rollout of the **Definition** tab. Choose the **OK** button to close the PropertyManager; the **Thermal3 (-X heat flux-)** node is added under the **Results** node and corresponding legend is displayed along with the model, refer to Figure 7-17.

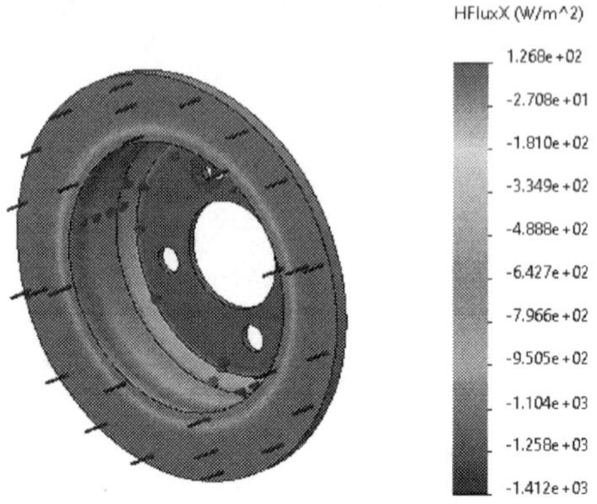

***Figure 7-17** Directional heat flux along the X-axis for the whole body*

Note

*In this tutorial, the directional heat flux is calculated along the X-axis. You can determine the directional heat flux along the Y or Z axis also by selecting the required option from the **Component** drop-down list in the **Display** rollout of the **Definition** tab of **Thermal plot PropertyManager**.*

You can toggle between different results generated from the thermal study.

5. Right-click on the name of the result under the **Results** node and choose **Show**; the resultant result is displayed in the Drawing area with its corresponding legend.

The table given next displays the results obtained for the total heat flux and Directional heat flux.

S. No.	Parameter	Max Value (W/m^2)	Min Value (W/m^2)
1	Total heat flux	1429.38	$3.587\ e^{-4}$
2	Directional heat flux (X Axis)	126.89	$-1.412e^{+3}$

It is obvious from the above results that the total heat flux is maximum in the region where the model is marked red and minimum where it is marked blue. Each color contour depicts a value and can be seen in the legend in the drawing area.

Tip

*The material used so far in this tutorial is Alloy Steel. However, you can also analyze a component by assigning a different material to it. You can do so by duplicating the **Thermal 1** study at bottom of the drawing area and adding different material to the model from the Simulation Tree and then solve the model.*

Saving the Project and Exiting SOLIDWORKS Simulation

After analyzing the results, save the entire project and exit SOLIDWORKS Simulation.

1. Choose the **Save** button to save the project at the following location:

   \Documents\SOLIDWORKS Simulation\c07

2. Choose **File > Close** from the SOLIDWORKS menus to close the document.

Tutorial 2

In this tutorial, you will create revolved feature of the Piston model, as shown in Figure 7-18. The sketch of the model and its dimensions are shown in Figure 7-19. You can also download this model from www.cadcim.com. The path of the file is as follows:

Textbooks > CAE Simulation > SOLIDWORKS Simulation > SOLIDWORKS Simulation 2022: A Tutorial Approach > Input Files > c07_simulation_2022_input.zip

Next, you will perform the transient thermal analysis on it. Before transient analysis, you will perform steady-state analysis using the initial temperature of the system. The initial temperature of the system before combustion is assume to be 25°C. The temperature of the model at the time of combustion is 2000°C for transient study and consider the system to be water cooled. You will evaluate temperature distribution on the body and also the total heat flux for the complete model.

(Expected time: 45 min)

Boundary Conditions:
Ambient temperature: 22^{0}C (295K)
Material of the Piston: Alloy Steel

Figure 7-18 *The model for Tutorial 2*

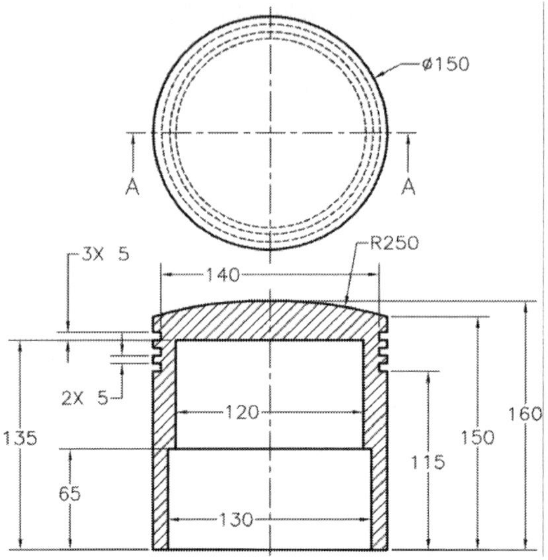

***Figure 7-19** Dimensions of the model for Tutorial 2*

The following steps are required to complete this tutorial:

a. Start SOLIDWORKS and create the model.
b. Invoke SOLIDWORKS Simulation and start a new study.
c. Apply material and steady state boundary conditions to the model.
d. Generate the mesh and run the study.
e. Start a new transient study.
f. Apply transient boundary conditions to the model.
g. Generate mesh on the model.
h. Analyze the results.
i. Display required results.
j. Save the model and exit SOLIDWORKS.

Starting SOLIDWORKS and Creating the Model

First, you need to start SOLIDWORKS and then create the model. For creating the model, set your units to **MMGS(millimeter, gram, second)**.

1. Start a new part file using the **New SOLIDWORKS Document** dialog box and invoke the sketching environment.

2. Choose **Front Plane** as the sketching plane and draw the sketch of the model. For dimensions, refer to Figure 7-19.

3. Choose the **Revolved Boss/Base** tool from the **Features CommandManager** and make a revolved feature to create the model, refer to Figure 7-18.

4. Save the model with the name **c07_sws_2022_tut02.**

Invoking SOLIDWORKS Simulation and Starting a New Study

After the model is created, you need to add the SOLIDWORKS Simulation add-in and start a new study.

1. Start SOLIDWORKS Simulation and choose **New Study** tool from the **Study Advisor** flyout in **Simulation CommandManager**; the **Study PropertyManager** is invoked, refer to Figure 7-20.

2. Choose the **Thermal** button from the **Advanced Simulation** rollout and enter the name of the study as **Steady-state** in the **Study name** edit box in the **Name** rollout.

3. Next, choose the **OK** button from the **Study PropertyManager;** a Simulation Tree is added below the **FeatureManager Design Tree** with **Steady-state** as the name of the study.

4. Right-click on the name of the study(**Steady-state**) and choose the **Properties** option from the shortcut menu; the **Thermal** dialog box is displayed.

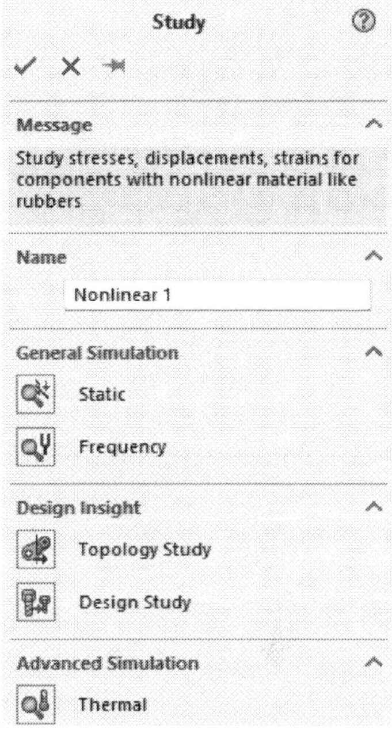

Figure 7-20** Partial view of the **Study PropertyManager

5. Select the **Steady state** radio button from the **Solution type** area in this dialog box, if it is not selected by default.

6. Click on the **OK** button to close this dialog box.

Applying Material and Steady State Thermal Boundary Conditions to the Model

After selecting the study types, you need to assign material to the model in Simulation Tree if not assigned in modeling environment. After that you need to apply boundary for the steady state.

1. Choose the **Apply Material** tool from **Simulation CommandManager** or right-click on the **Part** node in Simulation Tree; a shortcut menu is displayed; choose the **Apply/Edit Material** option, the **Material** dialog box is displayed.

2. Select **Alloy Steel** from the materials displayed and click on the **Apply** button. Next, choose the **Close** button to close this dialog box.

   After applying material to the model, you need to apply the boundary condition under which the thermal analysis will be performed. Note that you need to add temperature load on the cylindrical face of the model the one before combustion.

3. Right-click on the **Thermal Loads** node and choose **Temperature** or you can also select it from the **Thermal Loads** flyout in the **Simulation CommandManager**; the **Temperature PropertyManager** is displayed, refer Figure 7-21.

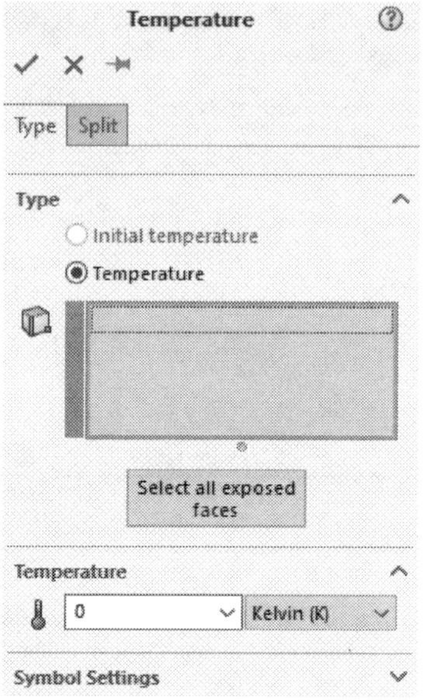

Figure 7-21** Partial view of the **Temperature PropertyManager

4. Select the top cylindrical face of the model, as shown in Figure 7-22, and enter **25⁰ C (or 298.15K)** as the value of temperature load in the **Temperature** edit box of the **Temperature** rollout of this PropertyManager and choose **OK** button.

5. Again, right-click on the **Thermal Loads** node in the Simulation Tree or choose **Convection** from the **Simulation CommandManager;** the **Convection PropertyManager** is displayed, refer to Figure 7-23.

   Now you need to select all the faces of the model except the top cylindrical face.

6. Choose the **Select all exposed faces** button and deselect the face not required, refer to Figure 7-24.

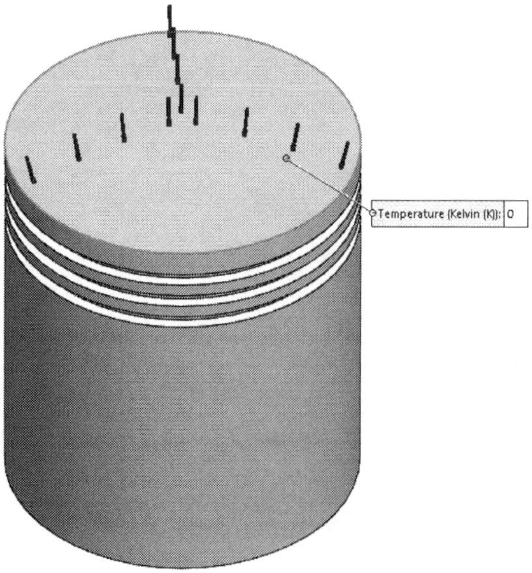

***Figure 7-22** Face selected for applying Temperature load*

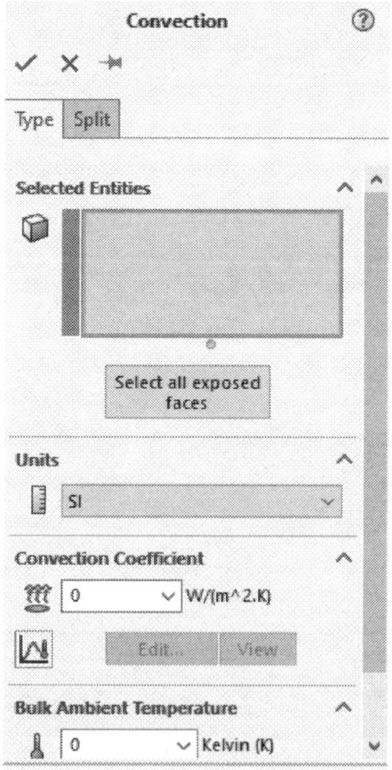

Figure 7-23** The **Convection PropertyManager

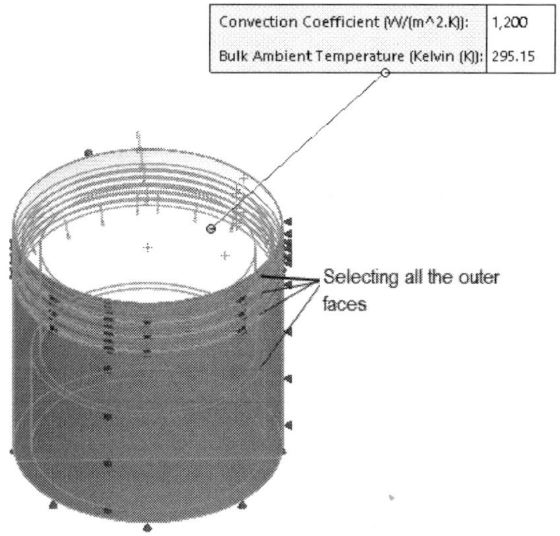

Figure 7-24 *Selecting all the outer faces for Convection load*

7. Enter **1200** in the **Convection Coefficient** edit box in the **Convection Coefficient** rollout and **295.15K** in the **Bulk Ambient Temperature** edit box in the **Bulk Ambient Temparature** rollout as the values of convection coefficient and ambient temperature, respectively.

8. Click the **OK** button to close this **PropertyManager**.

Generating Mesh on the Model and Running the Study

After applying the mentioned boundary conditions to the model, you need to generate mesh before running this study.

1. Choose the **Create Mesh** tool from the **Run This Study** flyout in the **Simulation CommandManager** or right-click on the **Mesh** node in the Simulation Tree; the **Mesh PropertyManager** is displayed, as shown in Figure 7-25.

2. Move the **Mesh Factor** slider in the **Mesh Density** rollout toward Fine; the number of elements increases and you get increased mesh accuracy.

3. Click **OK** to close this PropertyManager; the **Mesh** progress window is displayed and the mesh is generated with the defined settings, refer to Figure 7-26.

   Now, you need to know the number of elements created after meshing.

4. Right-click on the **Mesh** node in Simulation Tree and choose **Details** option from the shortcut menu; the **Mesh Details** window is displayed.

   In mesh details window, you get the information about study name, mesh type, mesher used, mesh quality, aspect ratio, total nodes, total elements, and so on. This is evident from **Mesh Details** window that the total elements are 69014 in thermal study.

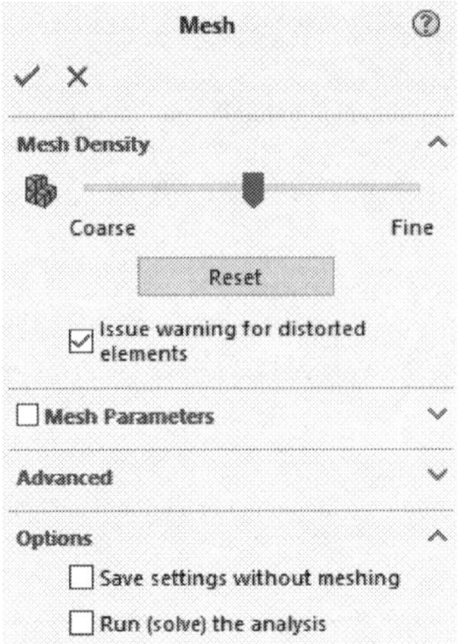

Figure 7-25 *The* ***Mesh PropertyManager***

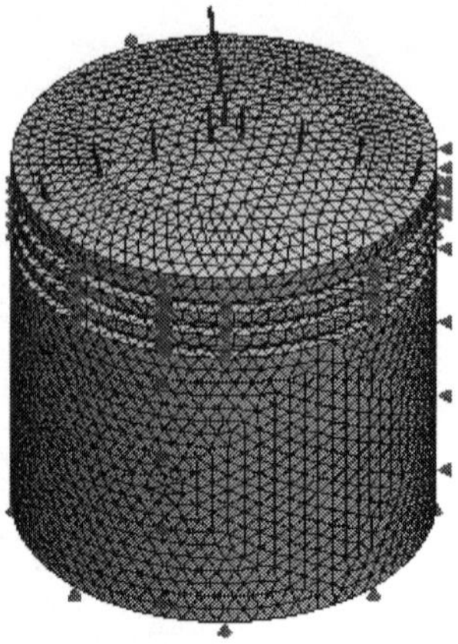

Figure 7-26 *Mesh generated on the model with changed settings*

Now, after the thermal boundary conditions are applied, it is now important to run this study with respect to the boundary conditions applied.

5. Click on the **Run This Study** tool from the **Simulation CommandManager** or right-click on the name of the study in Simulation Tree and choose **Run**: the Solver Status message box is displayed, refer to Figure 7-27, and the **Results** folder is added in the Simulation Tree indicating that the solution is done.

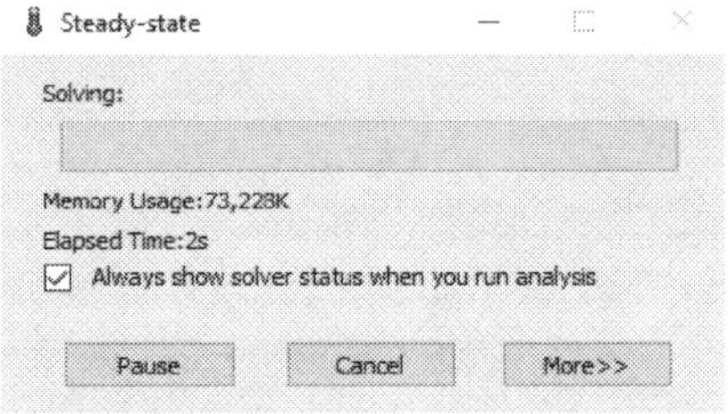

Figure 7-27 *The Solver Status message box*

Starting a New Transient Study

After running the steady state analysis you need to run the transient analysis on the model to get the final results.

1. Right-click on the **Steady-state** tab at the bottom of the Drawing area and choose **Create New Simulation Study**; the **Study PropertyManager** is invoked.

2. Choose the **Thermal** button from the **Advanced Simulation** rollout and enter the name of the study as **Transient** in the **Study name** edit box in the **Name** rollout. Choose the **OK** button from the **Study PropertyManager**; a Simulation Tree is added below the **FeatureManager Design Tree** with **Transient** as the name of the study.

3. Now, right-click on the name of the study **(Transient)** and choose the **Properties** option from the shortcut menu; the **Thermal** dialog box is displayed.

4. Select the **Transient** radio button from the **Solution type** area in this dialog box if not selected.

5. Select the **Initial temperatures from thermal study** check box; the name of the study performed earlier will be displayed in the drop-down list provided on the right of **Thermal study** in the **Solution type** area.

6. Click the **OK** button to close this dialog box.

Applying Transient Thermal Boundary Conditions to the Model

After starting the study you need to apply boundary conditions for the transient analysis on the model.

1. Right-click on the **Thermal Loads** node in the Simulation Tree and then choose the **Temperature** option from the shortcut menu or you can choose Temperature from the CommandManager; the **Temperature PropertyManager** is displayed.

2. Select **Temperature** from the **Type** rollout of the **Temperature PropertyManager** displayed.

3. Select the top cylindrical face of the model as selected in the steady state analysis and enter **2000^{0}C (or 2273.15K)** as the value of temperature load in the **Temperature** edit box of the **Temperature** rollout of this PropertyManager.

4. Choose the **OK** button to close this PropertyManager.

5. Right-click on the **Thermal Loads** node in the Simulation Tree and then choose **Convection** from the shortcut menu or you can choose **Convection** from the **Simulation CommandManager**; the **Convection PropertyManager** is displayed.

   Now you need to select all the outer faces of the model except the top cylindrical face.

6. Select all the required face (outer faces) under the **Faces for Convection** selection box in the **Selected Entities** rollout, refer to Figure 7-28.

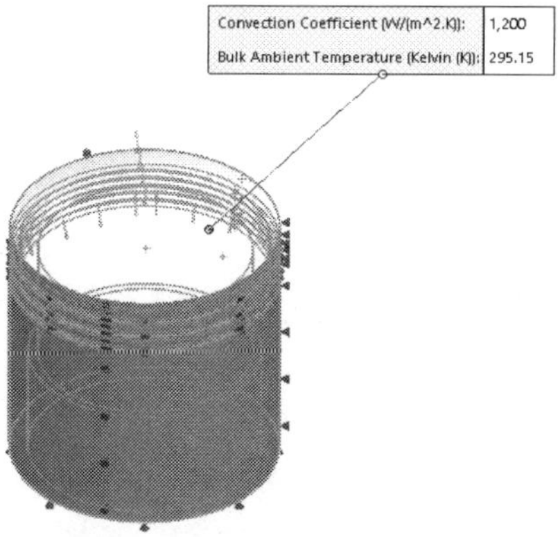

***Figure 7-28** Selecting all the outer faces for Convection load*

7. Enter **1200** in the **Convection Coefficient** edit box of the **Convection Coefficient** rollout and enter **295.15K** in the **Bulk Ambient Temperature** edit box of the **Bulk Ambient Temperature** rollout as values of convection coefficient and ambient temperature, respectively.

8. Choose the **OK** button to close this PropertyManager.

Generating Mesh on the Model

After applying the mentioned boundary conditions to the model, you need to generate mesh before running this study.

1. Right-click on the **Mesh** node in the Simulation Tree and choose **Create Mesh** or choose **Create Mesh** from the Run This Study flyout in the **Simulation CommandManager;** the **Mesh PropertyManager** is invoked.

2. Move the **Mesh Factor** slider toward the extreme right and choose the **OK** button; the **Mesh** progress window is displayed and the mesh is generated, refer to Figure 7-29.

   Note that you can view the number of elements generated. To do so, right-click on the **Mesh** node in the Simulation Tree and then select the **Details** option from the shortcut menu to invoke the **Mesh Details** window, refer to Figure 7-30. This dialog box shows the details of the mesh.

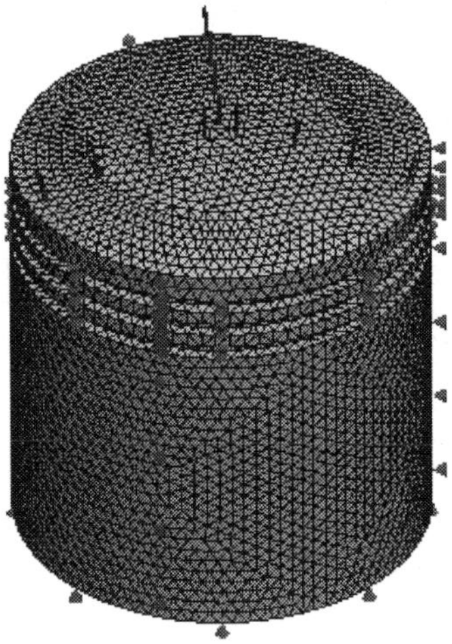

Figure 7-29 Mesh generated on the model

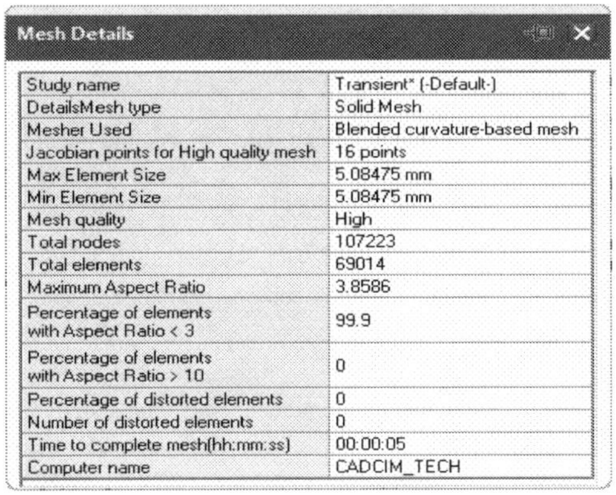

Mesh Details

Study name	Transient* (-Default-)
DetailsMesh type	Solid Mesh
Mesher Used	Blended curvature-based mesh
Jacobian points for High quality mesh	16 points
Max Element Size	5.08475 mm
Min Element Size	5.08475 mm
Mesh quality	High
Total nodes	107223
Total elements	69014
Maximum Aspect Ratio	3.8586
Percentage of elements with Aspect Ratio < 3	99.9
Percentage of elements with Aspect Ratio > 10	0
Percentage of distorted elements	0
Number of distorted elements	0
Time to complete mesh(hh:mm:ss)	00:00:05
Computer name	CADCIM_TECH

Figure 7-30 *The* ***Mesh Details*** *window*

Analyzing the Results

After applying the thermal boundary conditions and generating mesh, it is now important to analyze the behavior of the model with respect to the boundary conditions applied.

1. Click on the **Run This Study** tool from the **Simulation CommandManager** or right-click on the name of the study in Simulation Tree and choose **Run**; the Solver Status message box is displayed, refer to Figure 7-31, and **Results** node is added to the Simulation Tree indicating that the solution is done.

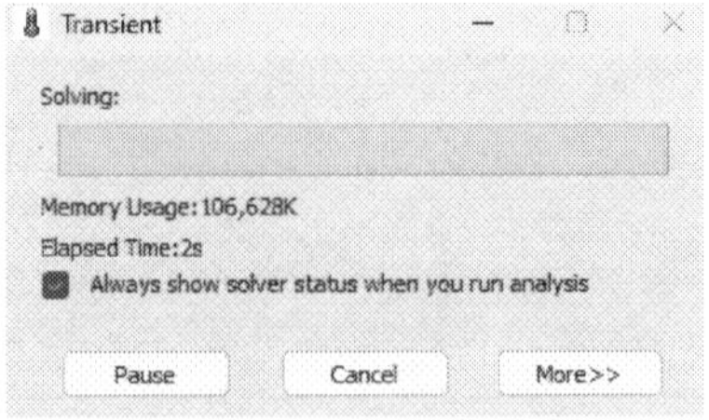

Figure 7-31 *The Solver Status message box*

   Under the **Results** node, the **Thermal1 (-Temperature-)** node is added. Also, the model is displayed in different color contours along with the Legend indicating the distribution of the temperature in the model, refer to Figure 7-32.

Figure 7-32 shows the color contours that indicate the distribution of temperature along various regions of the model. Red color contour displays the region with the maximum temperature whereas the blue color contour shows the region with minimum temperature. The other colors represent various temperatures that lie within the maximum (red contour) and minimum (blue contour) temperatures. The color contours and their respective values can be seen in the legend, refer to Figure 7-32. The temperature on the dome of the piston is maximum because the Temperature boundary condition is applied on it and the component has attained maximum temperature in this region. As you move farther the temperature of the piston decreases and will be minimum at the farthest point, refer to Figure 7-32. In this model, the maximum temperature is 2000 °C and the minimum temperature is 22.004 °C.

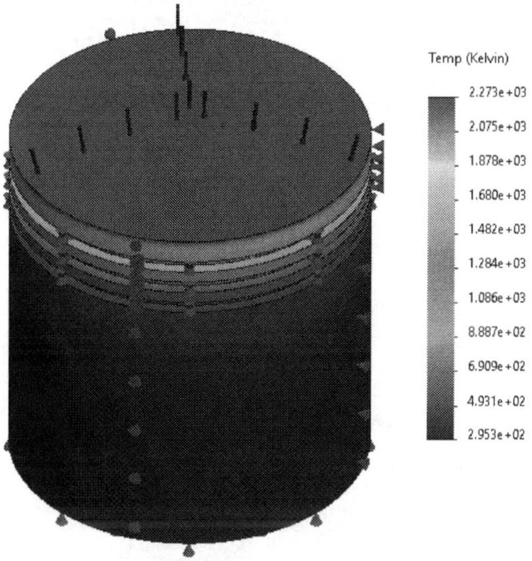

Figure 7-32 *The color contours displayed in the drawing area*

Displaying the Required Results

After solving the model, notice that only one result node is present. You can add more result nodes such as total heat flux and directional heat flux for the model.

1. Right-click on the **Results** node and choose the **Define Thermal Plot** option to invoke the **Thermal Plot PropertyManager**, refer to Figure 7-33.

2. Choose the **HFLUXN: Resultant heat flux** option from the **Component** drop-down list in the **Display** rollout of the **Definition** tab. Choose the **OK** button; the PropertyManager is closed, the **Thermal2 (-Res heat flux-)** node is added under the **Results** node and corresponding legend is displayed along with the model, refer to Figure 7-34.

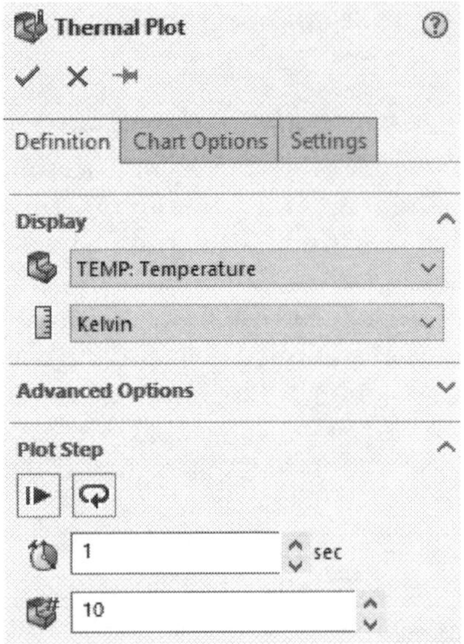

Figure 7-33 *The* ***Thermal Plot PropertyManager***

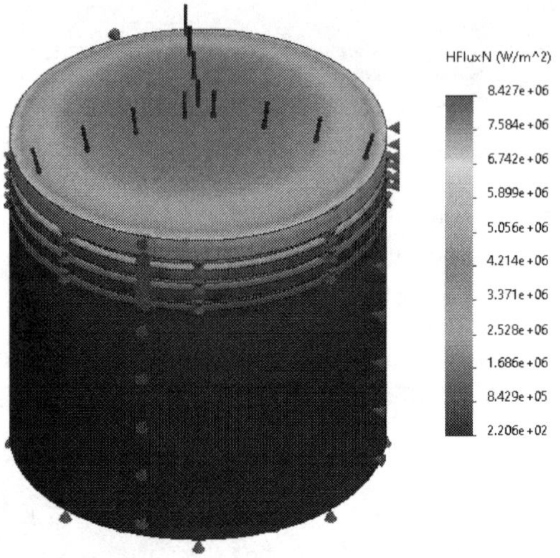

Figure 7-34 *The color contours of the total heat flux displayed on the model*

3. Choose **HFLUXY: Y Heat Flux** from the **Component** drop-down list in the **Display** rollout of the **Definition** tab. Next, choose the **OK** button; the PropertyManager is closed and **Thermal3 (-Y heat flux-)** node is added under the **Results** folder and corresponding legend is displayed along with the model, refer to Figure 7-35.

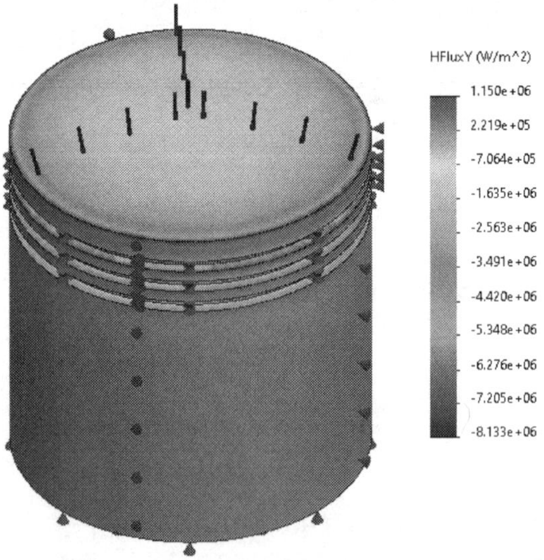

Figure 7-35 *Color contours of the directional heat flux displayed on the model*

Note that thermal analysis of a piston plays an important role in understanding the behavior of the component under given thermal loading conditions.

Saving the Project and Exiting SOLIDWORKS

Now you need to save the project.

1. Choose the **Save** button to save the project with the name *c07_sws_2022_tut02* at the following location:

   \Documents\SOLIDWORKS Simulation\c07

2. Choose **File > Close** from the SOLIDWORKS menus to close the document.

Tutorial 3 Transient Thermal

In this tutorial, you will create the assembly of the Heat Sink shown in Figure 7-36. The dimensions and views of the models are given in Figure 7-37 and 7-38. You can also download this model from *www.cadcim.com*. The path of the file is as follows:

Textbooks > CAE Simulation > SOLIDWORKS Simulation > SOLIDWORKS Simulation 2022 : A Tutorial Approach > Input Files > c07_simulation__2022_input.zip

Next, you will run a transient thermal analysis on it. You will evaluate temperature distribution and also total heat flux for the assembly on the basis of the boundary conditions given next.

(Expected time: 45 min)

Boundary Conditions:
Power generated by microchip: 20 W
Ambient temperature: 27^{0}C (300K)
Convection on radiator walls with film coefficient: 25 (W/m^2)/K
Thermal resistance: 0.001 (K.m^2)/W
Material of the microchip: Ceramic Porcelain
Material of the radiator: 1060 Alloy
Initial temperature: 27^{0}C (300K)

Figure 7-36 *View of the model*

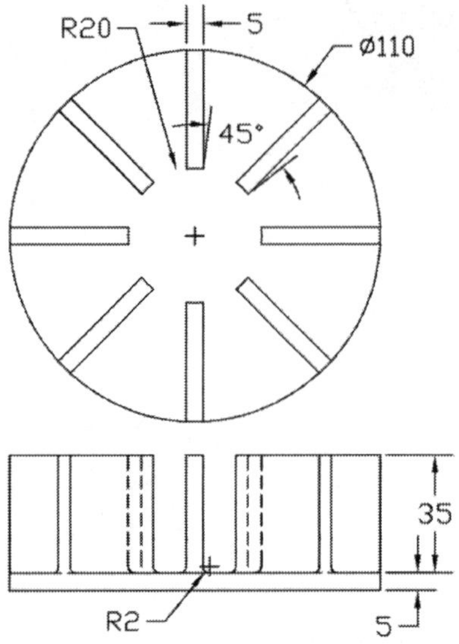

Figure 7-37 Top and front views of the radiator of heat sink with dimensions

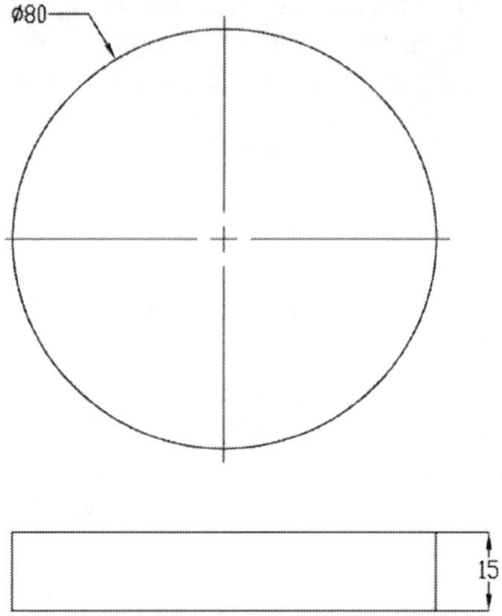

Figure 7-38 Top and front views of the microchip of heat sink with dimensions

The following steps are required to complete this tutorial:

a. Start SOLIDWORKS and create the parts and their assembly.
b. Invoke SOLIDWORKS Simulation and start a new study.
c. Apply material and boundary conditions to the model.
d. Generate mesh.
e. Analyze the results.
f. Save the project and exit SOLIDWORKS Simulation.

Starting SOLIDWORKS and Creating the Parts and their Assembly

First, you need to start SOLIDWORKS and then create the parts for the model. For creating the parts, set your drawing units to **MMGS (millimeter, gram, second).**

1. Start a new part file using the **New SOLIDWORKS Document** dialog box and invoke the sketching environment.

2. Choose **Top Plane** as the sketching plane and draw the sketch of the base of the radiator. For dimensions, refer to Figure 7-37.

3. Choose the **Extruded Boss/Base** tool from the **Features CommandManager** and make an extrude feature to create the base of the radiator, refer to Figure 7-39.

4. Create another extruded feature to complete the radiator, refer to Figure 7-40.

5. Similarly, create the microchip. For dimensions, refer to Figure 7-38.

Figure 7-39 The base feature of the model

Figure 7-40 Final model of the radiator

6. Save both the parts and create their assembly by placing microchip at center of the base feature of the radiator.

7. Save the assembly with the name **c07_sws_2022_tut03.**

Invoking SOLIDWORKS Simulation and Starting a New Study

After the model is created, you need to add the SOLIDWORKS Simulation add-in and start a new study.

1. Start SOLDWORKS Simulation and choose the **New Study** tool from the **Study Advisor** flyout in **Simulation CommandManager**; the **Study PropertyManager** is invoked, refer to Figure 7-41.

2. Choose the **Therma**l button from the **Advanced Simulation** rollout and enter the name of the study as **Heat Sink** in the **Study name** edit box in the **Name** rollout. Choose the **OK** button from the **Study PropertyManager**; a Simulation Tree is added below the **FeatureManager Design Tree** with Heat Sink as the name of the study.

3. Right-click on the study name and choose the **Properties** option from the shortcut menu; the **Thermal** dialog box is displayed.

4. Select the **Transient** radio button from the **Solution type** area in this dialog box, if not selected.

5. Next, enter **3000** in **Total time** and **20** in **Time increment** edit boxes, respectively of the **Solution Type** area of the **Thermal** dialog box.

6. Click on the **OK** button to close this dialog box.

Figure 7-41** Partial view of the* ***Study PropertyManager

Applying Material and Thermal Boundary Conditions to the Model

After starting new study, you need to apply material and boundary conditions under which thermal analysis will be performed to the parts of the assembly. Note that you need to add a thermal resistance contact between the contact faces of the microchip and the radiator.

1. Select **Microchip-1** under the **Parts** node in the Simulation Tree and then choose **Apply Material** from the **Simulation CommandManager** and apply **Ceramic Porcelain** as the material from the **Other Non-metals** category under the **SOLIDWORKS Materials** node in the **Material** dialog box.

2. Similarly, apply **1060 Alloy** from Aluminum alloy as the material to the radiator.

   Now, you need to apply thermal resistance between the contact surfaces of the microchip and the radiator.

3. Right-click on the **Connections** node and choose **Local Interactions** or choose **Local Interactions** from the **Connections Advisor** flyout in **Simulation Command Manager**; the **Local Interactions PropertyManager** is displayed, as shown in Figure 7-42.

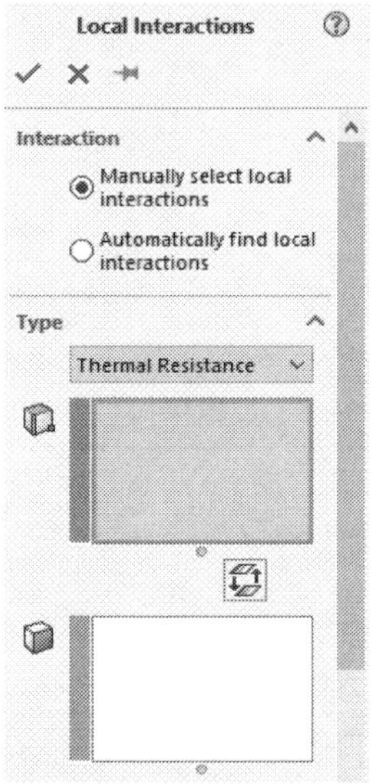

Figure 7-42** The Partial view of the **Local Interactions PropertyManager

4. Select the **Manually select local interactions** radio button in the **Contact** rollout if not selected by default.

5. Select the bottom face of the radiator and the top face of the microchip in **Faces, Edges, Vertices for Set 1** and **Faces for Set 2** selection boxes, respectively, refer to Figure 7-43.

6. Select the **Thermal Resistance** option in the **Type** drop-down list of the **Type** rollout. Also, select the **Thermal Resistance** check box and **Distributed** radio button in the **Thermal Resistance** rollout and specify **0.0001** (K.m^2)/W as the value in the edit box.

   Note that the value of the thermal resistance is obtained through tests.

7. Choose the **OK** button to close this PropertyManager.

8. Specify the heat power generated in the microchip. Right-click on the **Thermal Loads** node and choose **Heat Power** or select it from the **Thermal Loads** flyout in the **Simulation CommandManager**; the **Heat Power PropertyManager** is displayed.

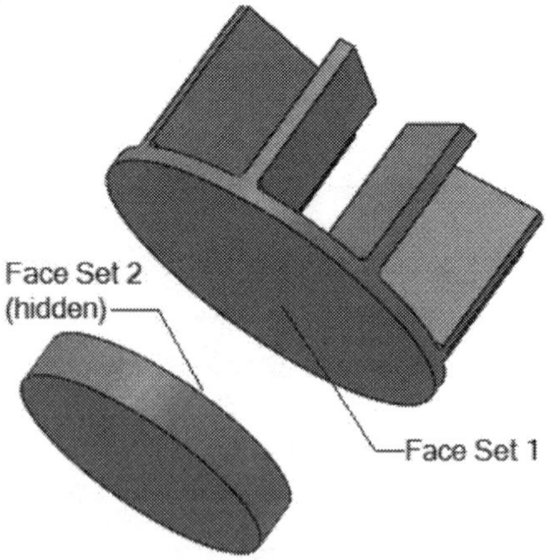

***Figure 7-43** Faces for applying thermal resistance contact*

9. Select **Microchip** from the **FeatureManager Design Tree** and enter **20** in the **Heat Power** edit box of the **Heat Power** rollout.

10. Click the **OK** button to close this PropertyManager.

11. Right-click on the **Thermal Loads** node and choose the **Convection** option from the shortcut menu or from the **Simulation CommandManager**; the **Convection PropertyManager** is displayed, refer to Figure 7-44.

12. Select all the faces of the radiator except the one touching the microchip by clicking on the **Select all exposed faces** button and deselecting the faces which are not required.

13. Enter **25** in the **Convection Coefficient** edit box of the **Convection Coefficient** rollout and **270C (300K)** in the **Bulk Ambient Temperature** edit box of the **Bulk Ambient Temperature** rollout as the values of convection coefficient and ambient temperature, respectively.

14. Click the **OK** button to close this PropertyManager.

15. Right-click on the **Thermal Loads** node in the Simulation Tree and choose the **Temperature** option from the shortcut menu or from the **Simulation CommandManager**; the **Temperature PropertyManager** is displayed.

16. Select the **Initial Temperature** radio button and then select the faces, refer to Figure 7-43.

17. Enter the **300** in the **Temperature** edit box and make sure the **Kelvin [K]** option is selected in the Unit drop-down list.

18. Choose the **OK** button to close the PropertyManager.

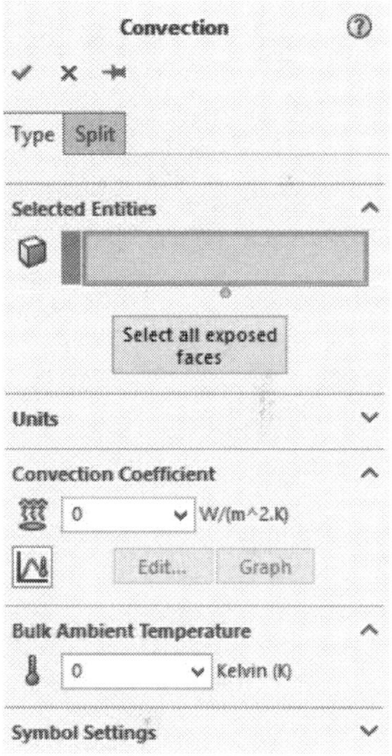

*Figure 7-44 The **Convection PropertyManager***

Generating Mesh on the Model

After applying the mentioned boundary conditions to the model, you need to generate mesh before running this study.

1. Right-click on the **Mesh** node in the **Simulation Tree** and choose **Create Mesh** or select it from the **Run This Study** flyout in the **Simulation CommandManager**; the **Mesh PropertyManager** is invoked.

2. Slide the **Mesh Factor** slider toward extreme right and choose the **OK** button; the **Mesh** progress window is displayed and the mesh is generated, refer to Figure 7-45.

   Note that you can view the number of elements generated by right-clicking **Mesh** and choosing **Details** to invoke the **Mesh Details** window with details, refer to Figure 7-46.

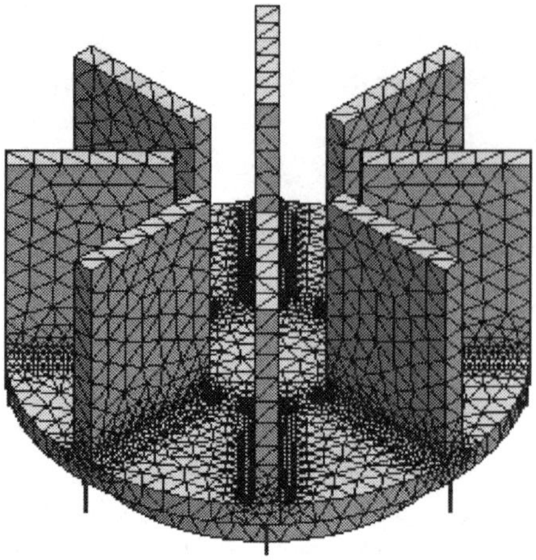

Figure 7-45 Model after generating mesh

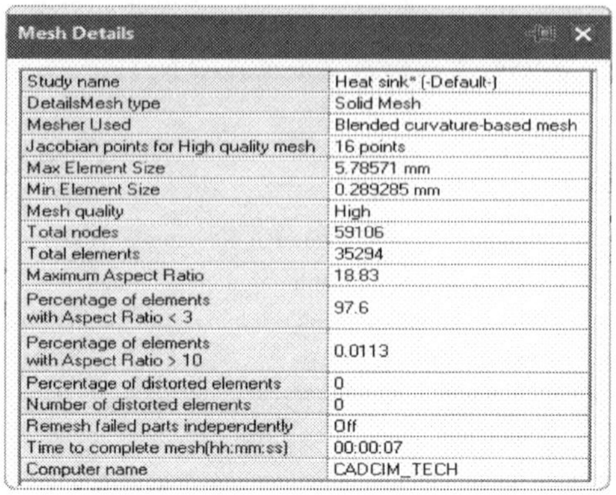

Mesh Details	
Study name	Heat sink* (-Default-)
DetailsMesh type	Solid Mesh
Mesher Used	Blended curvature-based mesh
Jacobian points for High quality mesh	16 points
Max Element Size	5.78571 mm
Min Element Size	0.289285 mm
Mesh quality	High
Total nodes	59106
Total elements	35294
Maximum Aspect Ratio	18.83
Percentage of elements with Aspect Ratio < 3	97.6
Percentage of elements with Aspect Ratio > 10	0.0113
Percentage of distorted elements	0
Number of distorted elements	0
Remesh failed parts independently	Off
Time to complete mesh(hh:mm:ss)	00:00:07
Computer name	CADCIM_TECH

Figure 7-46 The ***Mesh Details*** *window*

Analyzing the Results

After applying the thermal boundary conditions and generating mesh, it is now important to analyze the behavior of the model with respect to the boundary conditions applied.

1. Right-click on the name of the study **(Heat Sink)** and choose **Run** or choose the **Run This Study** tool from the **Run This Study** flyout in the **Simulation CommandManager**; the Solver status message box is displayed. After the model is solved; the **Thermal1 (-Temperature-)** is added under the **Results** folder with the corresponding legend in the drawing area. Also, the model is displayed in different color contours indicating the distribution of the temperature, refer to Figure 7-47.

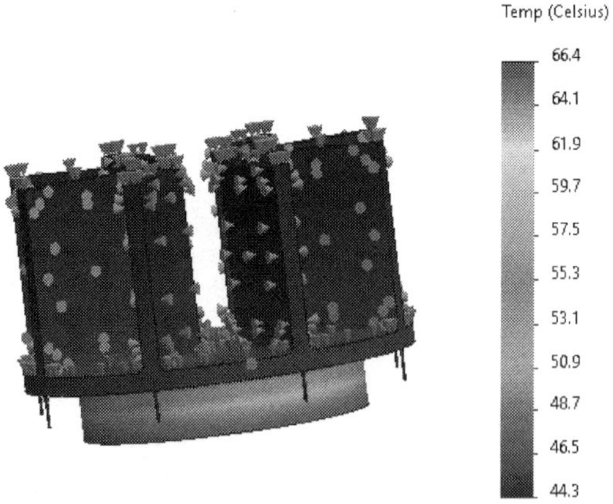

Figure 7-47 *Rotated view of the model with legends*

In the legend, the red color shows the maximum value of temperature attained in the model, which is 66.40 °C. Similarly, the blue color displays the minimum value of temperature in the model, which is 44.30 °C. The other colors display various temperatures attained in the model for the given boundary conditions.

For better information of the temperature distribution, set the model in the sectioned view. To do so, right-click on the **Thermal1 (-Temperature-)** node under the **Results** node and choose **Section Clipping** to invoke the **Section PropertyManager**, refer to Figure 7-48.

2. Choose the **Plane** button in the **Section1** rollout and select **Right Plane** from the **FeatureManager Design Tree** to display the name of respective planes in the **Reference entity** selection box. Next, choose the **OK** button; the sectioned view of the model is created, refer to Figure 7-49.

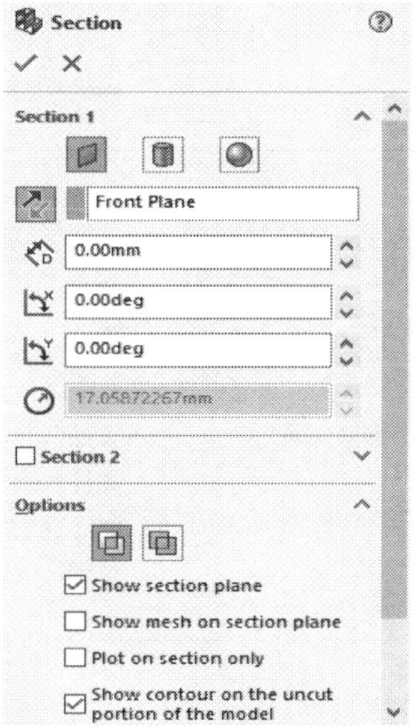

Figure 7-48 *The* ***Section PropertyManager***

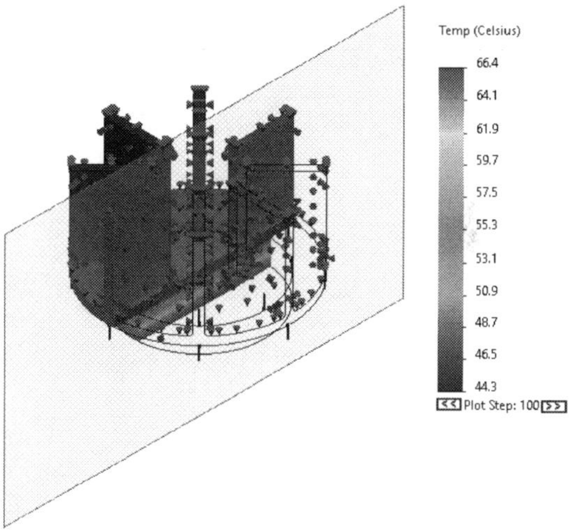

Figure 7-49 Sectioned view of the model along with the legend

3. Turn off the section view by using the **Clipping on / off** toggle button present at the bottom of the **Section PropertyManager.**

4. Right-click on the **Results** node and choose **Define Thermal Plot** to invoke the **Thermal Plot PropertyManager**, as shown in Figure 7-50.

5. Choose **HFLUXN: Resultant heat flux** from the **Component** drop-down list in the **Display** rollout of the **Definition** tab. Choose the **OK** button; the **Thermal2 (-Res heat flux-)** node is added under the **Results** node and corresponding legend is displayed along with the model in the drawing area, refer to Figure 7-51.

Since heat flux is a vector quantity. So to see the direction of the heat flux, you can set the model view to vector form.

6. Right-click the **Thermal2 (-Res heat flux-)** node and invoke the **Thermal Plot PropertyManager** by choosing **Edit Definition**.

7. Select the **Show as vector plot** check box under the **Advanced Options** rollout of the **Definition** tab in this PropertyManager.

8. Click the **OK** button; the PropertyManager is closed and the model is displayed, as shown in Figure 7-52. In this figure, the arrows indicate the direction of heat flux.

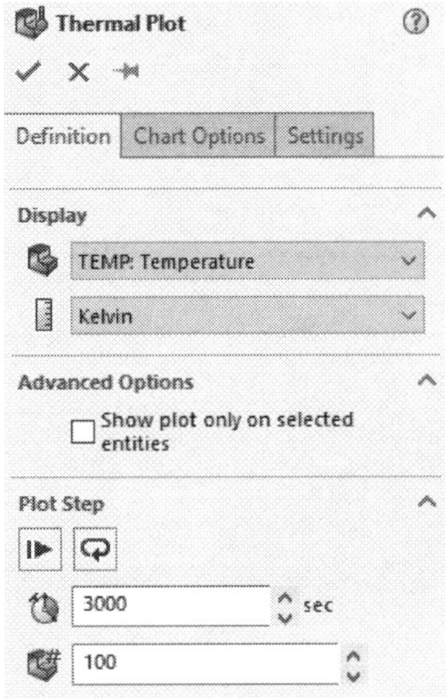

Figure 7-50** Partial view of the **Thermal Plot PropertyManager

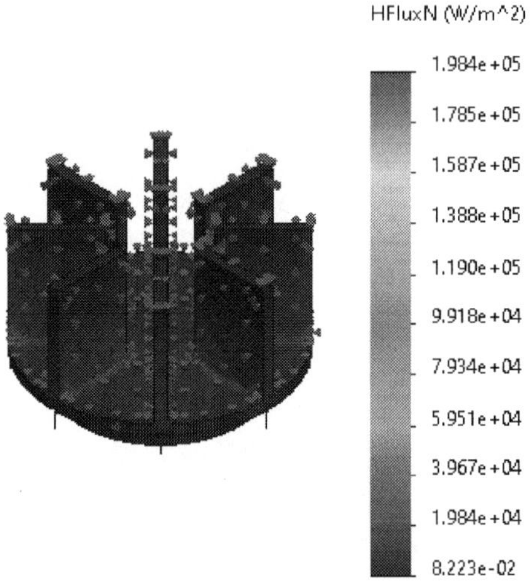

***Figure 7-51** Resultant heat flux legend along with the model*

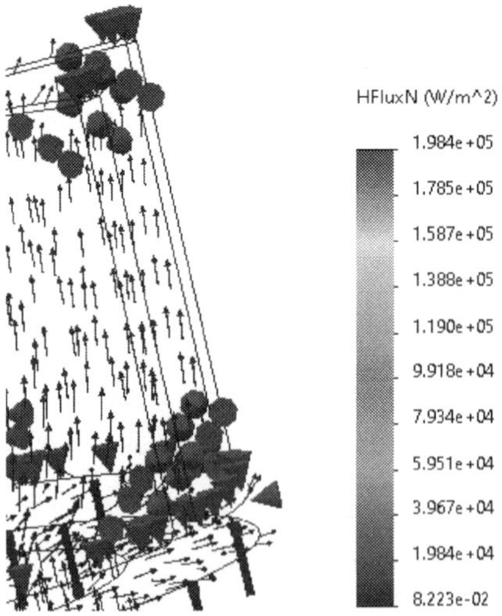

***Figure 7-52** Partial view of the vector plot of the Heat Flux*

For controlling the sizing and density of the arrows in vector form, right-click on the **Thermal2 (-Res heat flux-)** node and choose the **Vector Plot Options**; the **Vector plot options PropertyManager** is invoked. Now, you can define the settings as required in this PropertyManager. Click the **OK** button to close this PropertyManager.

Saving the Project and Exiting SOLIDWORKS

Now you need to save the project.

1. Choose the **Save** button to save the project with the name *c07_sws_2022_tut03* at the following location:

   \Documents\SOLIDWORKS Simulation\c07

2. Choose **File > Close** from the SOLIDWORKS Simulation menus to close the document.

Tutorial 4 Steady-State Thermal

In this tutorial, you will create the model shown in Figure 7-53. The sketch of the model and its dimensions are shown in Figure 7-54. You can also download this model from www.cadcim.com. The path of the file is as follows:

Textbooks > CAE Simulation > SOLIDWORKS Simulation > SOLIDWORKS Simulation 2022: A Tutorial Approach > Input Files > c07_simulation_2022_input.zip

Next, you will perform steady sate thermal analysis on it and find out the effect of thermal loading on the bracket attached to it. Consider a hot fluid of film coefficient 1200 W/m^2 K at 450 °C is running inside the pipe and find out the heat gained by the model from the fluid.

(Expected time: 45 min)

Boundary Conditions:
Convection coefficient of air: 20 W/m^2 K
Ambient temperature: 27^{0}C (300K)
Material of the Piston: Alloy Steel

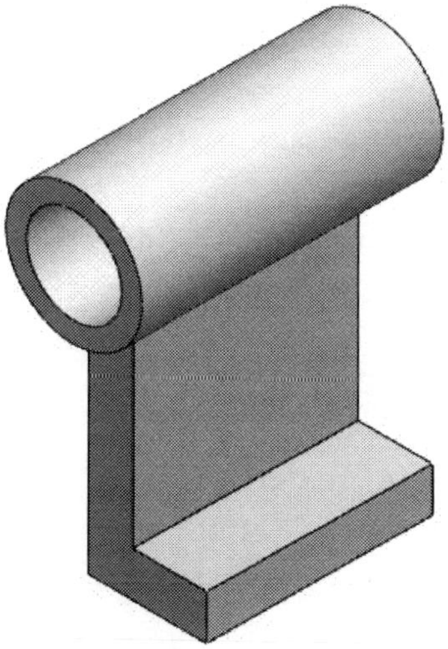

Figure 7-53 *Model for Tutorial 4*

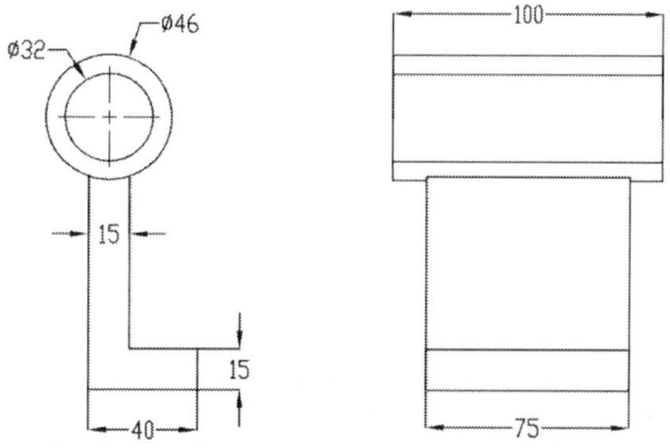

Figure 7-54 *The front and side views of the model with dimensions*

The following steps are required to complete this tutorial:

a. Start SOLIDWORKS and create the model.
b. Invoke SOLIDWORKS Simulation and start a new study.
c. Apply material and boundary conditions to the model.
d. Generate mesh.
e. Analyze the results.
f. Save the project and exit SOLIDWORKS Simulation.

Starting SOLIDWORKS and Creating the Model

First, you need to start SOLIDWORKS and then create the model. For creating the model set your drawing units to **MMGS (millimeter, gram, second)**.

1. Start a new part file using the **New SOLIDWORKS Document** dialog box and invoke the sketching environment.

2. Choose **Front Plane** as the sketching plane and draw the sketch of the model. For dimensions, refer to Figure 7-54.

3. Choose the **Contour Select** tool and extrude the sketch; you get the final model.

Invoking SOLIDWORKS Simulation and Starting a New Study

After the model is created, you need to add the SOLIDWORKS Simulation add-in and start a new study.

1. Start SOLIDWORKS Simulation and choose the **New Study** tool from the **Study Advisor** flyout in **Simulation CommandManager**; the **Study PropertyManager** is invoked, as shown in Figure 7-55.

3. Choose the **Thermal** button from the **Advanced Simulation** rollout and enter the name of the study as **Connector Pipe** in the **Study name** edit box in the **Name** rollout. Choose the **OK** button from the **Study PropertyManager**; a Simulation Tree is added below the **FeatureManager Design Tree** with **Connector Pipe** as the name of the study.

4. Right-click on the study name and choose the **Properties** option from the shortcut menu; the **Thermal** dialog box is displayed.

5. Select the **Steady state** radio button from the **Solution type** area in this dialog box, if it is not selected by default.

6. Click on the **OK** button to close the dialog box.

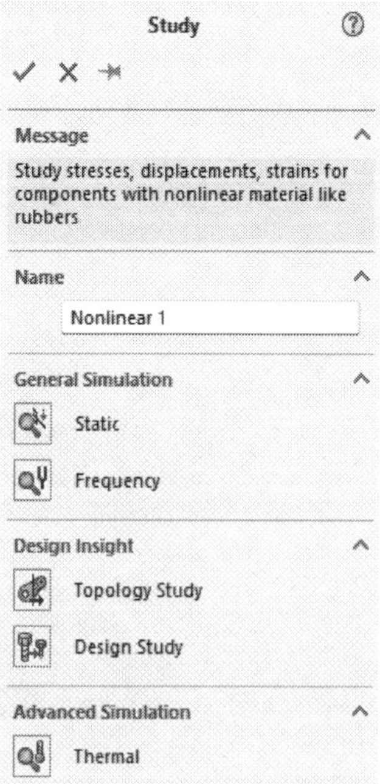

Figure 7-55** Partial view of the **Study PropertyManager

Applying Material and Thermal Boundary Conditions to the Model

After starting the new study, you need to apply material and boundary conditions under which thermal analysis will be performed to the parts of the assembly. In this study, you need to apply two convection loads on the model.

1. Choose **Apply Material** from the **Simulation CommandManager**; the **Material** dialog box is displayed.

2. Select **Alloy Steel** as the material and click on the **Apply** button and then close the dialog box; the material is applied on the model.

   After applying the material to the model, you need to apply boundary conditions also.

3. Right-click on the **Thermal Loads** node and choose **Convection** or choose **Convection** from the **Thermal Loads** flyout in the **Simulation CommandManager**; the **Convection PropertyManager** is displayed, refer Figure 7-56.

4. Select the inside face of the cylindrical feature for applying the convection load, refer to Figure 7-57; the name of the selected face is displayed in the **Faces for Convection** selection box.

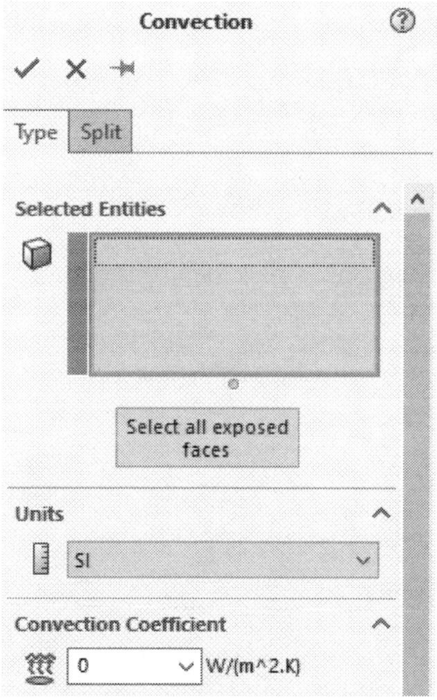

Figure 7-56** The **Convection PropertyManager

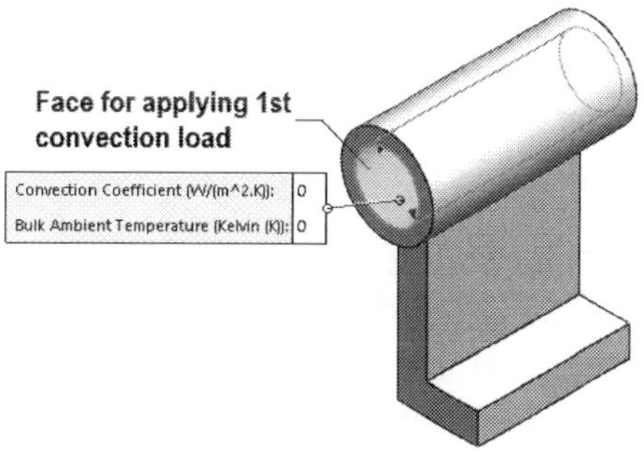

Figure 7-57** Face for applying the **1st Convection load

5. Enter **1200** in the **Convection Coefficient** edit box of the **Convection Coefficient** rollout and **450^{0}C (723.15K)** in the **Bulk Ambient Temperature** edit box of the **Bulk Ambient Temperature** rollout as the values of convection coefficient and ambient temperature, respectively

6. **Select** the **OK** button to close this PropertyManager.

7. Again, right-click on the **Thermal Loads** folder and choose **Convection** or choose **Convection** from the **Simulation CommandManager;** the **Convection PropertyManager** is displayed.

8. Select all the outer faces of the model excluding the inner cylindrical face.

   Since the model is a part of a longer section so you have to exclude the end faces of the pipe as well from the selection, refer to Figure 7-58.

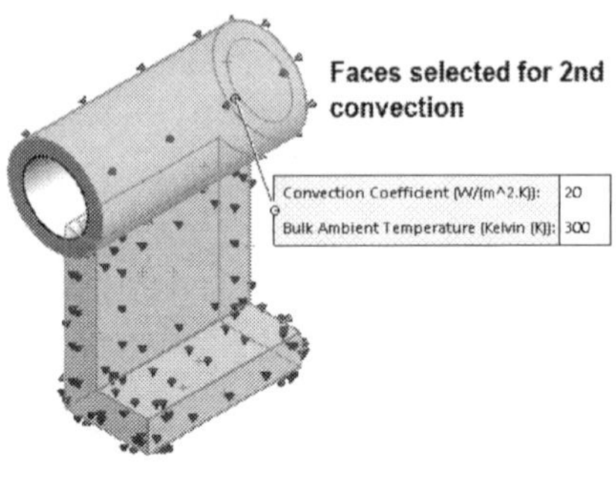

***Figure 7-58** Faces selected for applying the 2nd convection load*

9. Enter **20** in the **Convection Coefficient** edit box of the **Convection Coefficient** rollout and **27°C (300K)** in the **Bulk Ambient Temperature** edit box of the **Bulk Ambient Temperature** rollout as the values of convection coefficient and ambient temperature, respectively.

10. Click the **OK** button to close this PropertyManager.

Generating Mesh on the Model

After applying the mentioned boundary conditions to the model, you need to generate the mesh before running this study.

1. Right-click on the **Mesh** node in the **Simulation Tree** and choose **Create Mesh** or choose it from the **Run This Study** flyout in the **Simulation CommandManager**; the **Mesh PropertyManager** is invoked, refer to Figure 7-59.

2. Slide the **Mesh Factor** slider toward extreme right and choose the **OK** button; the **Mesh** progress window is displayed and the mesh is generated, refer to Figure 7-60.

   To view the number of elements generated, right-click on the **Mesh node** and choose **Details**; the **Mesh Details** window is displayed with details, refer to Figure 7-61.

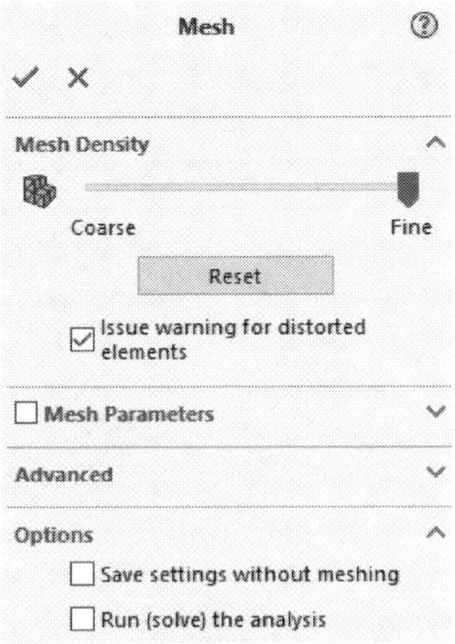

Figure 7-59** The **Mesh PropertyManager

Analyzing the Results

After applying the thermal boundary conditions and generating mesh, you have to analyze the behavior of the model with respect to the boundary conditions applied.

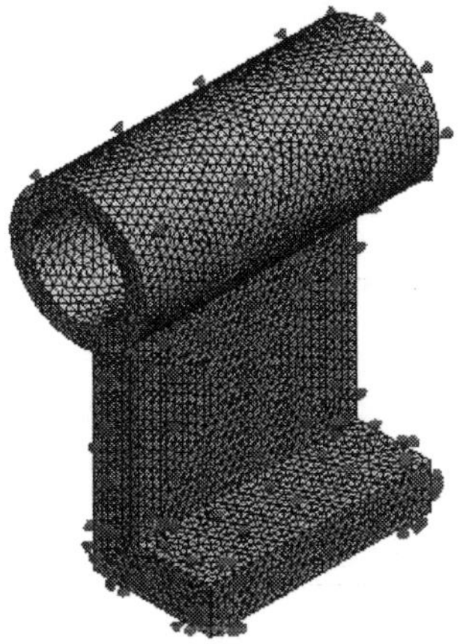

Figure 7-60 Model after generating mesh

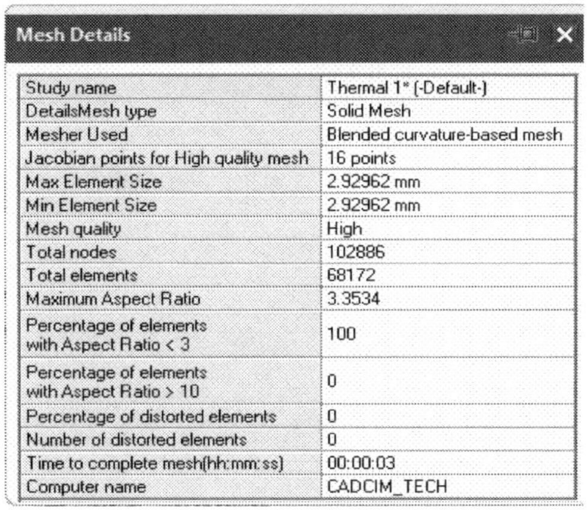

Mesh Details

Study name	Thermal 1* (-Default-)
DetailsMesh type	Solid Mesh
Mesher Used	Blended curvature-based mesh
Jacobian points for High quality mesh	16 points
Max Element Size	2.92962 mm
Min Element Size	2.92962 mm
Mesh quality	High
Total nodes	102886
Total elements	68172
Maximum Aspect Ratio	3.3534
Percentage of elements with Aspect Ratio < 3	100
Percentage of elements with Aspect Ratio > 10	0
Percentage of distorted elements	0
Number of distorted elements	0
Time to complete mesh(hh:mm:ss)	00:00:03
Computer name	CADCIM_TECH

***Figure 7-61** The **Mesh Details** window*

1. Right-click on the name of the study (**Connector Pipe**) and choose **Run** or choose the **Run This Study** tool from the **Run This Study** flyout in the **Simulation CommandManager**; the Solver status message box is displayed.

After the model is solved, the **Thermal1 (-Temperature-)** is added under the **Results** node with the corresponding legend in the drawing area. Also, the model is displayed in different color contours indicating the distribution of the temperature, refer to Figure 7-62.

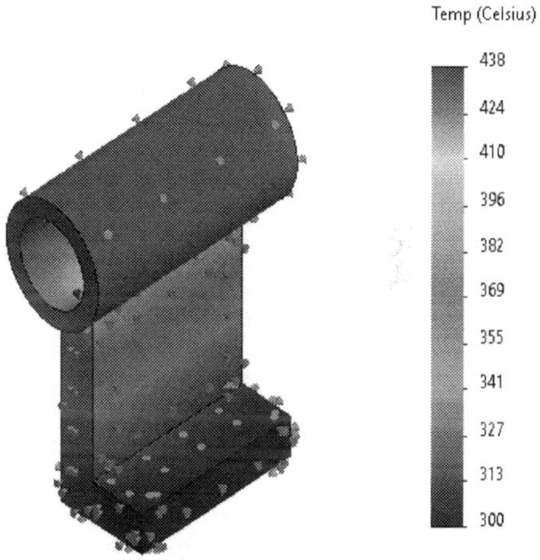

***Figure 7-62** Temperature distribution on the model with the legend*

In the legend, the red color shows the maximum value of temperature attained in the model, which is 438°C. Similarly, the blue color displays the minimum value of temperature in the model, which is 300°C. The other colors display various temperatures attained in the model for the given boundary conditions. For better information of the temperature distribution set the model in the sectioned view. To do so, right-click on the **Thermal1 (-Temperature-)** node under the **Results** folder and choose **Section Clipping** to invoke **Section PropertyManager**, refer to Figure 7-63.

2. Select the **Plane** button in the **Section1** rollout and select the **Right Plane** from the **FeatureManager Design Tree** to display in the **Reference entity** selection box.

3. Choose the **OK** button to close this PropertyManager. The sectioned view of the model is created, refer to Figure 7-64

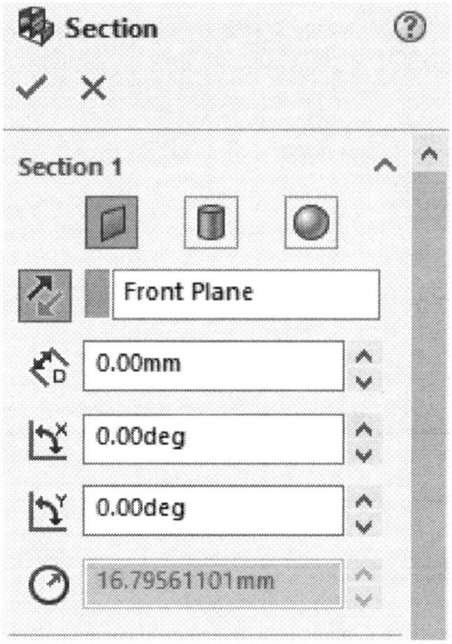

Figure 7-63 *The* ***Section PropertyManager***

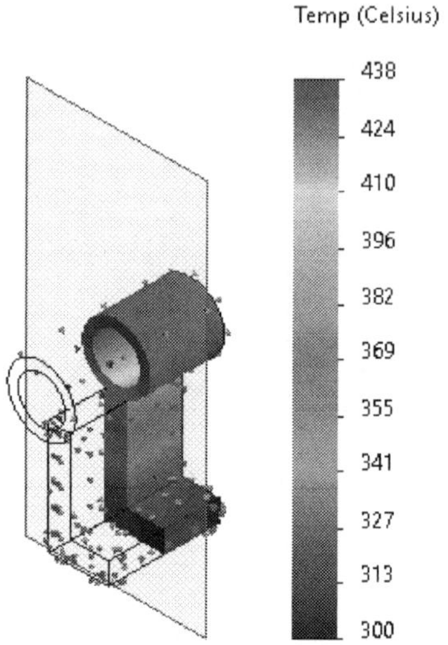

Figure 7-64 *Sectioned view of the model along with the legend*

4. Now, add more results for the model. Right-click on the **Results** folder and choose **Define Thermal Plot** to invoke the **Thermal Plot PropertyManager**, as shown in Figure 7-65.

5. Choose **HFLUXN: Resultant heat flux** from the **Component** drop-down list in the **Display** rollout of the **Definition** tab. Choose the **OK** button to close the PropertyManager; the **Thermal2 (-Res heat flux-) node** is added under the **Results** folder and corresponding **Legend** is displayed along with the model in the drawing area, refer to Figure 7-66.

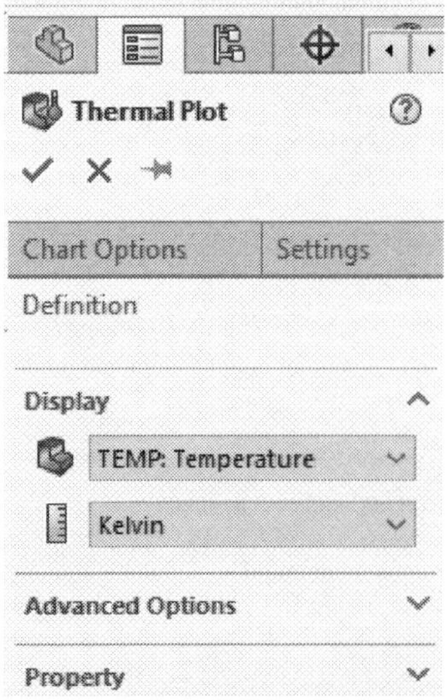

Figure 7-65 *The* ***Thermal Plot PropertyManager***

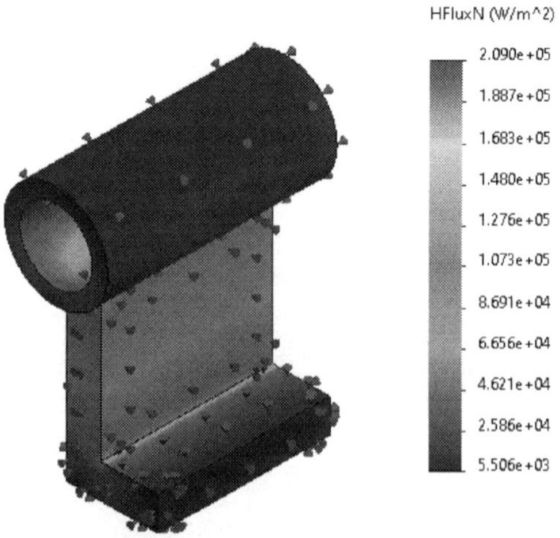

Figure 7-66 *Resultant heat flux legend along with the model*

Since, heat flux is a vector quantity so to see the direction of the heat flux, you can set the model view to vector form.

6. Right-click the **Thermal2 (-Res heat flux-**) node and invoke the **Thermal Plot PropertyManager** by choosing **Edit Definition.**

7. Select the **Show as vector plot** check box under the **Advanced Options** rollout of the **Definition** tab in this PropertyManager.

8. Choose the **OK** button to close this PropertyManager; the model is displayed with number of arrows inside it indicating the direction of heat flux, refer to Figure 7-67.

9. Right-click on the **Thermal2 (-Res heat flux-)** node and choose **Vector Plot Options**; the **Vector plot options PropertyManager** is invoked, refer to Figure 7-68. Note that to control the size and density of the arrows, respective result plot must be activated.

   Now, you can define the settings as required in this PropertyManager.

10. Choose the **OK** button to close this PropertyManager.

    Since the fluid flowing through the pipe is at very high temperature, therefore there will be some transfer of heat from the fluid to the model and hence the model will gain some heat power.

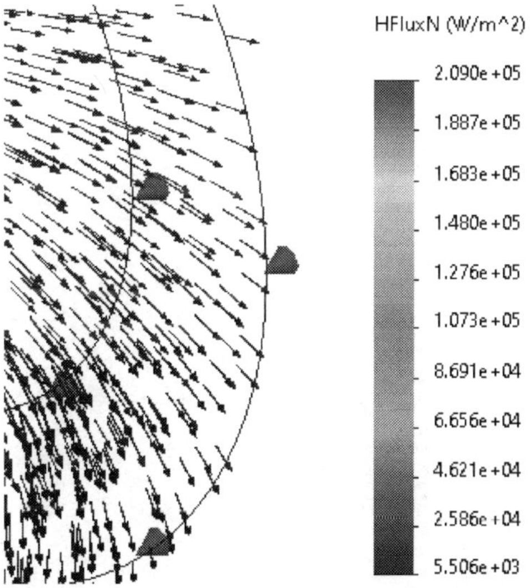

Figure 7-67 Partial view vector plot of the Heat Flux

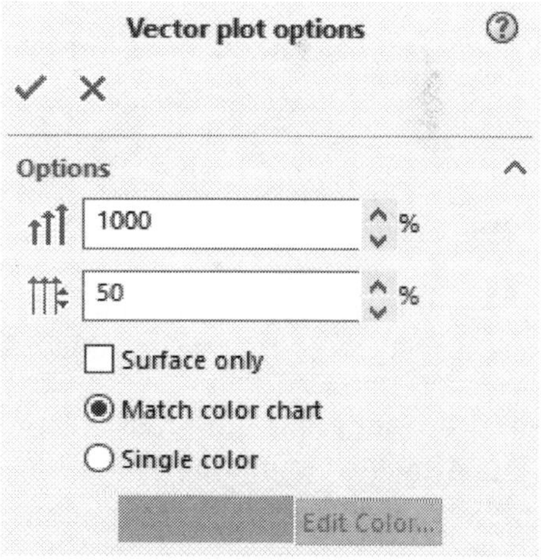

Figure 7-68** The **Vector plot options PropertyManager

11. Right-click on the **Results** node and choose **List Heat Power**; the **Heat Power PropertyManager** is displayed, as shown in Figure 7-69.

12. Select the inside cylindrical face of the model through which the fluid is flowing; the name of the selected face is displayed in the **Faces, Edges or Vertices** selection box in the **Selection** rollout.

13. Select **Watt** as the unit from the **Unit** drop-down list in the **Selection** rollout and click on the **Update** button; a callout with the value of heat power is displayed on the model, refer to Figure 7-70. Next, choose the **OK**.

    A positive value of heat power in the callout signifies a gain of heat by the selected face. Similarly, find heat power for other exposed faces.

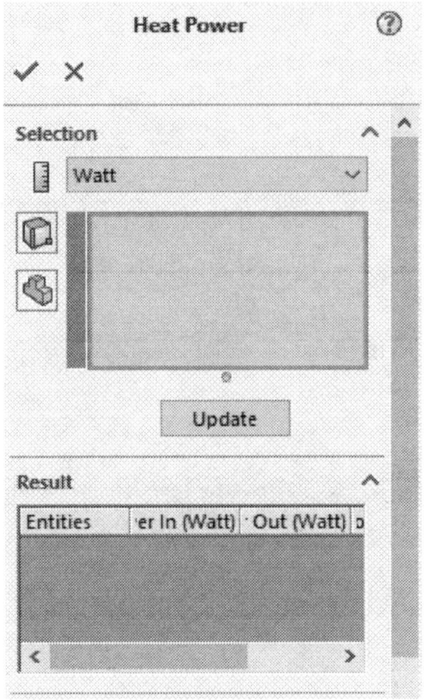

Figure 7-69 Partial view of the ***Heat Power PropertyManager***

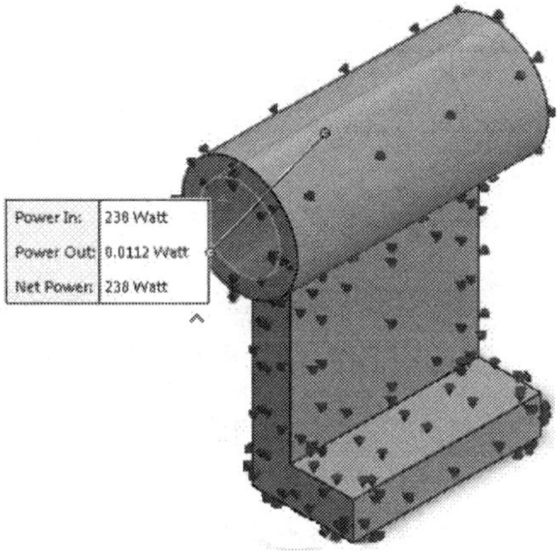

Figure 7-70 Model with the callout displaying value of heat power

14. Right-click on the **Results** node and choose **List Heat Power**; the **Heat Power PropertyManager** is displayed.

15. Select all the faces of the model except the inner cylindrical; the name of the selected face is displayed in the **Faces, Edges or Vertices** selection box in the **Selection** rollout.

16. Select **Watt** as the unit from the **Unit** drop-down list in the **Selection** rollout and click on the **Update** button; callouts for the selected faces with the value of heat power is displayed on the model, refer to Figure 7-71.

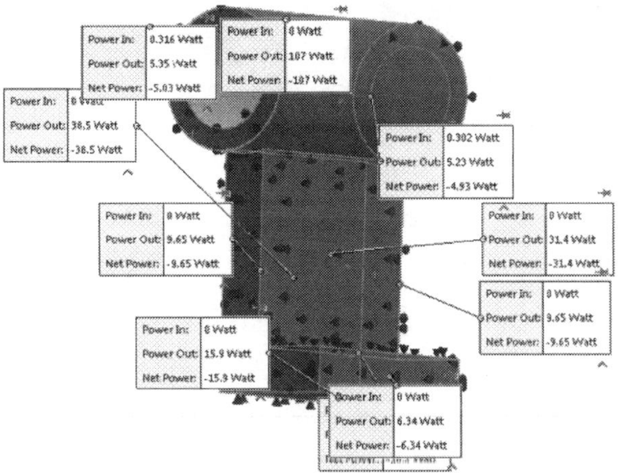

Figure 7-71 *Model with the callouts displaying value of heat power for selected faces*

All the faces selected shows that the value of heat power is negative, which implies heat loss through these faces.

Saving the Project and Exiting SOLIDWORKS

Now you need to save the project.

1. Choose the Save button to save the project with the name *c07_sws_2022_tut04* at the following location:
   \Documents\SOLIDWORKS Simulation\c07

2. Choose **File > Close** from the SOLIDWORKS menus to close the document.

Self-Evaluation Test

Answer the following questions and then compare them to those given at the end of this chapter:

1. You need to define the ________ in the **Convection PropertyManager** to apply convection on any surface.

2. You can insert temperature load by choosing the ____________ tool from the **Simulation** toolbar.

3. The unit of heat flow is ________.

4. In SOLIDWORKS Simulation, only the Steady-State Thermal analysis system is used to carry out all types of thermal analyses. (T/F)

5. The steady-state thermal analysis system is used to carry out a thermal analysis where conditions do not change over time. (T/F)

6. In thermal analysis, only thermal degrees of freedom of the elements are considered. (T/F)

7. You can insert the boundary conditions for a thermal analysis by using the tools available in the Simulation toolbar. (T/F)

8. A probe is used to evaluate results only along the edges of the geometry. (T/F)

9. You can apply different convection loads at different regions of the model. (T/F)

10. You can import various film coefficients in SOLIDWORKS. (T/F)

Review Questions

Answer the following questions:

1. By default, the _______ color in the legend represents the maximum value.

2. Right-click on the **Thermal Loads** node in the Simulation Tree and select the **Temperature** option to insert a Temperature load. (T/F)

3. Right-click on the **Solution** node and choose the **Temperature** option from the shortcut menu to view temperature distribution over a body or surface. (T/F)

4. The minimum and maximum values obtained from an analysis are displayed in a legend. (T/F)

5. In SOLIDWORKS, you can carry thermal analysis for time-dependent loads as well. (T/F)

6. Thermal analysis is used to find out temperature changes with respect to time. (T/F)

7. In Transient Thermal analysis, the loads are time-dependent. (T/F)

EXERCISES

Download the input files to perform the exercises of this chapter and then extract and save them at desired location. To download the files, path is given below:

Textbooks > CAE Simulation > SOLIDWORKS Simulation > SOLIDWORKS Simulation 2022: A Tutorial Approach > Input Files

Extract the downloaded zipped file at the location *C:\SOLIDWORKS Simulation\c07*.

Exercise 1

In this exercise, you will use *c07_exr01* file downloaded earlier to perform a transient thermal study on the Cylinder Head model of two-stroke automobile engine considering a hot fluid of film coefficient 1200 W/m^2 K at 450 °C is running inside the pipe. The objective to find out the effect of thermal loading with the given conditions on the Cylinder Head model. Figure 7-72 shows the model of Cylinder Head of two-stroke automobile engine for study.

(Expected time: 45 min)

Material: Aluminum alloys
Initial Temperature: 27°C (300K)

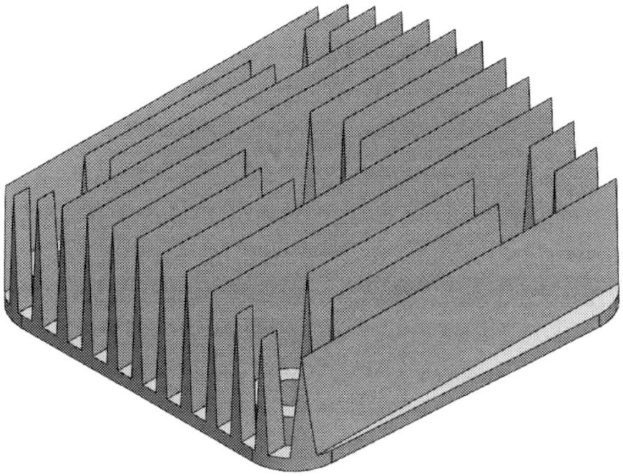

Figure 7-72 *Cylinder Head model*

Exercise 2

In this exercise, you will use *c07_exr02* file downloaded earlier to perform a thermal study on the Pipe Elbow model with boundary conditions as given in the Figure 7-73. You need to find out the effect of thermal loading such as temperature distribution in the given condition on the Pipe Elbow model. No convection coefficient is defined on the any surface of the model because heat can enter and leave the model only through the end faces of the Elbow. In the Steady State thermal Study, thermal conditions have been applied and it is assumed that all the parameters of the heat flow no longer vary with time. **(Expected time: 45 min)**

Material: Aluminum alloys
Initial Temperature: 27°C (300K)

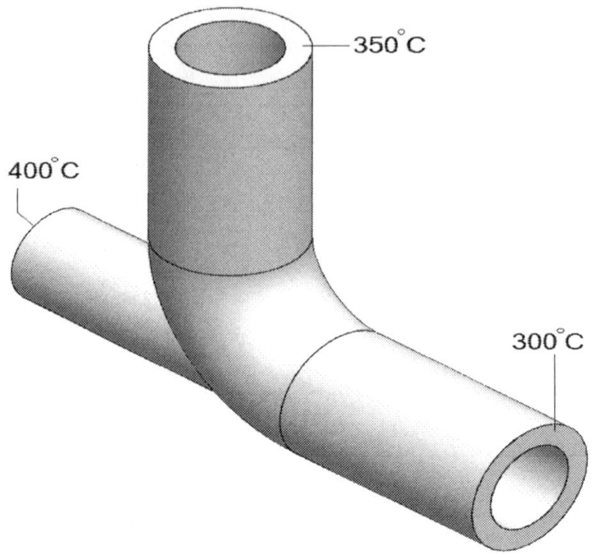

Figure 7-73 Model for Exercise 2

Exercise 3

A spherical interplanetary probe of 0.75 m diameter contains electronics that dissipate 250 W. If the probe surface has an emissivity of 0.8 and the probe does not receive radiation from other surfaces, as, for example, from the sun, what is its surface temperature?

(a) Use Simulation to find the surface temperature of the sphere.

(b) Use the Stefan-Boltzmann law to calculate the surface temperature of the sphere to validate the results obtained by using Simulation. If the values are not same, calculate the percentage error based on calculated temperature.

(Expected time: 45 min)

Exercise 4

Compare the temperature distribution along the length of three tubes with following specifications:

Length of the tubes:	1000 mm
Outer diameter of tubes:	100 mm
Inside diameter of tubes:	80 mm

Material

Tube 1:	1060 Alloy
Tube 2:	Aluminum Bronze

Tube 3:	Stainless Steel (Ferrite)
h of air:	5 w/m ^ 2 .K
Ambient Temperature:	300K
Temperature on left end of tubes:	600K

Design the simulation study, apply the boundary conditions and do the simulation. Note the temperature at various points along the length of the tube and plot the temperature versus length graph.

Note: Use Excel program to plot the graph of temperature distribution in three given tubes.

(Expected time: 45 min)

Exercise 5

The wall of an industrial furnace constructed from 0.25m thick fire clay brick having a thermal conductivity of 1.8 w/m.K. The temperature of inner and outer surfaces are 1400 and 1160K respectively.

(a) Use Simulation to find the rate of heat flow (Q) through a wall that is 0.5 m wide and 1.5 m long.

(b) Use the Fourier's Law [Q" = k (T_i – T_o)/X] to calculate the heat loss to validate the results obtained by using Simulation. If the values are not same, calculate the percentage error.

(c) The furnace is surrounded by air having convection coefficient h = 80 w/m2.K and ambient temperature of 300K, using Simulation find out the rate of heat flow (flux – Q").

(Expected time: 45 min)

Exercise 6

An oval shape handle is connected to a pan for easy handling, as shown in Figure 7-74. The temperature of the pan can reach as high as 200 C. The dimensions of the hollow oval shaped handle are: major diameter = 50 mm and minor diameter = 30 mm. Inside diameters are 40 mm and 20 mm, respectively. The handle is made of Wrought Stainless Steel with thermal conductivity (k) of 19 w/m.K and the pan is made of Alloy Steel. Ambient temperature is 27° C and heat transfer coefficient (h) of surrounding air is 20 w/m ^ 2.K.

(a) The user has decided to make the handle 300 mm long and the specification call for the temperature not to exceed 40 C at the end of the handle. Use Simulation to determine if the designer's decision to make the handle 300 mm long is the right decision.

(b) Draw the temperature profile, temperature vs. length, of the handle.

(Expected time: 45 min)

Figure 7-74 *Model for Exercise 6*

Answers to Self-Evaluation Test

1. **Convection Coefficient**, **2**. **Temperature**, **3**. Watt, **4**. F, **5**. T, **6**. T, **7**. T, **8**. F, **9**. T. **10**. T

Chapter 8

Nonlinear Analysis

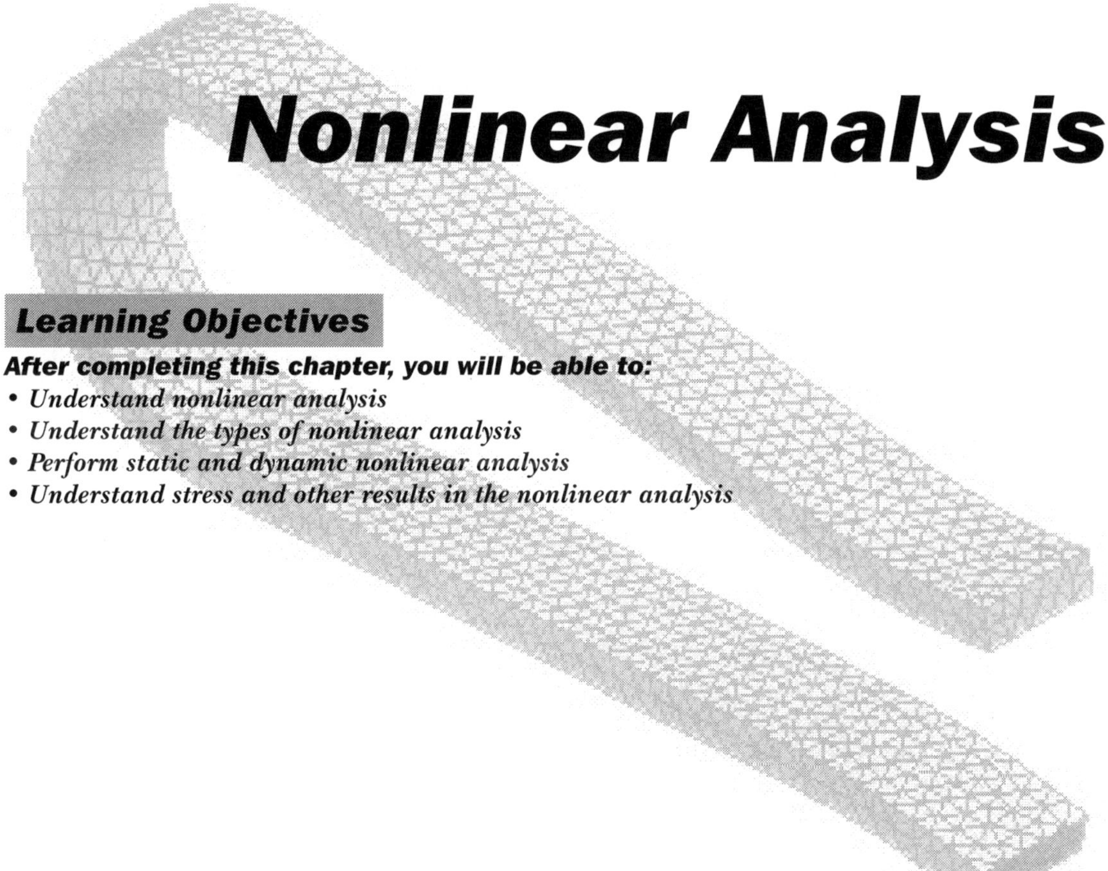

Learning Objectives

After completing this chapter, you will be able to:

- *Understand nonlinear analysis*
- *Understand the types of nonlinear analysis*
- *Perform static and dynamic nonlinear analysis*
- *Understand stress and other results in the nonlinear analysis*

NONLINEAR ANALYSIS

A nonlinear analysis is an analysis of a nonlinear relation existing between applied forces and displacements. Such nonlinear relations cause nonlinear effects which can originate due to geometrical nonlinearity (large deformations), material nonlinearity (elasto-plastic material), and contact nonlinearity. These effects result in a stiffness matrix which is not constant during load application. As a result, a different solving strategy is required for the nonlinear analysis and therefore a different solver.

Nonlinear Structural Analysis is a prediction of the response of nonlinear structures by model based analysis. Analysis involves a combination of mathematical modeling, discretization methods and numerical techniques. Finite element methods dominate the discretization step in this analysis.

In structural analysis, there are four sources of nonlinear behavior. The corresponding nonlinear effects are identified in terms of material, and boundary conditions (force and displacement). The source of nonlinearity can be attributed to multiple system properties, for example, materials, geometry, nonlinear loading, and constraints. Different types of nonlinearity that occur in the analysis are discussed next.

Geometric Nonlinearity

In finite element analyses, geometric nonlinearity involves changes in geometry. In this type of analysis, structural deformation are considered in formulating the constitutive and equilibrium equations. Many engineering applications such as metal forming, tire analysis, and medical device analysis require the use of large deformation analysis based on geometric nonlinearity. Small deformation analysis based on geometric nonlinearity is required for some applications, like analysis involving cables, arches, and shells.

Change in geometry is physical source of geometric nonlinearity as the structural deformation is taken into account in setting up the strain-displacement and equilibrium equations. There are many applications of geometric nonlinear analysis such as slender structures in aerospace, civil and mechanical engineering applications, and tensile structures such as cables, inflatable membranes, metal, and plastic forming.

Material Nonlinearity

Material nonlinearity involves the nonlinear behavior of a material based on a current deformation history, rate of deformation, temperature, pressure, and so on. There are many examples of material nonlinearity in the models such as large strain (visco), elasto-plasticity, and hyperelasticity (rubber and plastic materials)

There is a common misconception that nonlinear analysis is the analysis of nonlinear material only but material nonlinearity is just one of the types of the nonlinear analysis.

Material behavior depends on current deformation state and possibly past history of the deformation. There are so many factors such as pre-stress, temperature, time, moisture, electromagnetic fields which may be involved in material nonlinearity. Nonlinearity in engineering is considered in structures, which undego, nonlinear elasticity, plasticity, viscoelasticity, creep, or inelastic rate effects.

Contact (Boundary Conditions) Nonlinearity

Contact (constraint) nonlinearity can occur if kinematic constraints are present in the model. The kinematic degrees-of-freedom of a model can be constrained by imposing restrictions on its movement. Nonlinear analysis takes more time compared to linear analysis in solving the equations.

It is necessary to understand the basic mechanism of the problem to reduce the time involved in getting its solution. For example, if the problems contain nonlinearity then you first need to understand how they will affect the response of structures by doing finite element analysis. Contact nonlinearity depends upon the applied forces and its deformation in the analysis.

One of the important concerns of nonlinear analysis is to simulate the pressure loads of fluids. These loads include hydrostatic loads on submerged or container structures in aerodynamic and hydrodynamic loads caused by the motion of aeroform and hydroform fluids such as wind loads, wave loads, and drag forces. The Mathematical interest are gyroscopic and non-conservative follower forces but these forces are of interest only in a limited problems, mainly in aerospace engineering.

TUTORIALS

To perform the tutorials of this chapter, you need to download the input files from *www.cadcim.com*. The complete path for downloading the files is as follows:

Textbooks > CAE Simulation > SOLIDWORKS Simulation > SOLIDWORKS Simulation 2022: A Tutorial Approach > Input Files > c08_simulation_2022_input.zip

Extract the downloaded zipped file and save it.

Tutorial 1

In this tutorial, you will open *c08_tut01* file that was downloaded earlier and perform the nonlinear structural study on the model shown in Figure 8-1.

The model is made of **AISI 1035 Steel (SS)** material and the boundary conditions (Fixed support on circular face and 5000 N bearing load on semicircular face) applied are also shown in Figure 8-1. To understand the importance of geometry optimization, you will optimize the model and then evaluate the results. **(Expected time: 1 hr)**

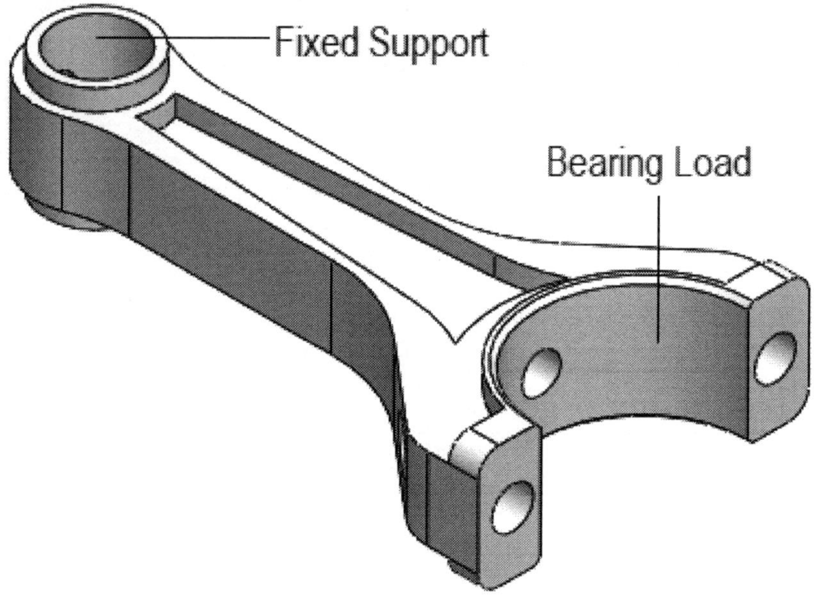

Figure 8-1 *Model for Tutorial 1*

Following steps are required to complete this tutorial:

a. Start SOLIDWORKS and open the model.
b. Generate mesh and apply structural boundary conditions to the model.
c. Analyze the results.
d. Save the model and exit SOLIDWORKS Simulation.

Starting SOLIDWORKS and Opening the Model

First, you need to start SOLIDWORKS and then open the model. After opening the model, you need to add the **SOLIDWORKS Simulation** add-in and start a new nonlinear study.

1. Open an input file named as *c08_tut01* and invoke the simulation environment.

2. Choose the **New Study** tool from the **Study Advisor** flyout in the **Simulation CommandManager**; the **Study PropertyManager** is invoked, refer to Figure 8-2.

3. Choose the **Nonlinear** button from the **Advanced Simulation** rollout and then the **Static** button. Next, enter **Nonlinear Analysis** as the name of the study in the **Study name** edit box of the **Name** rollout.

4. Choose the **OK** button from the **Study PropertyManager**. On doing so, a Simulation Tree is added below the Design Tree with Nonlinear Analysis as the name of the study.

5. Right-click on the name of the study node and choose the **Properties** option from the shortcut menu; the **Nonlinear - Static** dialog box is displayed.

6. Select the **Use large displacement formulation** check box in the **Geometry nonlinearity options** area in this dialog box if it is not selected by default, refer to Figure 8-3.

7. Choose the **OK** button to close this dialog box.

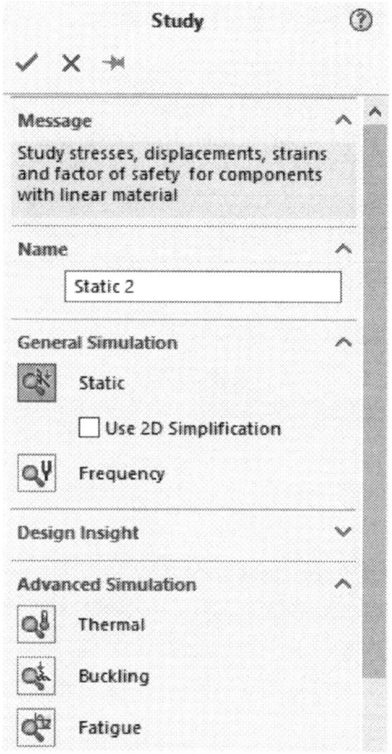

Figure 8-2 Partial view of the ***Study PropertyManager***

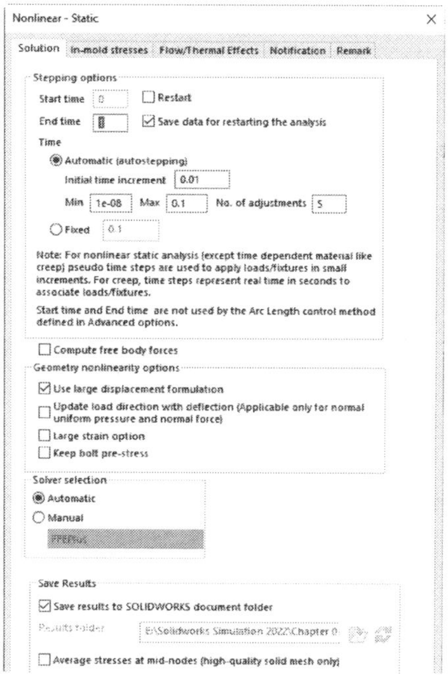

Figure 8-3 Partial view of the ***Nonlinear - Static*** *dialog box*

Assigning Material and Generating Mesh

After selecting the type of study, you need to assign material to the model in simulation environment if not assigned in the modeling environment. After that you need to mesh the model.

1. Choose the **Apply Material** tool from the **Simulation CommandManager**; the **Material** dialog box is displayed.

2. Select **AISI 1035 Steel (SS)** from the materials displayed in the **Material** dialog box under the **Steel** node.

3. Choose the **Apply** button and then the **Close** button to close the **Material** dialog box.

Alternatively, you can apply materials to the model by using shortcut menu. To do so, right-click on part in the Simulation Tree and then choose the **Apply/Edit Material** option from the shortcut menu; the **Material** dialog box is displayed.

After applying material, you need to mesh the model.

4. Choose the **Create Mesh** tool from the **Run This Study** flyout in the **Simulation CommandManager**; the **Mesh PropertyManager** is displayed, as shown in Figure 8-4.

Figure 8-4 *The* ***Mesh PropertyManager***

5. Choose the **OK** button from the PropertyManager; the mesh with default settings is generated, refer to Figure 8-5.

   Now, you need to know the number of elements created after meshing.

6. Right-click on the **Mesh** node in the Simulation Tree and choose the **Details** option from the shortcut menu; the **Mesh Details** window is displayed, as shown in Figure 8-6.

   The **Mesh Details** window provides you information such as study name, mesh type, mesher used, mesh quality, aspect ratio, total nodes, and total elements. This is evident from Figure 8-6 that the total number of elements and nodes displayed in the **Mesh Details** window are 19270 and 10998, respectively. Now, you need to improve the mesh quality to get more accurate results of nonlinear simulation study.

Figure 8-5 *Mesh generated with the default settings*

Mesh Details

Study name	Nonlinear Analy
DetailsMesh type	Solid Mesh
Mesher Used	Blended curvat
Jacobian points for High quality mesh	16 points
Max Element Size	3.17542 mm
Min Element Size	1.05846 mm
Mesh quality	High
Total nodes	19270
Total elements	10998
Maximum Aspect Ratio	14.31
Percentage of elements with Aspect Ratio < 3	94.7
Percentage of elements with Aspect Ratio > 10	0.291

Figure 8-6 *Mesh details with the default settings*

7. Choose the **Create Mesh** tool from the **Run This Study** flyout in the **Simulation CommandManager**; the **Mesh PropertyManager** is displayed.

8. Set the slider in the **Mesh Density** rollout on its right most extreme toward Fine; the number of elements increases and you get increased mesh accuracy.

9. Click **OK** to close this PropertyManager; the **Mesh** progress window is displayed and mesh is generated with modified settings, refer to Figure 8-7.

   Note that the number of elements in this mesh is 71415. This number is very large as compared to the element count that was achieved by the mesh generated with default mesh control settings.

Figure 8-7 *Mesh generated with the modified settings*

Applying Boundary Conditions to the Model

After the model is meshed, you need to apply boundary conditions for the study. You will apply a bearing load on the semi-cylindrical face and fixed support on the another cylindrical face of the model.

1. Choose the **Fixed Geometry** tool from the **Fixtures Adviser** flyout in the **Simulation CommandManager**; the **Fixture PropertyManager** is displayed, as shown in Figure 8-8.

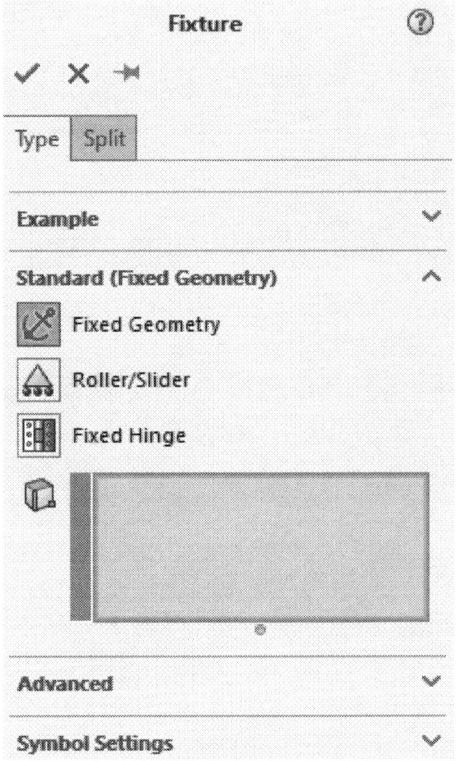

Figure 8-8** The **Fixture PropertyManager

2. Select the circular-cylindrical face of the model for applying the fixed support, refer to Figure 8-1; the name of the selected face is displayed in the **Faces, Edges, Vertices for Fixture** selection box.

3. Choose the **OK** button to close the PropertyManager.

4. Choose the **Bearing Load** tool from the **External Loads Adviser** flyout in the **Simulation CommandManager**; the **Bearing Load PropertyManager** is displayed.

5. Select the semi cylindrical faces of the model under the **Cylindrical Faces or Shell Circular Edges for Bearing Load** selection box in the PropertyManager; the names of the selected face get displayed in the selection box in the **Selected Entities** rollout and a callout is attached to it, refer to Figure 8-9.

6. Select the coordinate system created earlier from the Design Tree; the name of the coordinate system gets displayed in the **Select a Coordinate System** selection box.

   This coordinate system will be used to define the direction of the bearing load of the model.

7. Select the **SI** option in the **Unit** drop-down in the **Bearing Load** rollout of the PropertyManager, if not selected by default.

8. Select the **X-Direction** button and enter **5000** in the **X-Direction** edit box. If the direction of the load is not same, as shown in Figure 8-9, you need to select the **Reverse Direction** check box to reverse the direction of the bearing load.

9. Retain rest of the default values of the options in the **Bearing Load PropertyManager,** refer to Figure 8-10 and choose the **OK** button to close this PropertyManager.

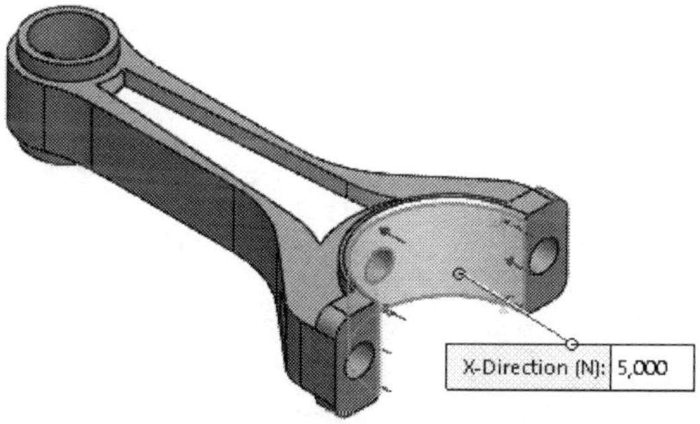

Figure 8-9 *Face to be selected for bearing load*

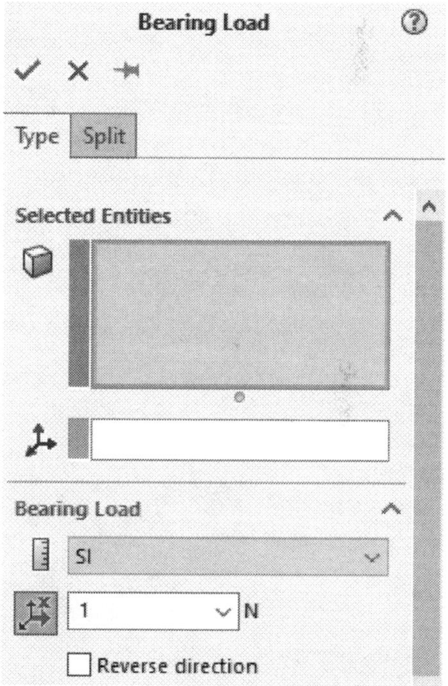

Figure 8-10 *The* ***Bearing Load PropertyManager*** *options*

Analyzing the Nonlinear Study

Now, you need to solve the nonlinear study and analyze the results for better functionality of the design.

1. Choose the **Run this Study** tool from the **Run this Study** flyout of the **Simulation CommandManager**; the **Nonlinear Analysis** window is displayed and solution process is started.

Note that error messages may be displayed during the solution of the study. You need to ignore them if they are not related to solver for the nonlinear study. Once the solution process is complete, the **Displacement** and **Stress** and **Strain** sub-nodes are added under the **Results** node in the Simulation Tree.

2. Select the **Displacement** and **Stress** results sub-nodes from the Simulation Tree to check the feasibility of the design. The results of the study are shown in Figures 8-11 and 8-12.

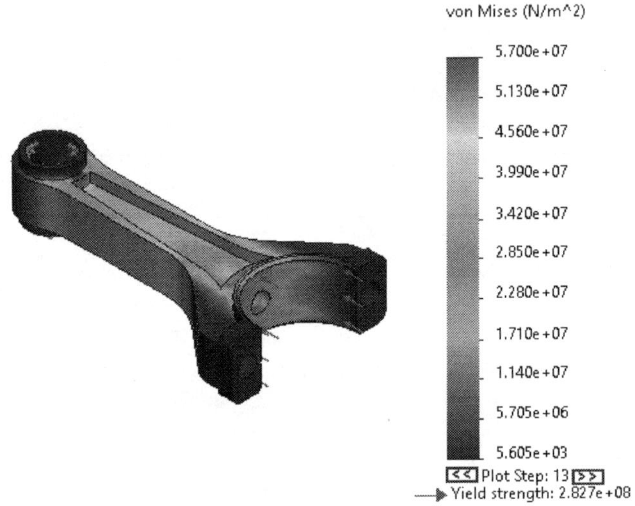

***Figure 8-11** vonMises Stress result for nonlinear study*

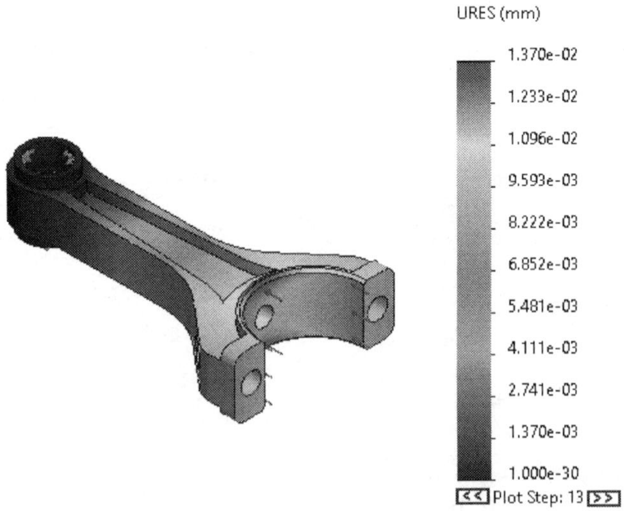

***Figure 8-12** Displacement result for nonlinear study*

Based on the maximum von Mises stress, you need to plot the factor of safety plot of the study at given boundary conditions.

3. Right-click on the **Results** node in the Simulation Tree; a shortcut menu is displayed.

4. Choose the **Define Factor of Safety Plot** option from the shortcut menu; the **Factor of Safety PropertyManager** is displayed.

5. Select the **Max von Mises Stress** option from the **Criterion** drop-down and then choose the **OK** button. The factor of safety plot is shown in Figure 8-13.

S. No.	Results Parameter	Max. Value	Min. Value
1	VonMises Stress	5.700e + 07	1.140e+07
2	Displacement	1.370e-01	1.000e-30

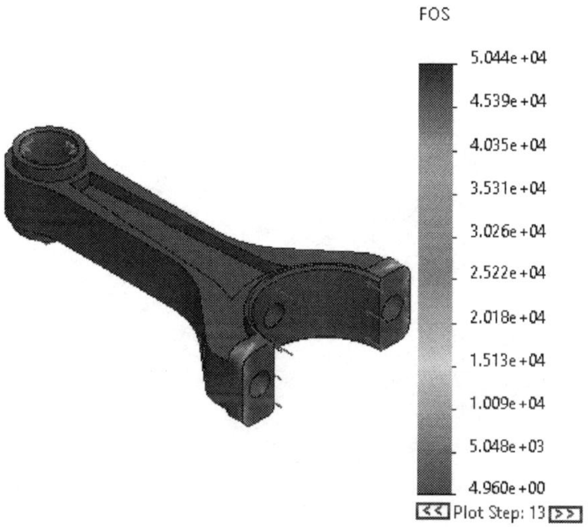

Figure 8-13 *FOS plot for nonlinear study*

Each color contour depicts a value and can be checked from the legend in the drawing area. It is obvious from the above results that the von Mises stress and displacement are maximum in the region where the model is marked red and minimum where it is marked blue. In linear static analysis, the stiffness matrix of the model is constant and the solving process is relatively shorter compared to nonlinear analysis on the same model. Therefore, for the first estimate, linear static analysis is often done prior to performing a full nonlinear analysis.

Saving the Study and Exiting SOLIDWORKS Simulation

After analyzing the results, save the entire study and exit SOLIDWORKS.

1. Choose the **Save** button to save the project at the following location:

   \Documents\SOLIDWORKS Simulation\c08

2. Choose **File > Close** from the SOLIDWORKS menus to close the document.

Tutorial 2

In this tutorial, you will open the *c08_tut02* file that was downloaded earlier and perform nonlinear structural study on the model shown in Figure 8-14. The model is made of **PE High Density** material and the boundary conditions (17 N force load and fixed support) applied are shown in Figure 8-14. To understand the importance of geometry optimization, you will optimize the model and then evaluate the results. **(Expected time: 1 hr)**

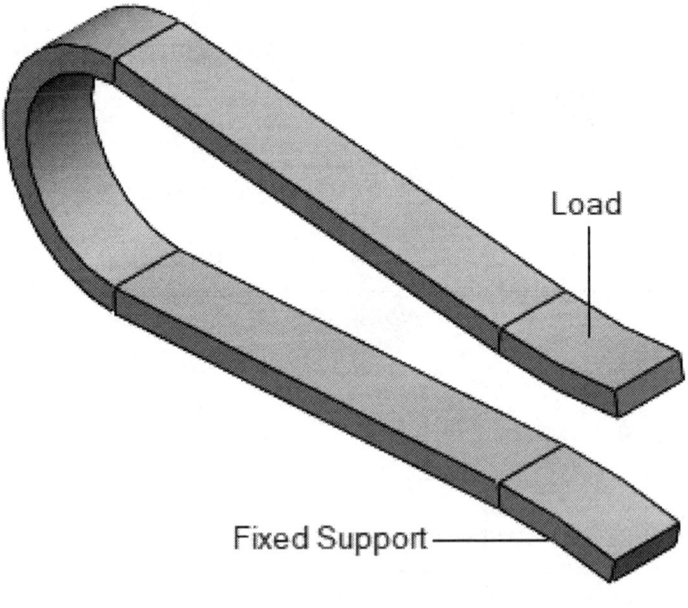

Figure 8-14 *Model for Tutorial 2*

Following steps are required to complete this tutorial:

a. Start SOLIDWORKS and open the model.
b. Generate mesh and apply structural boundary conditions to the model.
c. Analyze the results.
d. Save the model and exit SOLIDWORKS Simulation.

Starting SOLIDWORKS and Opening the Model

First, you need to start SOLIDWORKS and then open the model. After opening the model, you need to add the SOLIDWORKS Simulation add-in and start a new nonlinear study.

1. Open an input file named as *c08_tut02* and invoke the simulation environment.

2. Choose the **New Study** tool from the **Study Advisor** flyout in the **SimulationCommandManager**; the **Study PropertyManager** is invoked, refer to Figure 8-15.

3. Choose the **Nonlinear** button from the **Advanced Simulation** rollout and then select the **Static** button. Next, enter the name of the study as **Nonlinear Analysis** in the **Study name** edit box in the **Name** rollout.

4. Choose the **OK** button from the **Study PropertyManager**. On doing so, a Simulation Tree is added below the Design Tree with Nonlinear Analysis as the name of the study.

5. Right-click on the name of the study node and choose the **Properties** option from the shortcut menu; the **Nonlinear - Static** dialog box is displayed.

6. Select the **Use large displacement formulation** check box in the **Geometry nonlinearity options** area in this dialog box if not selected, refer to Figure 8-16.

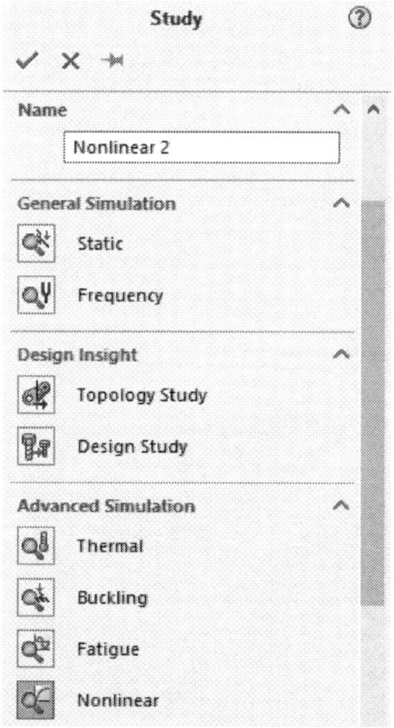

Figure 8-15** Partial view of the **Study PropertyManager

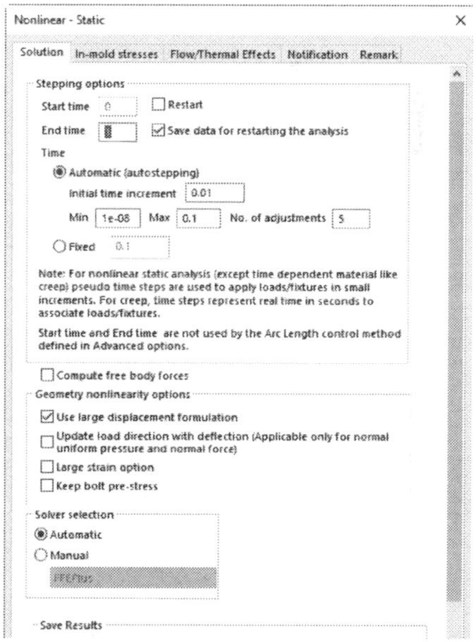

Figure 8-16** Partial view of the **Nonlinear PropertyManager

7. Choose the **OK** button to close this dialog box.

Assigning Material and Generating Mesh

After selecting the type of study, you need to assign material to the model in simulation environment if not assigned in the modeling environment. Then you need to mesh the model.

1. Choose the **Apply Material** tool from the **Simulation CommandManager**; the **Material** dialog box is displayed.

2. Select **PE High Density** from the **Plastics** materials displayed in the dialog box.

3. Choose the **Apply** button and then close the dialog box.

   After applying material, you need to mesh the model.

Figure 8-17 *The **Mesh PropertyManager***

4. Choose the **Create Mesh** tool from the **Run This Study** flyout in the **Simulation CommandManager**; the **Mesh PropertyManager** is displayed, as shown in Figure 8-17.

5. Choose **OK** from the PropertyManager; the **Mesh** progress window is displayed and mesh with default settings is generated, refer to Figure 8-18.

Figure 8-18 *Mesh generated with the default settings*

6. Right-click on the **Mesh** folder in the Simulation Tree; a shortcut menu is displayed. Select the **Details** option from the shortcut menu; the **Mesh Details** window is displayed, as shown in Figure 8-19.

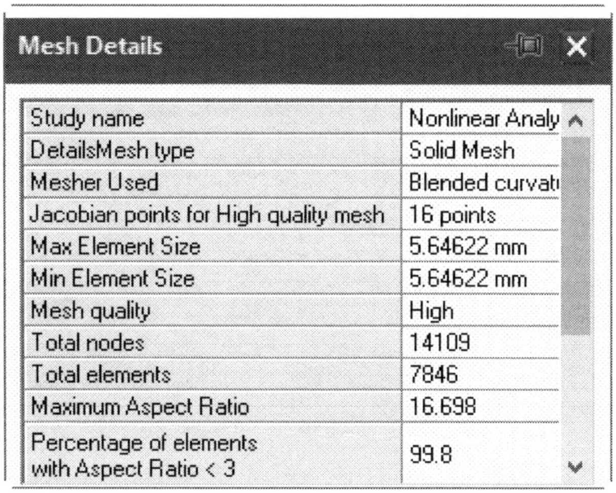

Figure 8-19 *Mesh details with the default settings*

It is evident from the **Mesh Details** window that the total elements in the study are 7846. Now, you need to improve the mesh quality to get more accurate results of nonlinear simulation study.

7. Select the **Mesh** node in the Simulation Tree and right-click; a shortcut menu is displayed.

8. Choose the **Apply Mesh Control** from the shortcut menu; the **Mesh Control PropertyManager** is displayed.

9. Select the model by using the selection filter; the model gets displayed in the **Faces, Edges, Vertices, Reference Points, Components for Mesh Control** selection box.

10. Next, enter **1.5** in the **Element Size** edit box in the **Mesh Parameters** rollout of the PropertyManager.

11. Choose the **Create Mesh** button from the PropertyManager to re-mesh the model; the **Simulation** dialog box is displayed.

12. Choose the **Yes** button from the dialog box; the **Mesh Progress** window is displayed and mesh is created.

13. Right-click on the **Mesh** node in the Simulation Tree; a shortcut menu is displayed. Choose the **Details** option from it; the **Mesh Details** window is displayed.

    Now, you notice that elements and nodes are increased in the study to get optimized results. The mesh details are shown in Figure 8-20.

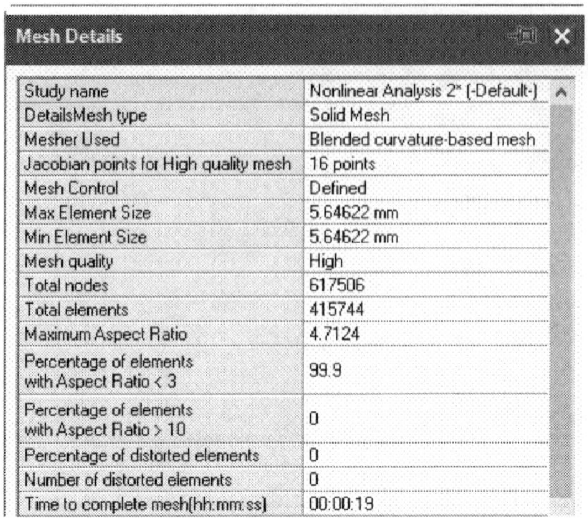

Mesh Details

Study name	Nonlinear Analysis 2* (-Default-)
DetailsMesh type	Solid Mesh
Mesher Used	Blended curvature-based mesh
Jacobian points for High quality mesh	16 points
Mesh Control	Defined
Max Element Size	5.64622 mm
Min Element Size	5.64622 mm
Mesh quality	High
Total nodes	617506
Total elements	415744
Maximum Aspect Ratio	4.7124
Percentage of elements with Aspect Ratio < 3	99.9
Percentage of elements with Aspect Ratio > 10	0
Percentage of distorted elements	0
Number of distorted elements	0
Time to complete mesh(hh:mm:ss)	00:00:19

Figure 8-20 *Mesh details with the modified settings*

Applying Boundary Conditions to the Model

After the model is meshed, you need to apply boundary conditions for the study. You will apply a load on the mentioned face and fixed support on the another face of the model.

1. Choose the **Fixed Geometry** tool from the **Fixtures Adviser** flyout in the **Simulation CommandManager**; the **Fixed PropertyManager** is displayed, as shown in Figure 8-21.

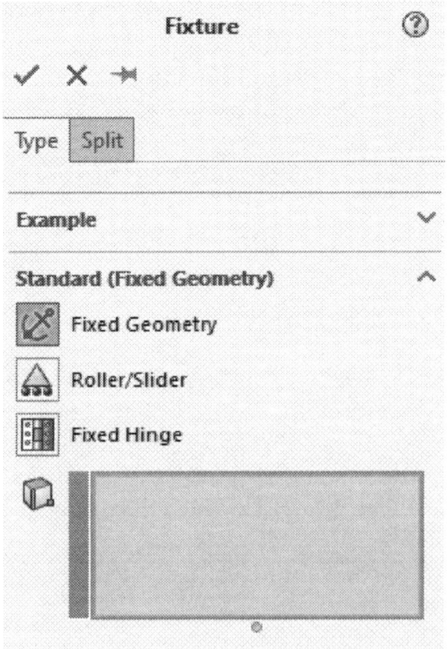

Figure 8-21 *The* ***Fixture PropertyManager***

2. Select the face of the model for applying the fixed support, refer to Figure 8-14; the name of the selected face is displayed in the **Faces, Edges, Vertices for Fixture** selection box.

3. Choose the **OK** button to close the PropertyManager.

4. Choose the **Force** tool from the **External Loads Advisor** flyout in the **Simulation CommandManager**; the **Force/Torque PropertyManager** is displayed.

5. Select the top faces of the model; the names of the selected face get displayed in the selection box in the **Force/Torque** rollout.

6. Select the **Force** and **Normal** radio buttons to apply the normal force to the selected face.

7. Select the **SI** option in the **Unit** drop-down in the **Force/Torque** rollout of the PropertyManager.

8. Enter **17** in the **Force Value** edit box. If the direction of the load is not same, as shown in Figure 8-22, you need to select the **Reverse Direction** check box to reverse the direction of the load.

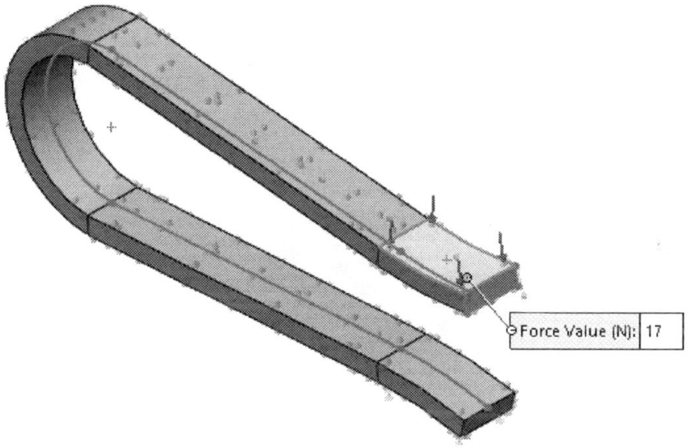

Figure 8-22 *Direction of the load for nonlinear study*

9. Select the **linear** radio button under the **Variation with Time** rollout if not selected.

10. Retain rest of the default values of the options in the **Force/Torque PropertyManager**, choose the **OK** button to close this PropertyManager.

11. Choose the **Connections** tool from the **Connection Advisor** flyout in the **Simulation CommandManager**; the **Local Interaction PropertyManager** is displayed.

12. Select the face opposite to the face on which the load is applied in **Faces, Edges, Vertices for Set 1** and select the corresponding face in **Faces for Set 2** and choose the **OK** button to close the PropertyManager.

Analyzing the Nonlinear Study

Now, you need to solve the nonlinear study and analyze the results for better functionality of the design.

1. Choose the **Run this Study** tool from the **Run this Study** flyout of the **Simulation CommandManager**; the **Nonlinear Analysis** window is displayed and the solution process is started.

   During the solution of the study, ignore the errors message unrelated to solver for the nonlinear study. Once the solution process is complete, the **Displacement**, **Stress** and **Strain** sub-nodes are added under the **Results** node in the Simulation Tree.

2. Double click on the displacement and stress results node in the Simulation Tree to check the feasibility of the design. Now, you need to compare this stress results with the analytical results of the study. The FEA results of the study are shown in Figures 8-23, 8-24, and 8-25.

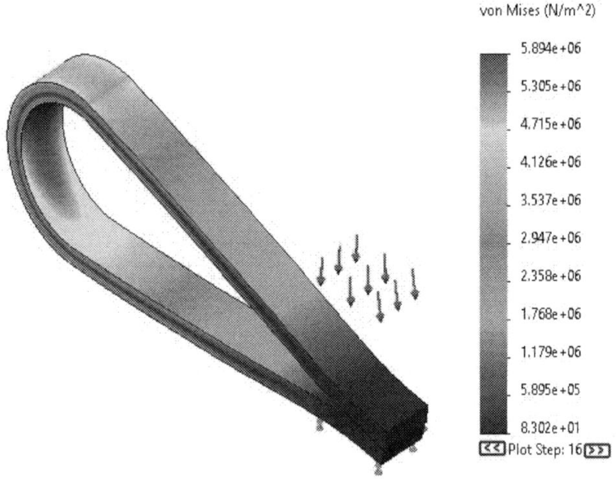

Figure 8-23 *von Mises Stress results for nonlinear study*

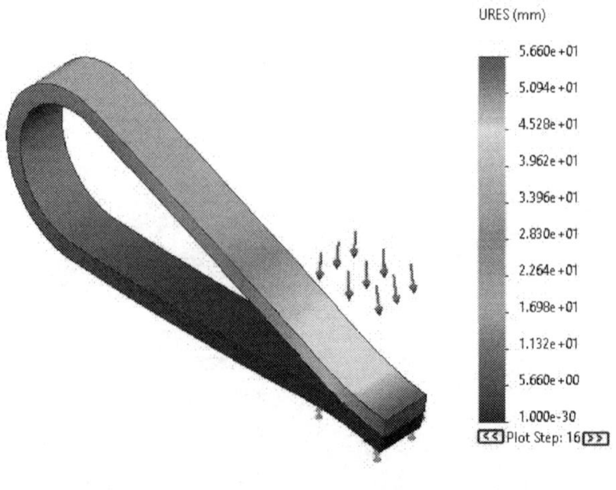

Figure 8-24 *Displacement results for nonlinear study*

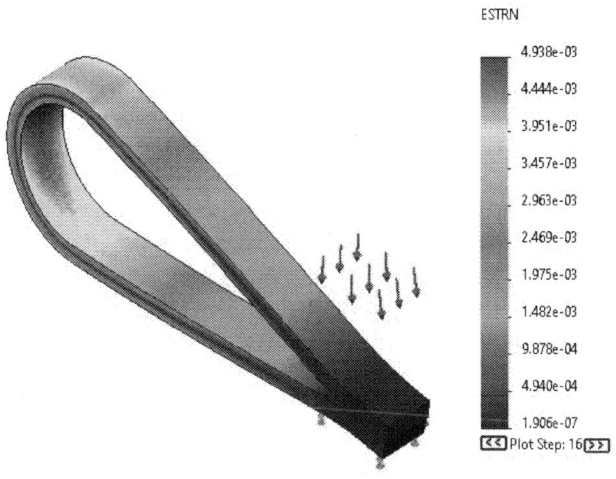

Figure 8-25 *Strain results for nonlinear study*

It is recommended that before proceeding for the nonlinear study, you should perform linear study and compare the FEA results with the analytical results.

Tutorial 3

In this tutorial, you will open the *c08_tut03* file that was downloaded earlier and perform the nonlinear structural study on the Retaining clip model, refer to Figure 8-26. The boundary conditions (100 lbf load and support) are also mentioned in the same figure and the model is made of **AISI 1035 Steel (SS)**. Also, validate the model and then evaluate the results.

(Expected time: 1 hr)

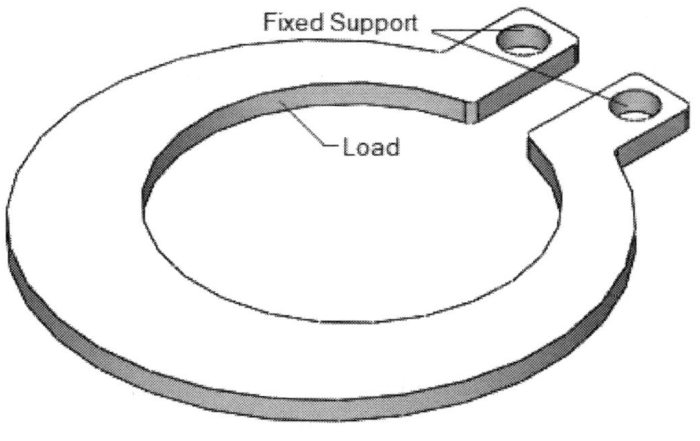

Figure 8-26 *Model for Tutorial 3*

Following steps are required to complete this tutorial:

a. Start SOLIDWORKS and open the model.
b. Apply the material to the model.
c. Generate mesh and apply structural boundary conditions to the model.
d. Analyze the results.
e. Save the model and exit SOLIDWORKS Simulation.

Starting SOLIDWORKS and Opening the Model

First, you need to start SOLIDWORKS and then open the model. After opening the model, you need to add the **SOLIDWORKS Simulation** add-in and start a new nonlinear study.

1. Open an input file named as *c08_tut03* and invoke the simulation environment.

2. Choose the **New Study** tool from the Study Advisor flyout in the **Simulation** the **Study CommandManager**; **PropertyManager** is invoked.

3. Choose the **Nonlinear** button and then the **Static** button from the **Advanced Simulation** rollout. Next, enter the name of the study as **Nonlinear** Analysis in the **Study name** edit box in the **Name** rollout. Next, choose the **OK** button from the **Study PropertyManager.** On doing so, a Simulation Tree is added below the Design Tree with **Nonlinear Analysis** as the name of the study.

4. Right-click on the name of the study node and choose the **Properties** option from the shortcut menu; the **Nonlinear - Static** dialog box is displayed.

5. Select the **Use large displacement formulation** check box in the **Geometry nonlinearity options** area in this dialog box if not selected by default, refer to Figure 8-27.

6. Choose the **OK** button to close this dialog box.

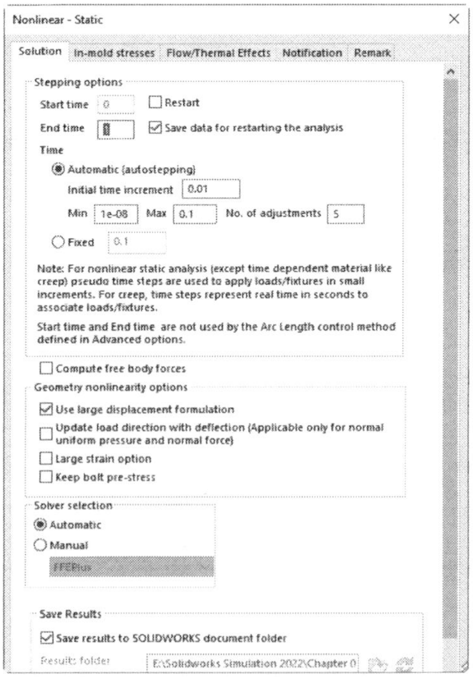

*Figure 8-27 Partial view of the **Nonlinear - Static** dialog box*

Assigning Material and Generating the Mesh

After selecting the type of study, you need to assign material to the model in simulation environment if not assigned in the modeling environment. Then, you need to mesh the model.

1. Choose the **Apply Material** tool from the **Simulation CommandManager;** the **Material** dialog box is displayed.

2. Select **AISI 1035 Steel (SS)** from the materials displayed in the dialog box.

3. Choose the **Apply** button and then the **Close** button to close this dialog box.

   After applying material, you need to mesh the model.

4. Choose the **Create Mesh** tool from the **Run This Study** flyout in the **Simulation CommandManager**; the **Mesh PropertyManager** is displayed, as shown in Figure 8-28.

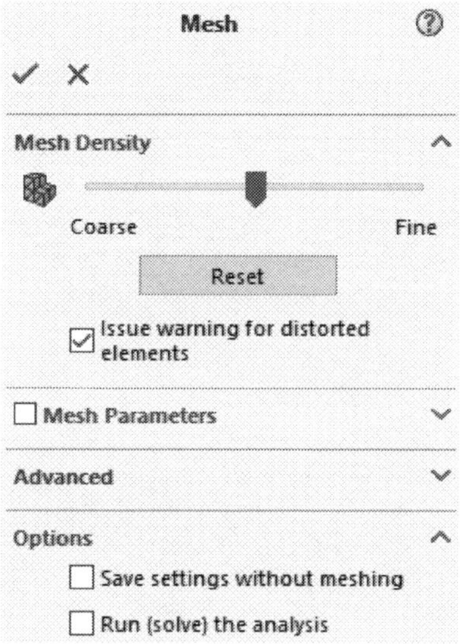

Figure 8-28** The **Mesh PropertyManager

5. Choose **OK** from the PropertyManager; the **Mesh Progress** window is displayed and mesh with default settings is generated, refer to Figure 8-29.

   Now, you need to know the number of elements created after meshing.

6. Right-click on the **Mesh** folder in Simulation Tree; a shortcut menu is displayed. Choose the **Details** option from the shortcut menu; the **Mesh Details** window is displayed. You get the information about mesh details such as nodes and elements from this window.

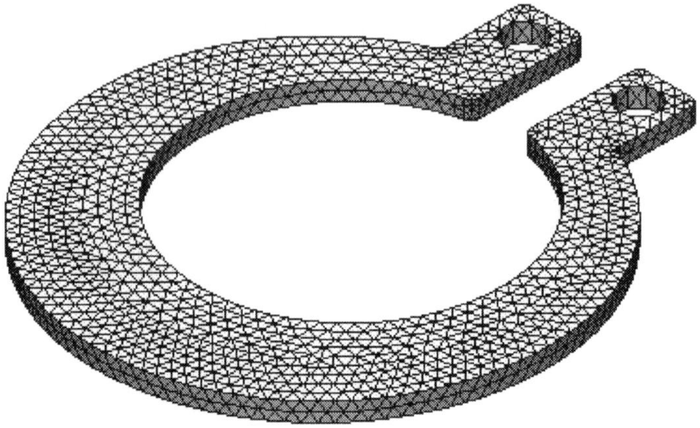

Figure 8-29 *Mesh generated with the default settings*

Applying Boundary Conditions to the Model

After the model is meshed, you need to apply boundary conditions for the study. You will apply a load on the circular face and fixed support on another cylindrical face of the model, as shown in Figure 8-26.

1. Choose the **Fixed Geometry** tool from the **Fixtures Adviser** flyout in the **Simulation CommandManager**; the **Fixture PropertyManager** is displayed.

2. Select the circular-cylindrical face of the model for applying the fixed support, refer to Figure 8-26; the name of the selected face is displayed in the **Faces, Edges, Vertices for Fixture** selection box.

3. Choose the **Fixed Geometry** button from the PropertyManager if not selected by default and then choose the **OK** button to close the PropertyManager.

4. Choose the **Force** tool from the **External Loads Adviser** flyout in the **Simulation CommandManager**; the **Force/Torque PropertyManager** is displayed.

5. Select the **Selected direction** radio button under the **Force/Torque** rollout of the PropertyManager; the **Edge, Face, Plan for Direction** selection box becomes available.

6. Select the edge in the direction of the arrows shown in Figure 8-30; the selected edge gets displayed in the selection box.

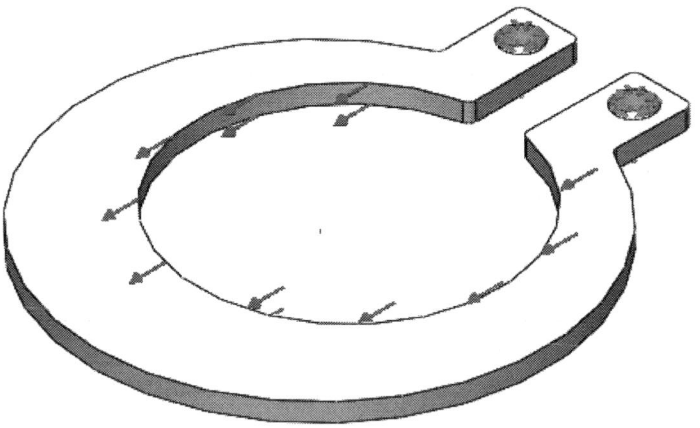

Figure 8-30 Direction of the load for nonlinear study

7. Select the **English (IPS)** option in the **Unit** drop-down list and make sure that the **Linear** radio button is selected in the **Variation with Time** rollout of the PropertyManager.

8. Enter **100** in the **Along Edge** edit box and then choose the **OK** button to close the PropertyManager.

Analyzing the Nonlinear Study

Now, you need to solve the nonlinear study and analyze the results for better functionality of the design.

1. Choose the **Run this Study** tool from the **Run this Study** flyout of the **Simulation CommandManager**; the **Nonlinear Analysis** window is displayed and the solution process is started.

   Error messages may occur during the solution of the study. Ignore them if they are not related to solver for the nonlinear study. Once the solution process is done, Displacement, Stress and Strain sub-nodes are added under the **Results** node in the Simulation Tree.

2. Double-click on the respective node to plot the displacement and stress results of the study to check the feasibility of the design. The results of the study are shown in Figures 8-31, 8-32, and 8-33.

   It is recommended that before proceeding the nonlinear study, you should perform linear study and compare the FEA results with the analytical results.

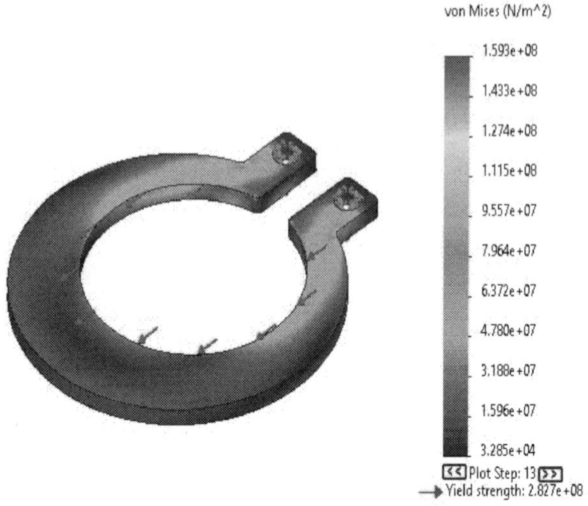

Figure 8-31 *von Mises stress results for nonlinear study*

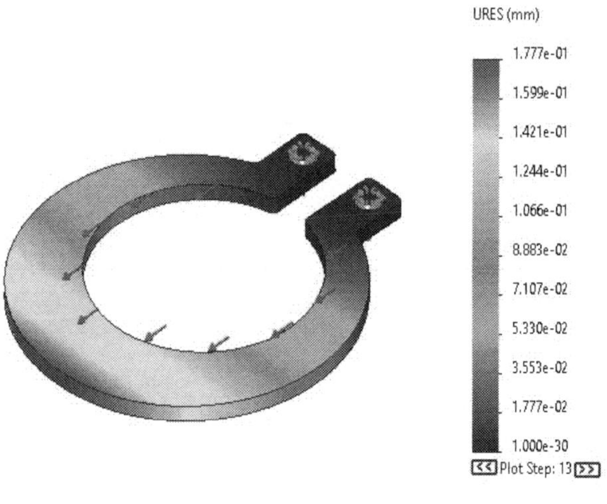

Figure 8-32 *Displacement results for nonlinear study*

Based on the maximum von Mises stress, you need to plot the factor of safety plot of the nonlinear study at given boundary conditions.

3. Right-click on the **Results** node in the Simulation Tree; a shortcut menu is displayed.

4. Choose the **Define Factor of Safety Plot** option from the shortcut menu; the **Factor of Safety PropertyManager** is displayed.

5. Select the **Max von Mises Stress** option in the **Criterion** drop-down list and than choose the OK button. The factor of safety plot is shown in Figure 8-33.

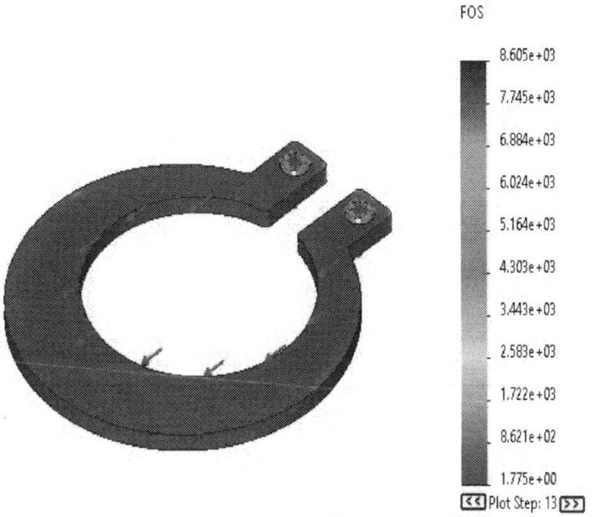

Figure 8-33 *Factor of Safety results for nonlinear study*

Tutorial 4

In this tutorial, you will open *c08_tut04_*assmbly file to perform a nonlinear analysis on a pipe bending machine, refer to Figure 8-34. 50000 N force is applied on the handle and the bottom of the wire is fixed, as shown in Figure 8-34. The model is made of Manganese Bronze and Alloy Steel. Also, to understand the importance of geometry optimization, you will optimize the model, apply contacts between the components, and then evaluate the results.

(Expected time: 2.5 hr)

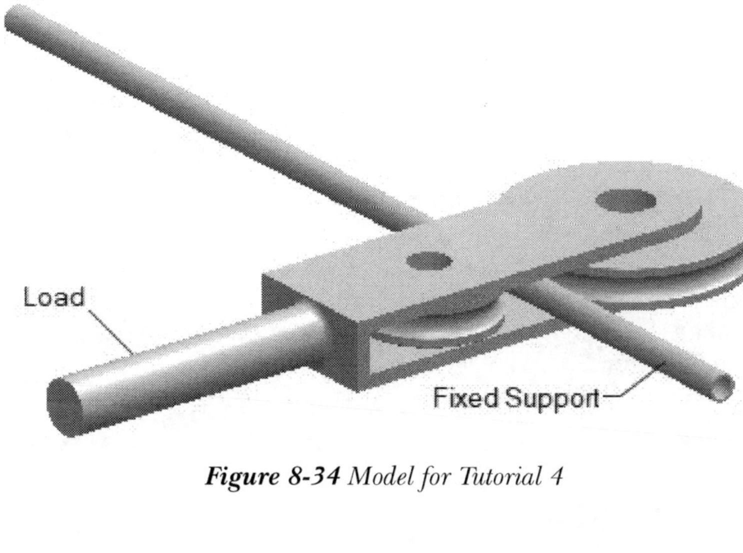

Figure 8-34 *Model for Tutorial 4*

Following steps are required to complete this tutorial:

a. Start SOLIDWORKS and open the model.
b. Generate mesh and apply structural boundary conditions to the model.
c. Analyze the results.
d. Save the model and exit SOLIDWORKS Simulation.

Starting SOLIDWORKS and Opening the Model

First, you need to start SOLIDWORKS and then open the model. After opening the model, you need to add the SOLIDWORKS Simulation add-in and start a new nonlinear study.

1. Open an input file named as *c08_tut04* and invoke the simulation environment.

2. Choose the **New Study** tool from the **Study Advisor** flyout in the **Simulation CommandManager**; the **Study PropertyManager** is invoked, as shown in Figure 8-35.

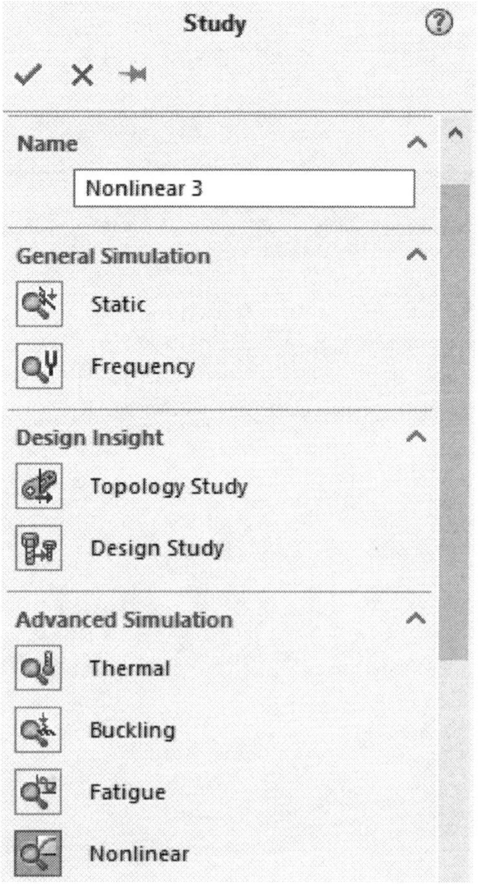

Figure 8-35 Partial view of the ***Study PropertyManager***

3. Choose the **Nonlinear** button from the **Advanced Simulation** rollout and then choose the **Dynamic** button. Next, enter **Nonlinear Analysis** as the name of the study in the **Study name** edit box in the **Name** rollout. Choose the **OK** button from the **Study PropertyManager**. On doing so, a Simulation Tree is added below the Design Tree with Nonlinear Analysis as the name of the study.

4. Right-click on the name of the study node and choose the **Properties** option from the shortcut menu; the **Nonlinear - Dynamic** dialog box is displayed.

5. Select the **Use large displacement formulation** check box in the **Geometry nonlinearity options** area in this dialog box if not selected by default, refer to Figure 8-36. Next, choose the **OK** button.

Assigning Material and Generating Mesh

After selecting the type of study, you need to assign material to the model in the simulation environment if not assigned in the modeling environment. After that you need to mesh the model

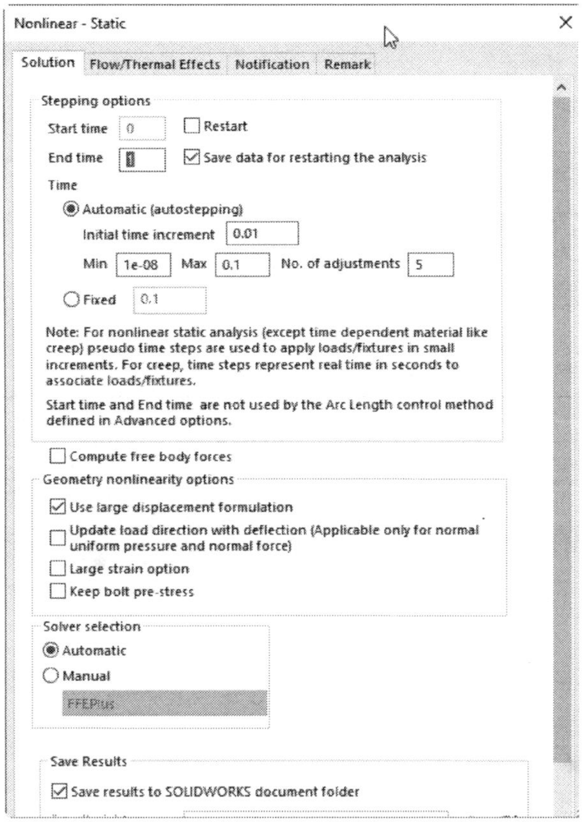

***Figure 8-36** Partial view of the **Nonlinear - Dynamic** dialog box*

1. Choose the **Apply Material** tool from the **Simulation CommandManager**; the **Material** dialog box is displayed as shown in Figure 8-37.

2. Choose **Manganese Bronze** from **Copper Alloys** for **Wire-1** from the parts node of Simulation Tree and choose secondary material **Alloy Steel** from **Steel** for **Handle-1, Larger Roller-1, Smaller- 1** from the **Material** dialog box.

3. Choose the **Apply** button and then the **Close** button to close this dialog box.

   After applying material, you need to mesh the model.

4. Choose the **Create Mesh** tool from the **Run This Study** flyout in the **Simulation CommandManager**; the **Mesh PropertyManager** is displayed, as shown in Figure 8-38.

5. Choose the **OK** button; the **Mesh Progress** window is displayed and mesh with default settings is generated, as shown in Figure 8-39.

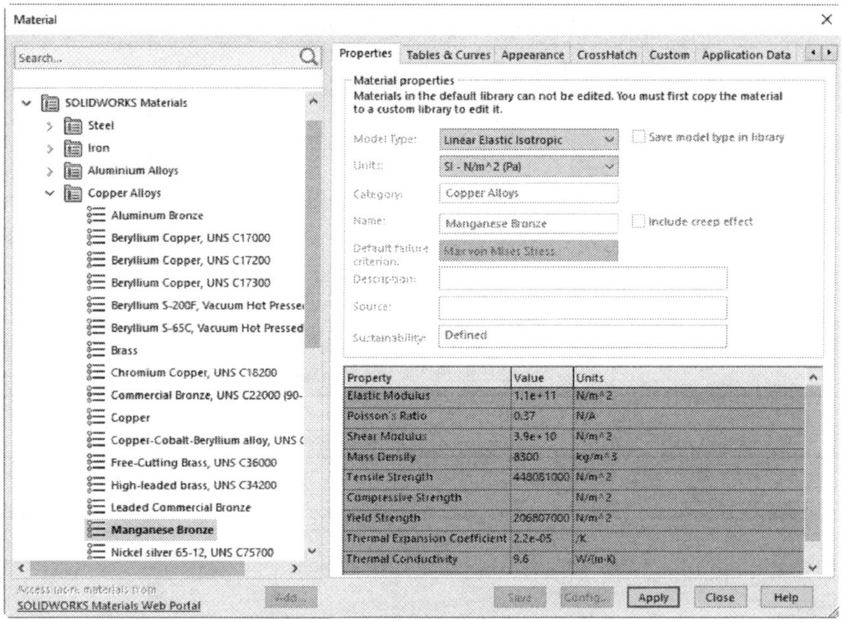

***Figure 8-37** The **Material** dialog box*

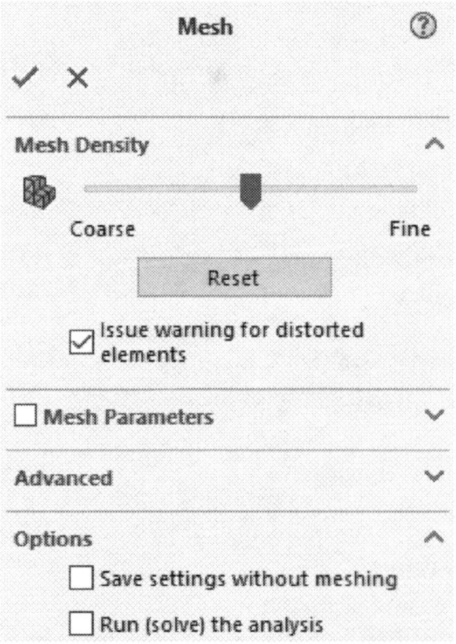

Figure 8-38** The **Mesh Property Manager

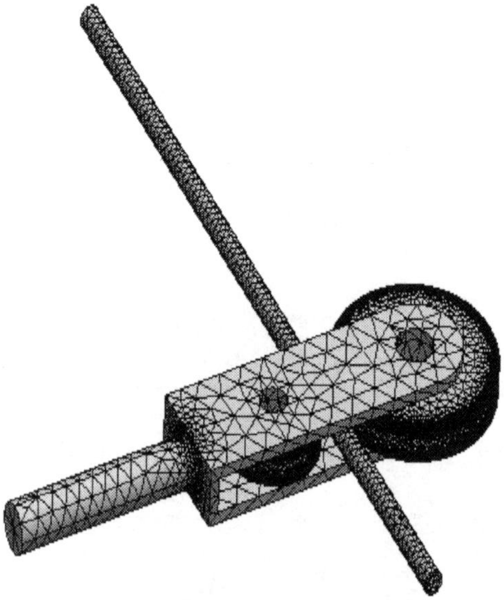

***Figure 8-39** Mesh generated with the default settings*

6. Right-click on the **Mesh** folder in the Simulation Tree; a shortcut menu is displayed. Choose the **Details** option from the shortcut menu; the **Mesh Details** window providing you mesh details such as nodes and elements is displayed. The number of elements in this mesh is 127171.

Now, you need to improve the mesh quality to get more accurate results of nonlinear simulation study.

7. Choose the **Create Mesh** tool from the **Run This Study** flyout in the **Simulation CommandManager**; the **Mesh PropertyManager** is displayed.

8. Move the slider corresponding to Mesh Factor in the **Mesh Density** rollout towards Fine; the number of elements increases and you get increased mesh accuracy.

9. Choose the **OK** button; the **Mesh Progress** window is displayed and mesh is generated with modified settings, refer to Figure 8-40.

   Note that the number of elements in this mesh is 313937. This number is very large as compared to the element count that was achieved by the mesh generated with the default mesh settings.

Applying Fixed Supports to the Model

After the model is meshed, you need to apply boundary conditions for the study. You will apply a linear load on the handle, fixed support on the rod, and hinged support on the larger wheel.

1. Choose the **Fixed Geometry** tool from the **Fixtures Adviser** flyout in the **Simulation CommandManager**; the **Fixture PropertyManager** is displayed, as shown in Figure 8-41.

Figure 8-40 *Mesh generated with the modified settings*

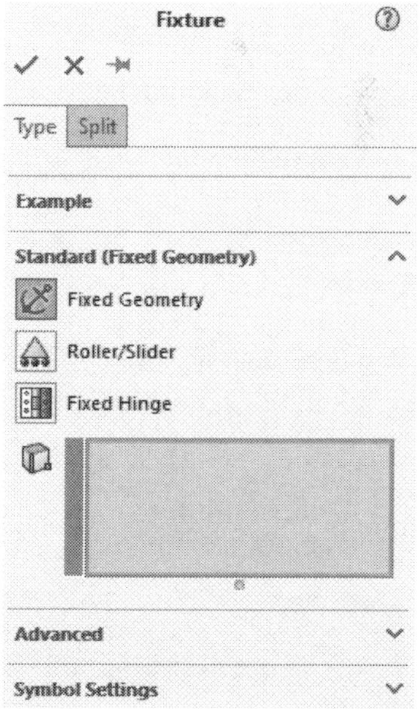

Figure 8-41 *The* ***Fixture PropertyManager***

2. Select the bottom cylindrical face of the Wire for applying the fixed geometry support, refer to Figure 8-42; the name of the selected face is displayed in the **Faces, Edges, Vertices for Fixture** selection box. Next, choose the **Fixed Geometry** button from the **Standard Fixed Geometry** rollout.

3. Choose the **OK** button.

4. Choose the **Fixed Hinge** tool from the **Fixtures Adviser** flyout in the **Simulation CommandManager**; the **Fixture PropertyManager** is displayed.

5. Select the inner cylindrical face of the Larger Roller to apply the Fixed Hinge support, as shown in Figure 8-43; the name of the selected face is displayed in the **Cylindrical Faces for Fixture** selection box. Choose the **OK** button.

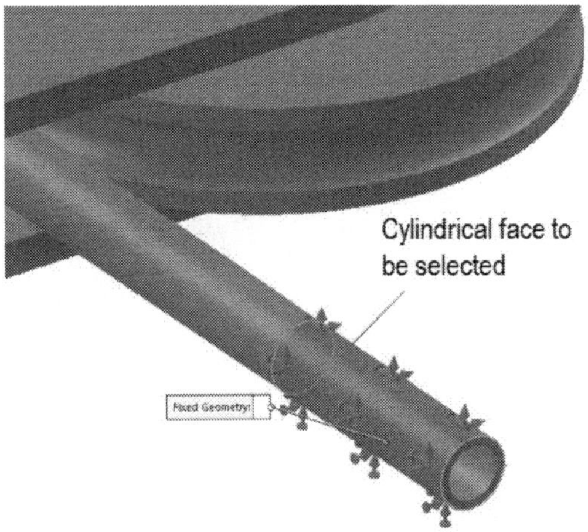

Figure 8-42 *Cylindrical face to be selected*

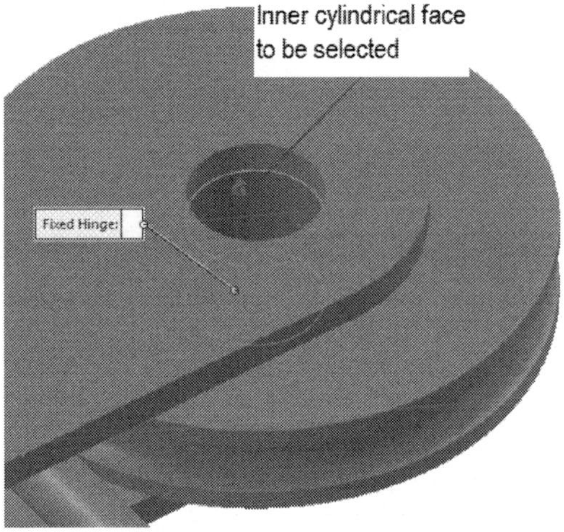

Figure 8-43 *Inner cylindrical face to be selected*

Applying Load to the Model

Now, you need to apply load on the handle part of the model.

1. Choose the **Force** tool from the **External Loads Advisor** flyout in the **Simulation CommandManager**; the **Force/Torque PropertyManager** is displayed, as shown in Figure 8-44.

2. Select the top cylindrical face of the Handle to apply the force; the names of the selected face get displayed in the **Faces and Shell Edges for Normal Force** selection box, refer to Figure 8-45.

3. Choose the **Selected direction** button in this PropertyManager to apply force in specific direction. Next, select the straight edge near the surface. The name of the edge gets displayed in the **Face, Edge, Plane for Direction** selection box, refer to Figure 8-46.

4. Select the **SI** option in the **Unit** drop-down list under the **Units** area of this PropertyManager.

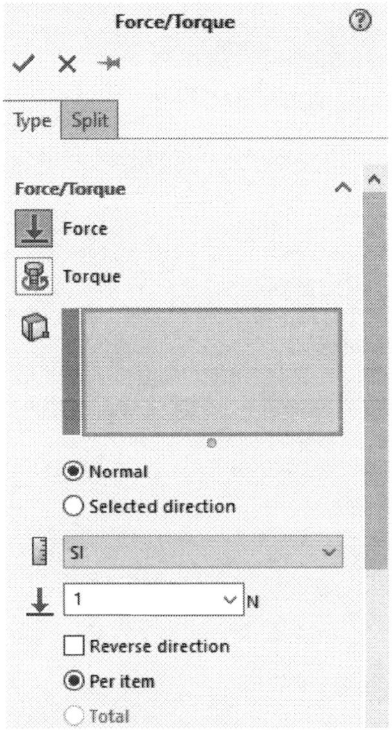

Figure 8-44** The **Force/Torque PropertyManager

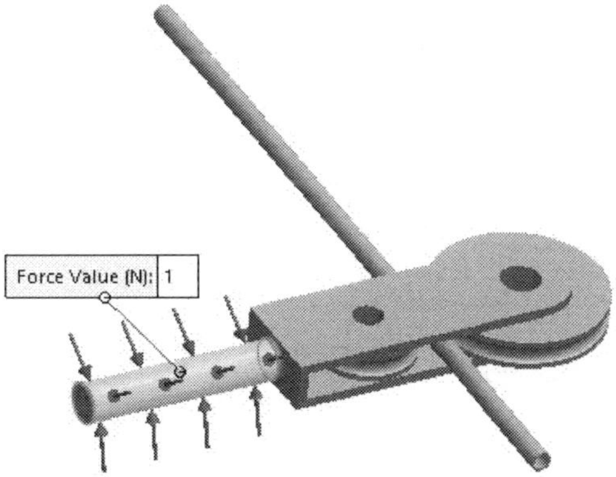

***Figure 8-45** Cylindrical face to be selected*

5. Enter **50000** in the **Along Edge** edit box under the **Force** area. If the direction of the force is not same, as shown in Figure 8-46, you need to select the **Reverse Direction** check box to reverse the direction of the force.

6. Retain rest of the default settings in the Property Manager. Choose the **OK** button.

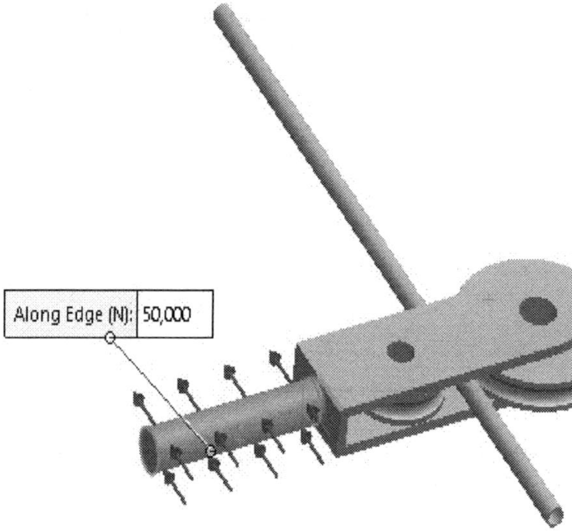

Figure 8-46 Edge to be selected

Applying Connectors on the Joints of the Model

Now, you need apply virtual connectors on the joints of the model by which the model can move properly during the application of load.

1. Choose the **Local Interactions** tool from the **Connections Advisor** flyout to invoke the **Local Interactions PropertyManager**, as shown in Figure 8-47.

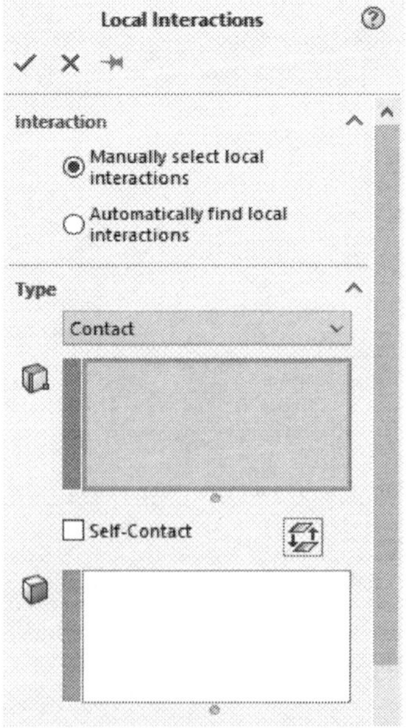

Figure 8-47** Partial view the **Local Interactions PropertyManaager

2. Select the **Contact** option under the **Type** drop-down list of this PropertyManager if not selected by default.

3. Select the outer cylindrical face of the Wire, refer to Figure 8-48; the name of selected face is displayed in the **Faces, Edges, Vertices for Set 1** selection box.

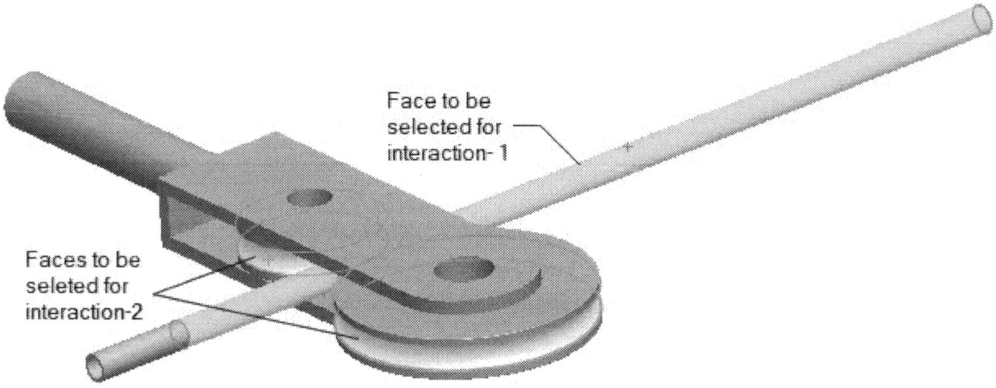

Figure 8-48 *Faces to be selected for set-1 and set-2 selection boxes*

4. Select the outer cylindrical faces of both small and large rollers, refer to Figure 8-48; under the **Faces for Set 2** selection box and choose the **OK** button.

   Next, you have to apply the pin connectors between the parts of the model.

5. Choose the **Pin** tool from the **Connections Advisor** flyout in the **Simulation CommandManager**; the **Connectors PropertyManager** is displayed, as shown in Figure 8-49.

6. Select the **Pin** option in the **Type** drop-down list from the **Type** rollout of this PropertyManager

7. Select the top inner cylindrical face of the Handle and inner cylindrical face of the Larger Roller, as shown in Figure 8-50; the selected faces get displayed in the first and second **Cylindrical Faces/Edges** selection boxes, respectively. Choose the **OK** button.

Similarly, apply the pin connector to the other side of the handle and to the inner cylindrical face of the larger roller.

Similarly, apply the pin connectors to the smaller roller and between both faces of the handle. After applying all the connectors, total number of the pin connectors must be four in the Simulation Tree. For better visibility, all the pin connectors are made hidden as soon as they created.

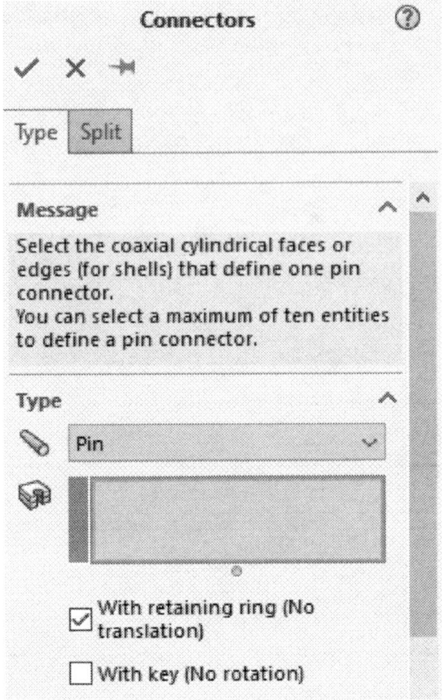

Figure 8-49 *Partial view of the* ***Connectors PropertyManager***

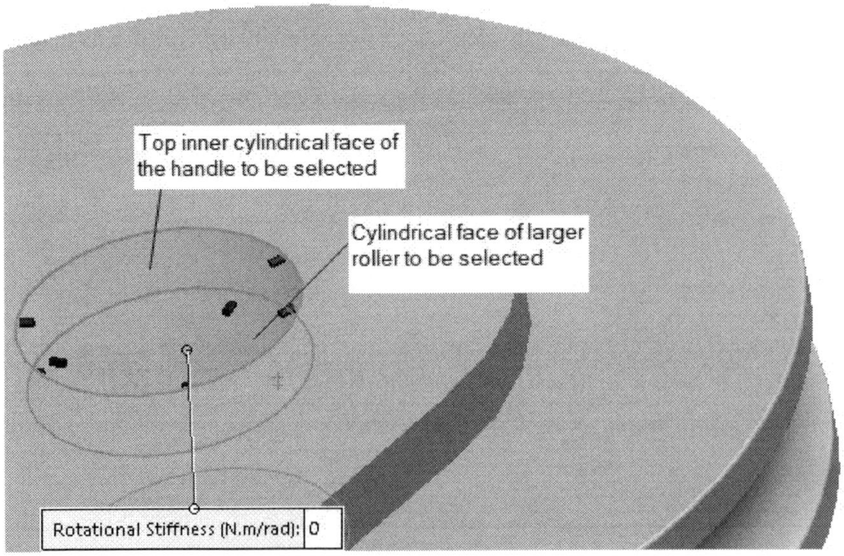

Figure 8-50 *Faces to be selected for Pin Connector-1*

Running the Simulation

After applying the fixtures, load, and connectors, now you have to run the simulation to check the durability and nonlinear behaviour of the model under the applied conditions. First, you need to remesh the model to make changes in the study.

1. Choose the **Run This Study** tool from the **Run This Study** flyout; the **Mesh progress** window is displayed. The model is meshed again and the **Simulation** message box is displayed, as shown in Figure 8-51, informing about the initial contact conditions.

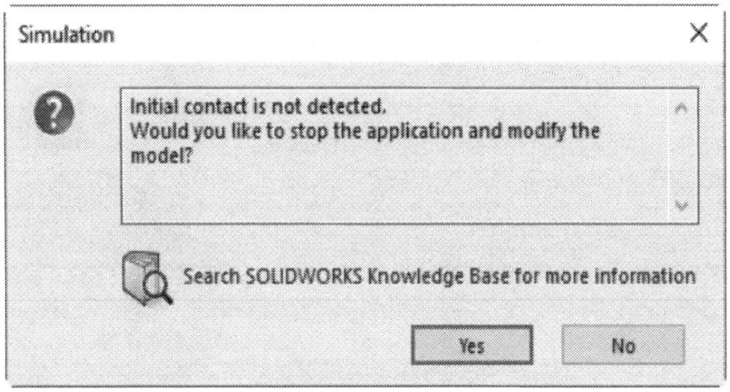

***Figure 8-51** The **Simulation** dialog box*

2. Choose **No** from this message box; the **Solver Status** window is displayed with name of the analysis (Nonlinear Analysis), as shown in Figure 8-52. The window shows the status of the solution.

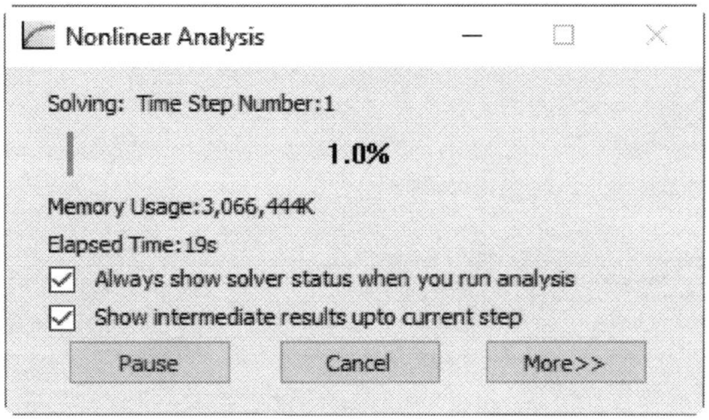

***Figure 8-52** The solver status window*

3. Once solution is completed, check the stress, displacement and strain plots by double-clicking on their names under the **Results** node in the Simulation Tree. By default, stress distribution is displayed. Figure 8-53 and Figure 8-54 show stress and displacement distribution when 50000 N load is applied to the study.

Animating Analysis Results

After completion of the analysis, you can create its animation.

1. Select the **Animate** option from the **Plot Tools** drop-down list; the **Animation PropertyManager** is displayed, as shown in Figure 8-55.

2. Choose the **Stop** button under the **Basics** area.

3. Select the **Save as AVI** file check box; the **Browse** button is displayed, refer to Figure 8-56.

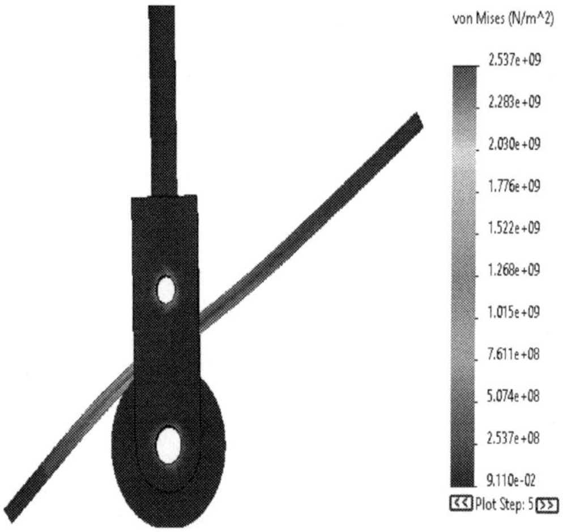

Figure 8-53 *Stress distribution on 50000N load application*

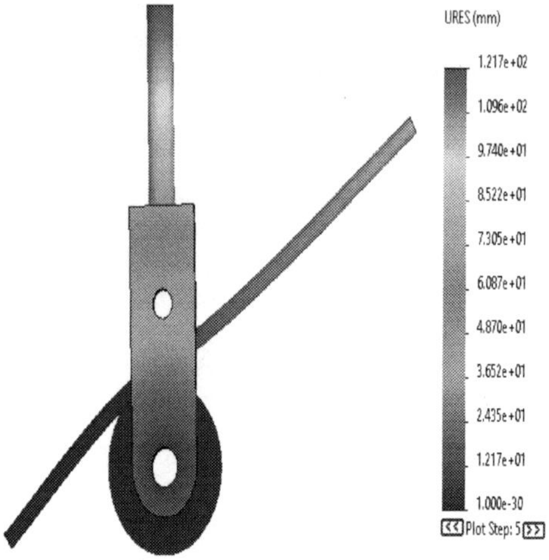

Figure 8-54 *Displacement distribution on 50000N load application*

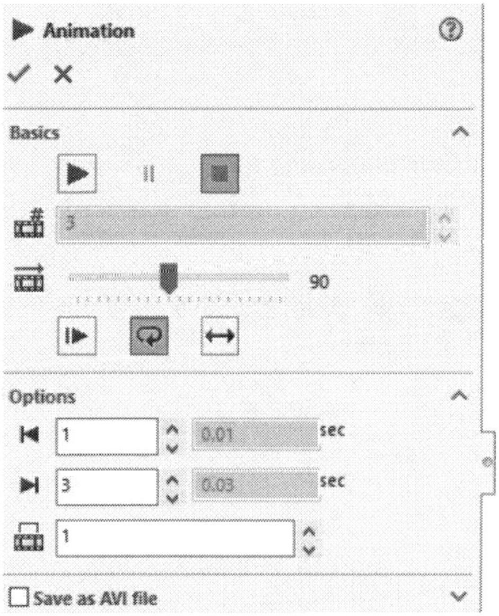

Figure 8-55 *The* ***Animation PropertyManager***

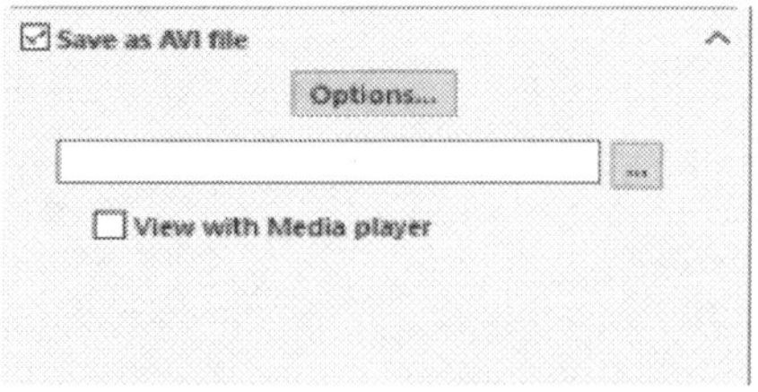

Figure 8-56 *The* ***Save as AVI file*** *check box selected in the* ***Animation PropertyManager***

4. Choose the **Browse** button; the **Save as** dialog box is displayed, as shown in Figure 8-57. Save the animation at desired location.

5. Enter the **Animation** in **File name** edit box. Select **Movie clip(*.AVI)** in the **Save as type** drop-down list and choose the **Save** button.

6. Choose the **Play** button under the **Basics** area; the animation is now saved. Close the **Animation PropertyManager.**

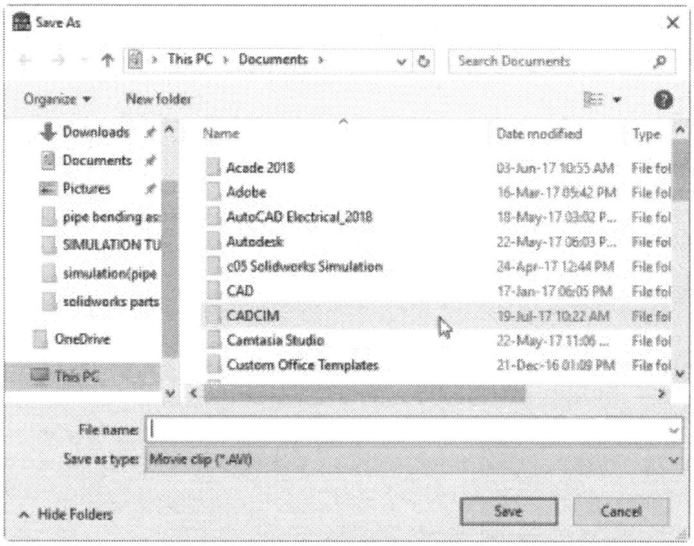

Figure 8-57 *The* ***Save as*** *dialog box*

Saving the analysis and Exiting SOLIDWORKS Simulation

After analyzing the results, save the entire study and exit SOLIDWORKS.

1. Choose the **Save** button to save the project at the following location:

   \Documents\SOLIDWORKS Simulation\c08.

2. Choose **File > Close** from the SOLIDWORKS menus to close the document.

Self-Evaluation Test

Answer the following questions and then compare them to those given at the end of this chapter:

1. How many independent equations will be satisfied for static equilibrium in a space structure?

   a) 2 b) 3
   c) 4 d) 6

2. In moment distribution method, what will be the sum of distribution factors of all the members meeting at any joint?

   a) zero b) less than 1
   c) 1 d) greater than 1

3. Bending moment at any section in a conjugate beam gives __________ in the actual beam. (T/F)

4. In SOLIDWORKS Simulation, once nonlinear static study is complete, you can create a factor of safety plot based on the defined failure criterion. (T/F)

5. Right-click on **Results** and choose the **Define Factor of Safety Plot** option from the shortcut menu. (T/F)

6. In SOLIDWORKS Simulation, Automatic (autostepping) option gives the solver freedom to adjust the step as needed in convergence. (T/F)

7. You can assign material to the model in simulation environment as well as in the modeling environment. (T/F)

8. If you use a material from the SOLIDWORKS database, the changes that can be brought about in material properties of linear and nonlinear materials are limited. (T/F)

9. One of the most crucial tasks in creating nonlinear simulations is the proper settings of the time stepping options. (T/F)

10. If Singularity Elimination Factor (SEF) is less than 1, the singularity of the structural stiffness gets reduced. (T/F)

Review Questions

Answer the following questions:

1. A Singularity Elimination Factor (SEF) value of less than 1 can help the non-linear solution proceed further to completion and the most effective SEF values are set between 0.5 to 0 (0 for maximum effect). (T/F)

2. The Factor of Safety plot helps you determine whether the design is safe or need further modifications. (T/F)

3. The von Mises criterion is commonly used to check the failure of ductile materials. (T/F)

4. The minimum and maximum values obtained from an analysis are displayed in a plot. (T/F)

5. A simulation study defines the analysis type, materials, fixtures, and loads. (T/F)

6. If the results of a nonlinear study in a properly meshed model are not as desired, you can change the material, loads, or fixtures to get the desired results and do not need to mesh the model again. (T/F)

7. In SOLIDWORKS Simulation, nonlinear static analysis calculates displacements, strains, stresses, and reaction forces. (T/F)

EXERCISES

Download the input files to perform the exercises of this chapter and then extract and save them at desired location. To download the files, the path is given below:

Textbooks > CAE Simulation > SOLIDWORKS Simulation > SOLIDWORKS Simulation 2022: A Tutorial Approach > Input Files

Extract the downloaded zipped file at the location *C:\SOLIDWORKS Simulation\c08.*

Exercise 1

In this exercise, you will perform analysis on a rectangular beam of 1.5 inch x 0.5 inch x 0.04 inch which is restrained at one end. A load of 315 lbf is applied on the other end of the beam. The beam has a tangential modulus of 33e6 psi and is made up of plain carbon steel. Find the von Mises stress and compare the FEM result with analytical result. **(Expected time: 1 hr)**

Exercise 2

In this exercise, you will perform analysis on the model shown in Figure 8-58. A load of 1000 N is applied on one side of the rod. The model has a tangential modulus of 33e6 psi and is made up of aluminum alloys. Find the von Mises stress and optimize the results for better functionality. **(Expected time: 1 hr)**

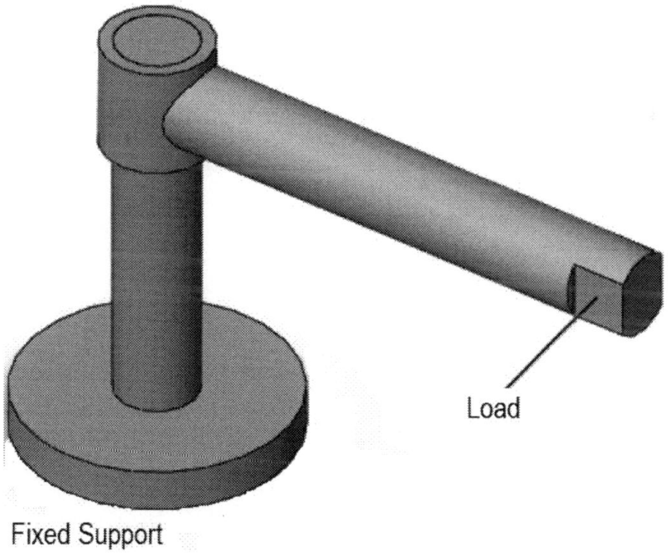

Figure 8-58 *Model for Exercise 2*

Exercise 3

In this exercise, you will determine vertical displacement on a beam caused by the load for the nonlinear load case. The load applied is 5000 lb and the length of the beam is 105 inch. The dimensions of the beam section are 1.0 inch x 2.0 inch. The Young's modulus is 30 x 106 and the Poisson's Ratio is 0.3. **(Expected time: 1 hr)**

Answers to Self-Evaluation Test

1. d, 2. c, 3. Deflection, **4. T, 5. T, 6. T, 7. T, 8. T, 9. T, 10. T**

Chapter 9

Implementation of FEA

Learning Objectives

After completing this chapter, you will be able to:

- *Verify and validate results*
- *Understand the FEA project management*
- *Understand project check points*
- *Generate FEA reports*

VERIFICATION AND VALIDATION OF THE RESULTS

In any analysis, the verification and validation of results is a very important process to make final conclusions about a product's life. Verification is a typical checking process which is performed while developing an FEA model. The validation process is used to check the FEA model based on available test data. Figure 9-1 shows the virtual product validation flow in the system.

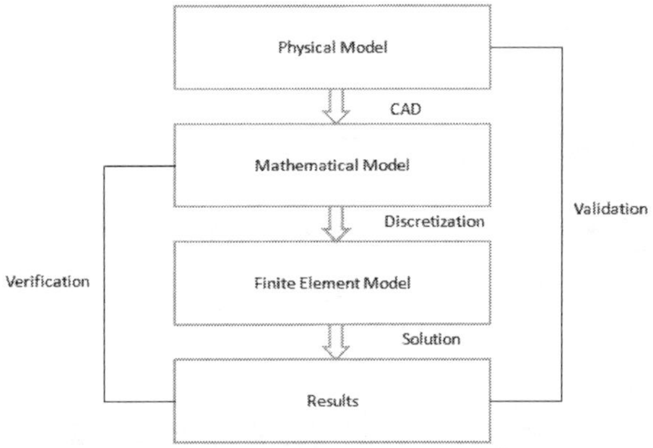

Figure 9-1 *Virtual product validation flow*

Why Verification Is important?

The finite element analysis has two key factors: the results obtained are approximate and the results are not robust. Small errors in modeling, data input, and boundary conditions can lead to very large errors in the results. These errors even can be relatively small and difficult to identify but, still have significant impact on performance or service life. For example, using the incorrect reference temperature for thermal strain in a structural model may only affect the stresses by a less percent and this error may not be large when comparing FEA results to a manual calculation. However, the error may have significant impact on the fatigue life of the model. These types of mistakes are hard to catch unless someone reviews the analysis thoroughly. Errors are likely to occur in finite element model. The more complex the model, the higher the probability that errors will occur. Catching these errors before they cause any type of damage to the model is called the verification process.

Verification Process

The verification process can take many different forms depending on the type of analysis being conducted, the parts being analyzed, the accuracy required, the level of risk involved. The analysis must begin with clearly defined goals, accuracy requirements, and key assumptions. For ease of use, repeatability, and ultimate success, it is better to create a verification template. Verification is a gated process which is used for performing FEA.

After checking the credibility of the mathematical (discretized mesh) model and the errors, you can compare the design with FEA answers to be assured that they are within the acceptable range. If the model was not sufficient to represent the problem then you need to adjust the mesh and other numerical inputs in the model. Once the verification process is complete, you can validate the FEA model with the physical model to make a decision on the feasibility of the model.

How to Interpret FEA Results?

Results of structural FEA are provided in the form of displacements and stresses. But, how you can decide if a design passes or fails? What constitutes the failure? To answer these questions, you need to establish some criteria to interpret FEA results, which may include maximum acceptable displacements, maximum stress, or lowest acceptable natural frequency. The displacement and frequency criteria are quite obvious and easy to establish while stress criterion is not. You need to conduct a stress analysis in order to ensure that stress is within an acceptable range. To understand stress results, you need to understand the mechanism of potential failure. SOLIDWORKS Simulation can present stress results in any desired form, but you need to decide which stress measures should be used to analyze the results.

FINITE ELEMENT ANALYSIS PROJECT MANAGEMENT

Before conducting analysis, you should be familiar with the basic concepts and terminologies of analysis which are discussed next.

Requirement of Finite Element Analysis (FEA)

The finite element analysis (FEA) is a computing technique that is used to obtain approximate solution to boundary value problems. FEA is essential to speed up the production process by cutting down the number of iterative cycles. Thus, by using FEA, you can improve the safety, efficiency, and cost-effectiveness, so that the production time is reduced automatically.

Durability of a part is also one of the main factors to be checked. Early FEA packages were more rudiment and used to offer linear static analysis abilities only, whereas, current FEA packages offer advance options such as non-linear stress analysis, fluid flow, thermal flow, fatigue due to vibration, and so on. Due to pre-loaded material properties, such as stress/strain, deformation, and natural frequencies, the modeled part for durability can be easily tested which was not possible with physical prototypes in early stages.

FEA is also used to troubleshoot structural issues in a new design. Since, most CAE capabilities are integrated well with CAD capabilities, required modifications can be implemented seamlessly, thus saving time and money.

FEA plays a major role in development projects and also helps in slashing expensive prototype testing. Figure 9-2 shows various steps and processes in FEA system.

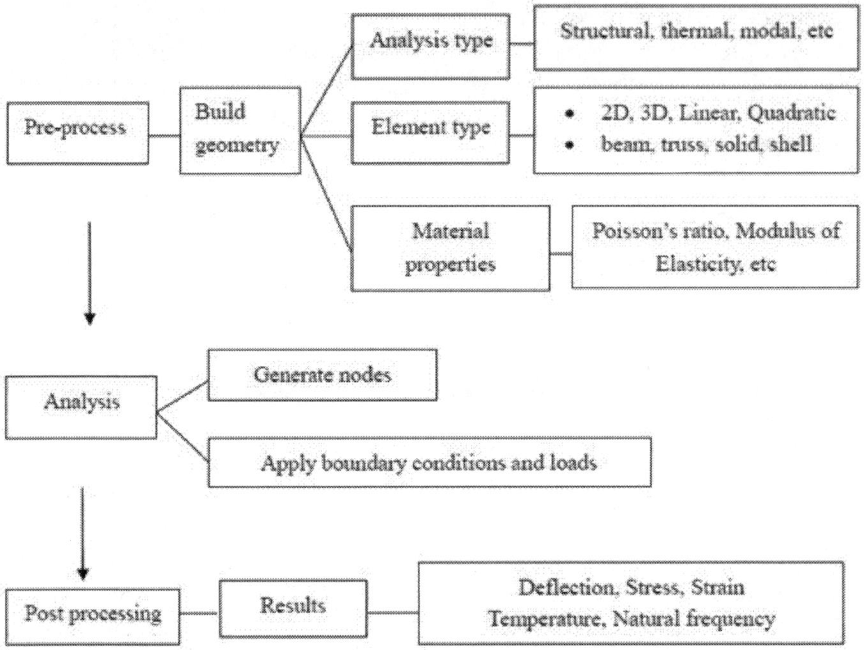

***Figure 9-2** Virtual product validation flow*

Finite Element Analysis (FEA) Report

The FEA report should be self-explanatory and contains enough information to duplicate, evaluate, and verify analysis results. A good report with backup, provides sufficient detail for rerunning the analysis without any additional instructions. Following are the main heads which should be included in the report.

Title Page

The title page of the report contains report number and title. Title should be specific, unique, and concise. Acronyms should not be used in the title without definition. A revision block should be included in this page for revision of report.

Table of Content

Table of content should be included in the report listing each section with applicable page number.

Executive Summary

The report should contain an executive summary describing how FEA was used to analyze the design, FEA model used, the results of FEA, and conclusion related to FEA results supporting the design decision. As part of the design registration submission, FEA documentation is subject to review and acceptance.

Introduction

Introduction should contain the detailed description about the scope of the FEA analysis related to the design, the justification for using FEA to support the design calculations, the FEA software used for the analysis, the type of FEA analysis (static, dynamic, elastic, plastic, small deformations, large deformations, and so on), and a complete description of the material properties used in the analysis.

Introduction should explain the purpose or objective of the report. This should contain a list of the parts being analyzed and should also include any historical information that would be helpful to further understand the context of the analysis. If the structure is a component of a larger assembly, this section should clearly describe how the component being analyzed fits into rest of the assembly.

Geometry and Material

This section of report describes complete description of FEA model used for the analysis. The description will include dimensional information and the model geometry. Simplification of geometry must be explained and justified appropriately. This section also contains all the allowable materials that are used in the analysis to calculate the structure margins of safety. All materials used in the analysis should be clearly identified (alloy, heat treat, and composite type) along with the applicable material properties.

Loads and Restraints

This section shows with illustrations the points where and how the load is applied such as force, pressure, or heat source as boundary conditions. The supports, loads, and forces should be clearly described and shown in this section.

Mesh

The mesh and its types (h, p, 2D, or 3D), shape, and order (2nd order or above) of the elements used must be described in this section. If different types of elements are used, a description of how the different elements are connected together is required. When using shell elements, describe the top or bottom orientation with plots of the elements and indicate that they are thick or thin elements.

Analysis type

This section of report describes various analysis types that are performed on the model.

Solver

In the solver, take all the input data, process them using specific algorithms, and give an output that corresponds to the result of the analysis.

Results of Analysis

A report contains the results such as reaction forces, displacements plot, deformed shape with un-deformed shape, stress plot with mesh, plot with element stress, and compared nodal (average) stress vs. element (non-averaged) stress.

When plots or figures are presented in the report, there must be description related to each and every figure explaining the purpose and importance of the figure.

Conclusions
The results of FEA should be related to the defined allowable stresses of the Code of construction using the appropriate ASME methodology.

Project Documentation
The project documentation is the documentation of the results calculated on the basis of numerical, prototype, and physical or field results of the project.

Follow-up
In this section, all the calculations and results should be crosschecked to get the final report.

CHECKPOINTS FOR AN FEA PROJECT
Proper execution of an FEA project is the Figure 9-3 FEA project check points major criteria for its success. The results of the FEA project cannot be achieved only by reviewing the analysis results but also depend upon the execution of project. Figure 9-3 shows the graphical representation of the FEA project check points.

Following are few check points which should be considered for FEA projects:

1. Can loads, supports, and modeling approach be accepted?
2. Are the mesh and elements appropriate?
3. Is the error value within specified criteria?
4. Do results agree with the analysis method?

DEVELOPMENT OF MECHANICAL PRODUCTS
Product development covers all aspects of product design cycle such as manufacturing, usage, servicing, and disposal of products. Product development also helps in the creation of the product different conditions such as design and functionality.

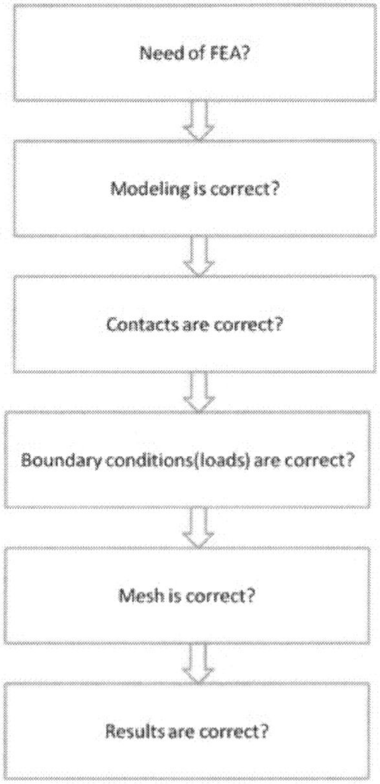

***Figure 9-3** FEA project check points*

Figure 9-4 shows various stages and aspects of the product development cycle.

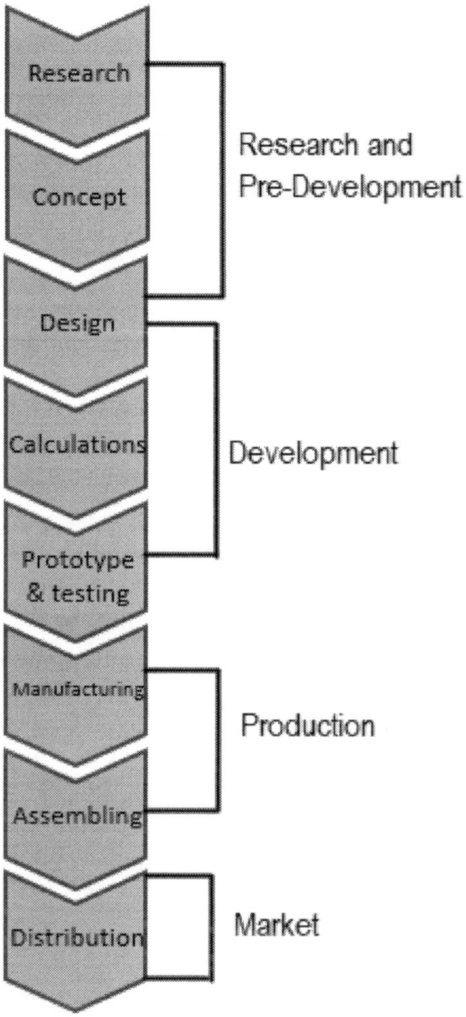

Figure 9-4 *Stages of the product development cycle*

In first phase, the main characteristics of a new product are defined and evaluated. The planning stage is often embedded in the concept phase. After the concept phase, the series development includes styling, design, and detail engineering phase. The extent of the product testing stage depends on the product type. In case of automotive engineering processes, the testing stage consists of field test and optimization work. Next, production related processes are developed and implemented. After the starting of production, the product manufacturing phase represents the last stage in the product life cycle.

Virtual Product Development

Virtual product development includes all computer aided technology support and virtual product modeling processes for the generation of a new product. Virtual product models are used to perform optimization and testing procedures in a virtual environment with the goal of saving development time and costs, while simultaneously increasing the product quality. Depending upon the categories of development applied, there are different types of virtual models. They can include market relevant or business case relevant data (economic models), workflow oriented information (process models), technical characteristics and descriptions (functional models), and geometry information (design models).

Depending upon the different requirements and characteristics of diverse disciplines, both virtual models and results generated can differ significantly. In general, the disciplines of virtual product development can be classified into following categories:

Computer-Aided Design (CAD)

Computer-aided design (CAD) is the process which uses the computer technology for designing and design documentations. It is the replacement of manual drafting with the automated process.

Computer-Aided Styling (CAS)

Computer Aided Styling (CAS) focuses on creating the exterior, interior, and styling or visual appearance of an automobile components. CAS users design freeform surfaces which are also known as Class A surfaces that combine form and functionality together.

Computer-Aided Engineering (CAE)

Computer-aided engineering (CAE) is the broad usage of computer software to aid in the engineering analysis tasks. It includes finite element analysis (FEA), computational fluid dynamics (CFD), multi-body dynamics (MBD), and optimization of product material and design.

Digital Mock-up (DMU)

Digital Mock-up or DMU is a concept that allows the description of a product for its entire life cycle which is usually in 3D. Digital Mock-up is enriched by all the activities that contribute in describing a product and its life cycle.

Computer-Aided Manufacturing (CAM)

Computer-aided manufacturing (CAM) is the process which includes the use of software to control machine tools and related ones in the manufacturing of the work pieces.

Computer-Aided Quality Assurance (CAQ)

Computer-aided quality assurance (CAQ) is the engineering application of computers and computer-controlled machines for the definition and inspection of the quality of the products.

Computer-Aided Testing (CAT)

Computer-aided testing (CAT) is the application of computers to control either analog or digital test techniques in order to evaluate the quality of components and products. It is used to check the parts, subassemblies, and full systems within the specified tolerances and specifications.

Depending on the type of product to be developed, several additional disciplines are applied during the product life cycle management, such as computer-aided software engineering (CASE), which are used in the development of IT applications or mechatronic products.

Efficient virtual product development is based on an effective interaction and integration of the various systems applied to enable a close cooperation with all participating departments and development partners. Virtual product development in the automotive industry uses a wide range of product models which are connected by global data management systems. Data management is organized in different structures, depending upon the requirement of the specific stages in the product development and product life cycle. Different terms are used to describe data management related processes and functionalities. Figure 9-5 shows the virtual product development revolution in the industries.

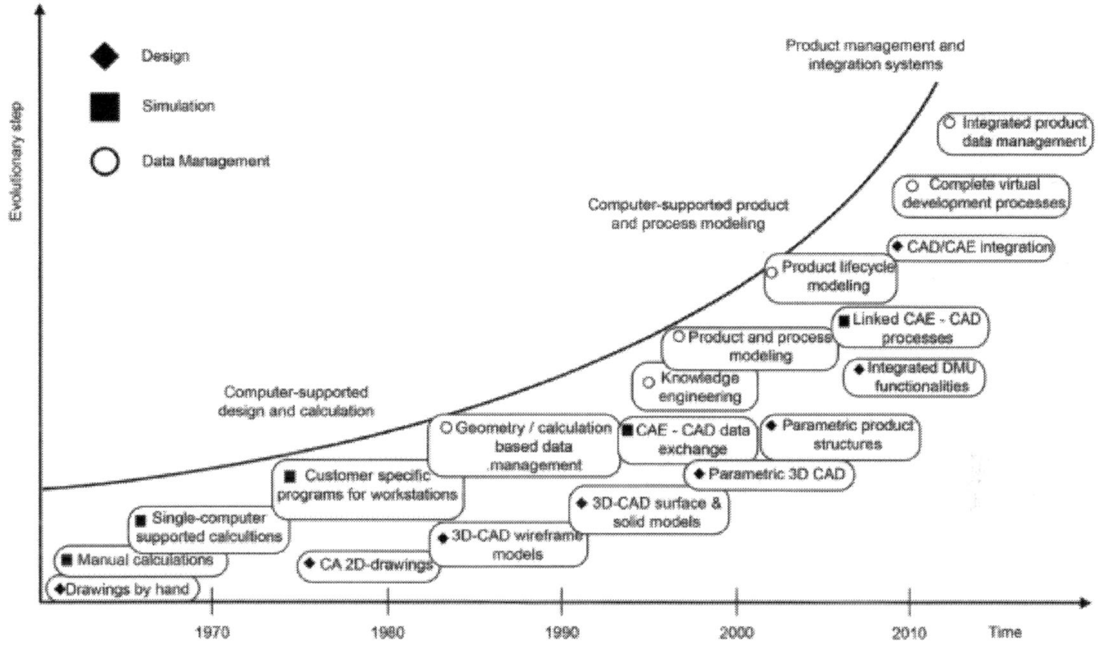

Figure 9-5 *Virtual product development revolution in the industries*

FAILURE CRITERIA IN DESIGN AND DEVELOPMENT

Important part of the design and analysis process is to define an upper bound on the stress state of a structure thereby defining its point of failure. This point becomes difficult to define when the structure is subjected to multi-axial loading. Failure criterion that addresses a multi-axial stress state including maximum shear stress theory, maximum distortion-energy theory, maximum normal stress theory, and the strain energy release rate approach. The former two criteria are best applied to ductile materials for which failure is defined by yield initiation. The latter two describe brittle failures characterized by fracture of the material.

Failure Theories for Ductile Materials

You have a more complicated situation at a point in the case of multidimensional stress. Since, it is impractical to test every material and every combination of stresses σ1, σ2, and σ3, a failure theory is needed for making predictions on the basis of a material's performance on the tensile test of how strong it will be under any other conditions of static loading. The theory behind various failures is that whatever is responsible for failure in the standard tensile test will also be responsible for failure under all other conditions of static loading.

In ductile material, it is understood that failure occurs due to relative sliding of materials atoms within the lattice structure. This sliding is caused by shear stresses and is accompanied by distortion of the shape of the part. Thus, the yield strength in shear is strength parameter of the ductile material used to design. Generally, following failure theories are used for ductile materials.

Maximum Shear Stress Criterion

The maximum shear stress criterion, also known as Tresca yield criterion, is based on the Maximum Shear stress theory. This theory predicts that the failure of a material will occur when the absolute maximum shear stress (σ_{max}) reaches the stress that causes the material to yield in a simple tension test. The maximum shear stress criterion is used for ductile materials.

At yielding in an uni-axial test, the principal stresses are:

$$\sigma 1 = S_{y,}\ \sigma 2 = 0 \text{ and } \sigma 3 = 0$$

Therefore, the shear strength at yielding:

Ssy =[1-(σ2 or σ3 =0)]/2. Therefore Ssy = Sy/2

$$\sigma_{max} >= \sigma_{limit}/2$$

FOS = limit/(2 * max)

To use this theory for either two or three-dimensional static stress in homogeneous, isotopic ductile materials, first compute the three principal stresses (σ1, σ2, σ3) and the maximum shear stress $\tau 13$ or τ_{max} as:

$$\tau_{max} = \frac{(\sigma_1 - \sigma_2)}{2} = \frac{(\sigma_{pmax} - \sigma_{pmin})}{2}$$

Then compare the maximum shear stress to the failure criterion.

$$\tau_{max} \leq S_{sy} = \frac{(\sigma_{pmax} - \sigma_{pmin})}{2} \leq S_{sy}$$

The safety factor for the maximum shear-stress theory is given by:

$$N = \frac{S_{sy}}{\tau_{max_1}}$$

Maximum von Mises Stress Criterion

Von Mises stress is a non-negative, scalar stress measure. Von Mises stress is commonly used to present results because the structural safety for many engineering materials showing elasto-plastic properties (for example, steel or aluminum alloy) can be evaluated using von Mises stress.

The maximum von Mises stress criterion is based on the von Mises-Hencky theory, also known as the Shear-energy theory or the Maximum distortion energy theory. The theory states that a ductile material starts to yield at a location when the von Mises stress becomes equal to the stress limit. In most cases, the yield strength is used as the stress limit.

In terms of the principal stresses (s1, s2, s3) the von Mises stress is expressed as:

$$\sigma_{vonMises} = \{[(s1 - s2)2 + (s2 - s3)2 + (s1 - s3)2] / 2\}1/2$$

The theory states that a ductile material starts to yield at a location when the von Mises stress becomes equal to the stress limit. In most cases, the yield strength is used as the stress limit. However, the software allows you to use the ultimate tensile stress limit or set your own stress limit.

$$\sigma_{vonMises >=} \sigma_{limit}$$

Yield strength is a temperature-dependent property. This specified value of the yield strength should consider the temperature of the component. The factor of safety at a location is calculated as:

$$\text{Factor of Safety (FOS)} = \sigma_{limit}/\sigma_{vonMises}$$

Note that in the case of pure shear, at the onset of yielding, the maximum shear stress in pure shear is √3 times lower than the yield stress in the case of simple tension.

Failure Theories for Brittle Materials

The fractures in the brittle material generally occur due to normal tensile stress alone and thus, the maximum normal-stress theory becomes applicable. Also, brittle fracture in compression is due to combination of normal compressive stress and shear stress. The Mohr-Coulomb and Maximum normal stress theories are discussed next.

Mohr-Coulomb Stress Criterion

The Mohr-Coulomb stress criterion is based on the Mohr-Coulomb theory also known as the Internal Friction theory. This criterion is used for brittle materials with different tensile and compressive properties. Brittle materials do not have a specific yield point and hence it is not recommended to use the yield strength to define the limit stress for this criterion. The theory predicts that the failure will occur when the combination of the maximum and minimum principal stress exceeds their respective stress limits.

Maximum Normal Stress Criterion

The maximum normal stress criterion, also known as Coulomb's criterion, is based on the Maximum normal stress theory. According to this theory, failure occurs when the maximum principal stress reaches the ultimate strength of the material for simple tension. This criterion is used for brittle materials. It assumes that the ultimate strength of the material in tension and compression is the same. This assumption is not valid in all cases. For example, cracks decrease the strength of the material in tension, while in compression, the effect of strength is smaller because the cracks tend to close.

Brittle materials do not have a specific yield point and hence it is not recommended to use the yield strength to define the limit stress for this criterion.

This theory predicts that the failure will occur when:

$$\sigma_1 >= \sigma_{limit}, \text{ where } \sigma_1 \text{ is the maximum principal stress.}$$

The Factor of Safety (FOS) is calculated as:

$$FOS = \sigma_{limit} / \sigma_1$$

Maximum Stress Failure Criterion

The maximum stress criterion applies to composite shells. According to this criterion, the failure occurs when the stress in one of the principal material directions exceeds the strength.

Note that the maximum stress criterion does not consider the interactions between different stress components.

The characteristic equations and failure applications of various failure criteria are shown in following table.

Failure Criteria	Characteristic Equation	Material/Failure
Tresca yield Criterion	$\tau_{max} = \frac{\sigma_Y}{2}$	Ductile/Yield
Von Mises-Hencky Theory	$\sigma_1^2 - \sigma_1\sigma_2 + \sigma_2^2 = \sigma_Y^2$	Ductile/Yield
Maximum-Normal-Stress Theory	$\lvert\sigma_1\rvert = \sigma_{ult}$ $\lvert\sigma_2\rvert = \sigma_{ult}$	Brittle/Fracture
Strain Energy Release Rate	$\frac{P}{2t}\frac{\partial u}{\partial a} \geq \frac{1}{E}K_{Ic}^2$	Anisotropic/Crack Propagation

Comparing von Mises and Tresca Stress Criteria

The maximum shear stress criterion is more conservative than the von Mises stress criterion since the hexagon representing the shear stress criterion is enclosed within the ellipse representing the von Mises stress criterion. Note that in case of pure shear, von Mises stress criterion predicts failure at 0.577*yield strength, whereas the shear stress criterion predicts failure at 0.5 yield strength. Actual torsion tests have shown that the von Mises stress criterion gives more accurate results than the maximum shear stress.

ERRORS OCCUR IN FINITE ELEMENT ANALYSIS (FEA)

Errors associated with Finite Element Analysis can be classified in three groups: Modeling errors, Discretization errors, and Solution errors.

Modelling errors are introduced via simplifying assumptions made in the formulation of the mathematical (usually simplified) representation. Discretization errors occur in mathematical model get discretized continuously into a FE model. In the same manner, solution errors occur due to the process of numerical solution of the FE equations.

Modelling errors

Suppose, you wish to analyze a support bracket. The questions that come to mind include: What do you want to find: the Maximum stress, Maximum deflections, the first few natural frequencies, the bending stiffness, temperature distribution, Does the bracket remain in the elastic range, How many loading conditions are critical, How should we represent support conditions?

The two major aspects to create the model geometry are: clear objectives and the understanding of limitations inherent in the theory being used. At times, this geometry may be similar to the CAD geometry but quite often we find it necessary to modify the topological description to simplify meshing. At this stage, decision making is an important part of modeling process which includes modeling thin walls with shell elements, taking advantage of symmetry or asymmetry or both, including or neglecting fillets, and deleting nonessential features. For example, deciding to use shell elements rather than solid elements means one makes an important decision concerning the mathematical model and hence the kind of operations the FEA software will perform.

After idealizing the topological description, one still needs to idealize material properties such as linear elastic, elastic-plastic, loads, and supports. These decisions further define the mathematical problem, which supposedly represents the bracket with respect to the data of interest. Important modeling decisions are sometimes made without much concern for their implications. It often happens that the formulated model is conceptually wrong. An indicator of a poor model comes when the strain energy corresponding to the exact solution is infinite or trivially zero. Another indicator is when the data of interest, corresponding to the solution of the mathematical model, is not defined. And, when the solution is defined, error is entirely discretization (mesh) dependent.

Many analysts think that an efficient mesh generator reduces modeling error with a high quality mesh. But, this is not correct. Modeling assumptions are made before meshing. Consequently, even the best mesh cannot fix an improperly defined mathematical model. You can check a brief check list of modeling errors to ensure the data of interest are accurate or not:

- Check the model properly conceived with respect to the data of interest.
- Ensure the strain energy of the exact solution is finite and not zero.
- Ensure the data of interest finite.
- Check the estimated relative error in energy norm (not the strain energy) reasonably small, about 5% or less.

- Check the data of interest converging to a limit as the number of degrees of freedom increases.
- Check where stress maximums are of interest, are the stress contours smooth in the highly stressed areas.

The first step in a finite-element analysis selects a mathematical model to represent the object being analyzed. The term mathematical model refers to a theory such as the theory of elasticity, the Reissner theory of plates, or deformation theory of plasticity, and also provide information that defines the problem geometric descriptions, material properties, as well as constraints and loading.

The analysis goal is to compute information from the exact solution and then calculate information such as the maximum von Mises stress, which could be F(u_EX). The function F(u_EX) depends only on the definition of the mathematical model and not on the method used for finding an approximate solution. Therefore, function does not depend on mesh quality, type, and size of elements. The difference between F(u_EX) and the physical property is called the modeling error.

Discretization Errors

The approximation error is the estimation of exact solution in the finite element method. This involves selecting a mesh and elements such as 2D plates, eight-node bricks, p-elements, shell elements, and so on. The mesh and elements define the finite element discretization.

In FEA, the system is assumed to be a mathematical model partitioned into regions called elements and can individually be solved with simple linear equations or by using differential equations. This partitioning creates a series of linear approximations that strive to map, as closely as possible, the true continuous solution. To solve the structural stress-strain problem, the FEA software takes the mesh approximation and solves for the displacements at each node given the loading and constraints on the model. Next, as a secondary operation, the software approximates the stress contour in each element by looking at the relative displacement of the nodes of that element. In this manner, a stress contour is determined for each element. Given that this contour is approximated based only on the displacement of nodes of the element in question, the stress contour will not necessarily be continuous from one element to the next.

Essentially, this discontinuity of the stress contour from one element to another is called discretization error. As the size of each element is reduced towards zero, this error also reduces to zero as the increasing number of partitions approach to true continuous system. To check the elements continuity, the convergence process is used. This process adds degrees of freedom to the FEM model to see how results change. Degrees of freedom are added either by using more elements (mesh refinement, called h convergence), or by using higher-element orders (p-convergence). Convergence should demonstrate that results converge to a finite value and, therefore, are not significantly dependent on the choice of discretization. If the process shows that results are significantly discretization dependent then the results are unreliable.

Solution Errors

Solution errors occur due to the process of numerical solution of the FE equations. The solution errors are round off errors accumulated by the analysis solver.

Self-Evaluation Test

Answer the following questions and then compare them to those given at the end of this chapter:

1. Which of the following processes is used to validate a model and determine the accuracy of results?

   (a) validation (b) designing
   (c) verification (d) modeling

2. An FEA solver may not be the cause of error all the times, the errors could be due to both verification and _________ in results.

3. Force method and displacement or stiffness method are generally associated with the finite element analysis. (T/F)

4. Point collocation, sub-domain collocation, least square, and galerkin's are all weighted residual methods. (T/F)

5. Local axes are established in an element and they change with the change in orientation of the element encountered in finite element analysis. (T/F)

6. Global axes are defined for the entire system and they are same in direction for all the elements even though the elements are differently oriented. (T/F)

7. Body force is the distributed force that acts on every elemental volume of the body. (T/F)

8. Traction force is defined as the distributed force acting on the surface of a body such as frictional resistance, viscous drag, and surface shear. (T/F)

9. In structural problems, displacement at each nodal point is obtained and then using the displacement solutions, stress, and strain on each element can be calculated. (T/F)

10. In p version, the number of elements is maintained constant and the order of polynomial approximation of element is increased. (T/F)

11. What will be the value of the sum-of-forces calculation done on a cylinder with uniform pressure applied?

Review Questions

Answer the following questions:

1. Verification is a checking process to ________ an FEA model.

2. Product development is the process of creating or developing a product under different ________________ .

3. Virtual product models are used to perform ____________and testing procedures in a virtual environment with the goal of saving development time and costs, while simultaneously increasing the product quality.

4. Results of structural FEA are provided in the form of displacements and stresses. (T/F)

5. FEA is also used before a new design has been developed to troubleshoot for any possible structural issues that might be difficult to carry out with simple hand calculations. (T/F)

6. Maximum shear stress criterion, also known as Tresca yield criterion, is based on the maximum shear stress theory. (T/F)

7. Modelling errors are introduced via simplifying assumptions made in the formulation of the mathematical (usually simplified) representation. (T/F)

8. In a thermal structural problem, the temperature or fluid pressure at each nodal point is obtained and then using these values, the properties such as heat flow, fluid flow for each element can be calculated. (T/F)

PROJECTS

FEA Project 1

Create component of the wheel design and optimize the design which focus on reducing the mass of the current design. New designs include reducing the thickness of the legs, modifying the fillet radius at the intersection of the legs and the hub, and modifying the radius of the fillet on the back face near the top of each leg. Create an FEA report of your project and conclude the FEA test. The design criteria of the project is given in the table below and the wheel is shown in Figure 9-6. The dimensions and model of the wheel center are shown in the Figure 9-7 to Figure 9-10. **(Expected time: 3 hr)**

Design criteria:

Yield Strength (MPa)	Deflection (mm)	Safety Factor
276 (AL 6061)	1.25 mm	2.0

Figure 9-6 *Wheel model for project*

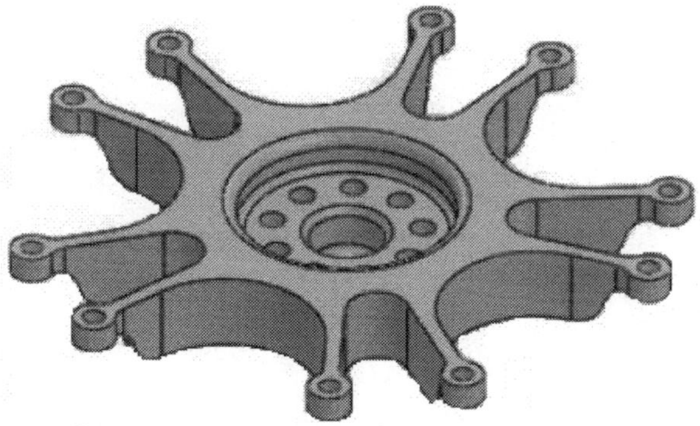

Figure 9-7 *Wheel center component model for project*

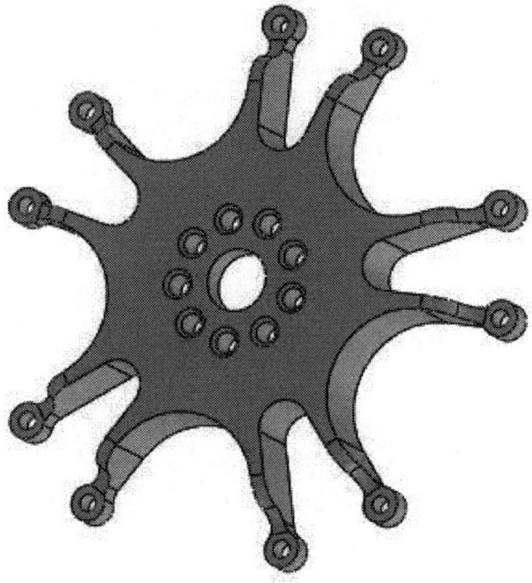

Figure 9-8 Wheel center component model for project

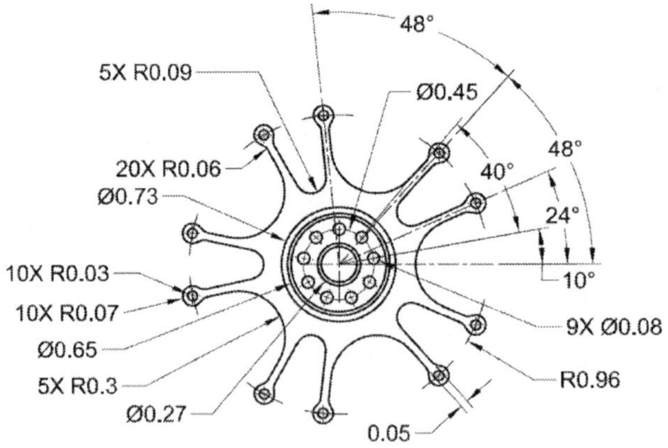

Figure 9-9 The dimensions of the model

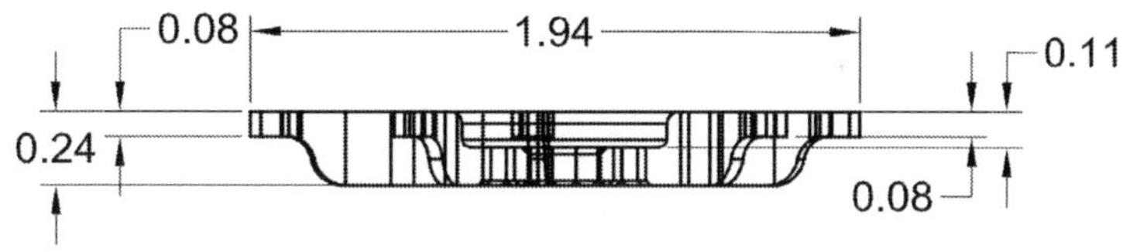

Figure 9-10 *The dimensions of the model*

FEA Project 2

Create the components of the truck chassis, as shown in Figure 9-11 and validate the model. Perform static structural analysis of the truck chassis such that structural systems of the chassis can be easily analyzed. Next, create the project report and conclude the FEA test. The design criteria of the project is given below and the chassis design is shown in Figure 9-11. The dimensions of the components are shown in Figure 9-12 to Figure 9-15.

Design criteria:

Maximum shear stress: 100 MPa.
Maximum displacement of analytical calculation: 3.00 mm

Material Properties:

Tensile Strength (N/mm^2)	Yield Strength (N/mm^2)	Modulus of Elasticity (N/mm^2)	Poisson Ratio
520 (ST 52)	360 (ST 52)	2×10^5	0.3

Loading condition:
The magnitude of force on the upper side of chassis is 117720 N and on one side bar is 58860 N. The fixed supports act on the front and rear overhang of the chassis.

(Expected time: 3 hr)

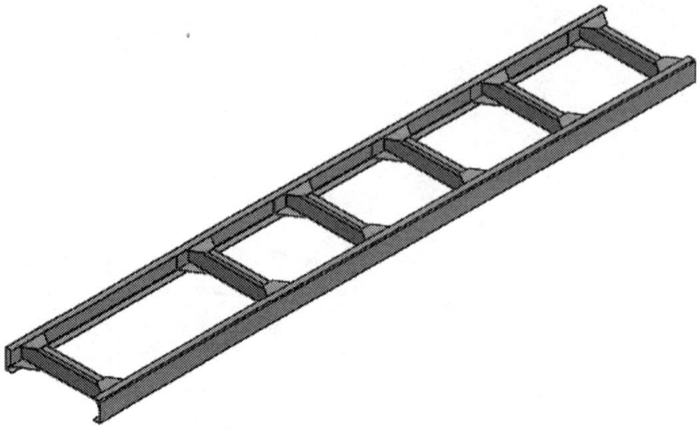

Figure 9-11 *The model for the project*

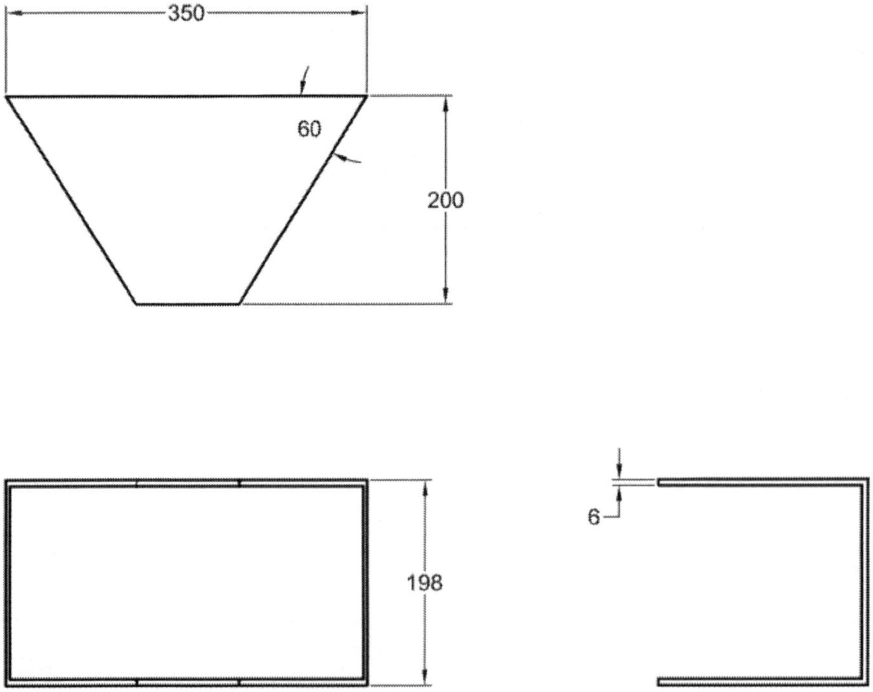

Figure 9-12 *The dimensions for the brackets*

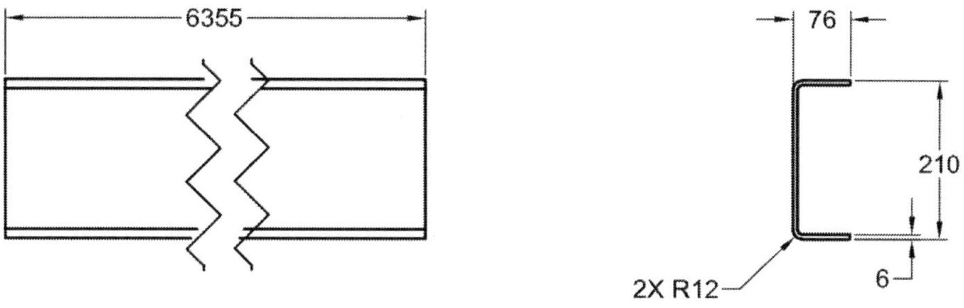

Figure 9-13 *The dimension for the c-section*

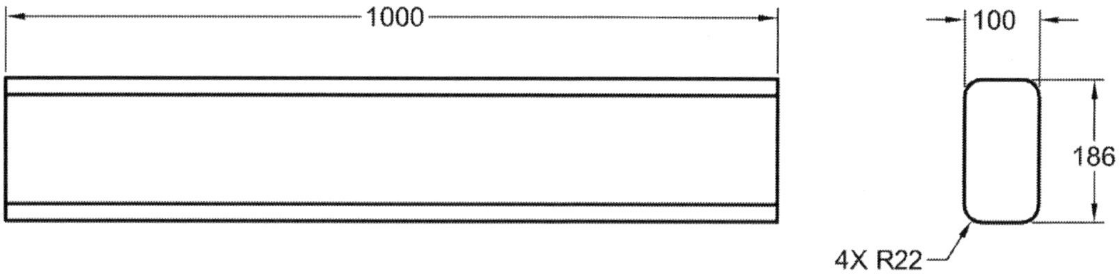

Figure 9-14 *The dimension for the lad*

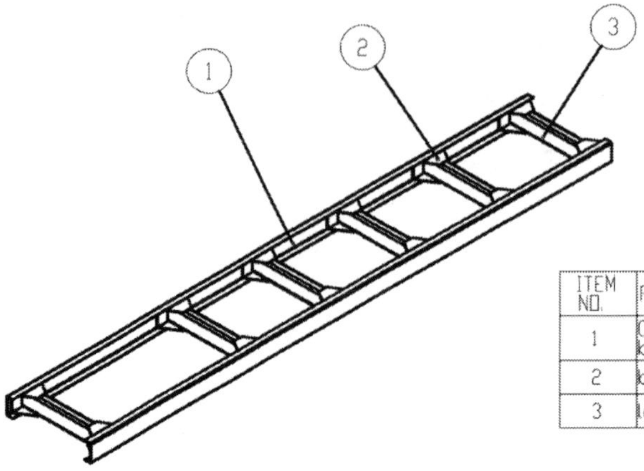

ITEM NO.	PART NUMBER	DESCRIPTION	QTY.
1	C-section beam		2
2	brackets		12
3	lad		6

Figure 9-15 *The assembly and BOM of the model*

FEA Project 3

In this project, create the model of the crank hook, as shown in Figure 9-16 and validate it for the different materials such as alloy steel, galvanized steel and aluminium alloys (1060 Alloy). Perform static structural analysis of the model such that structural systems of the crank hook can be easily analyzed. Next, create the project report and conclude the FEA tests. The design criteria of the project is given below and the model is shown in Figure 9-16. The dimensions of the model is shown in Figure 9-17.

Design criteria:

Factor of Safety: 1.5 min
Factor of Safety: 3.0 max **(Expected time: 2.5 hr)**

Figure 9-16 *The model of the crank hook*

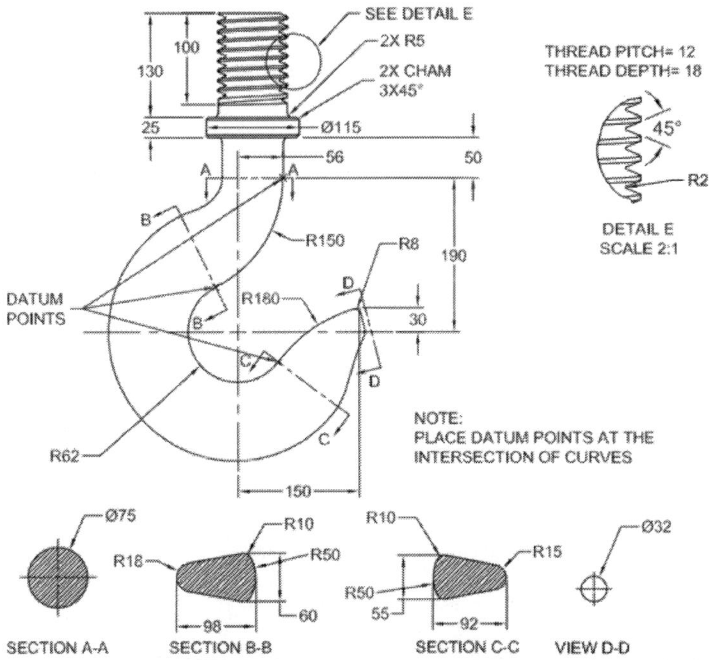

Figure 9-17 *The dimensions of the model*

FEA Project 4

In this project, create the model of the Pulley Support assembly, as shown in Figure 9-18 and Figure 9-19. You need to validate the design for better functioning and safety purpose. The bottom face of the assembly is in contact with ground and pulley rotates with 500 RPM which generate 1Nm torque. The material details are given in Figure 9-24. Perform the structural analysis on the model such that structural systems of the model can be easily analyzed. Next, create the project report and conclude the FEA test. The design criteria of the project is given below. The dimensions of the model are shown in Figure 9-20 to Figure 9-23.

Design criteria:
Factor of Safety: 2.0 Min **(Expected time: 3 hr)**

Figure 9-18 *The model of the Pulley Support assembly*

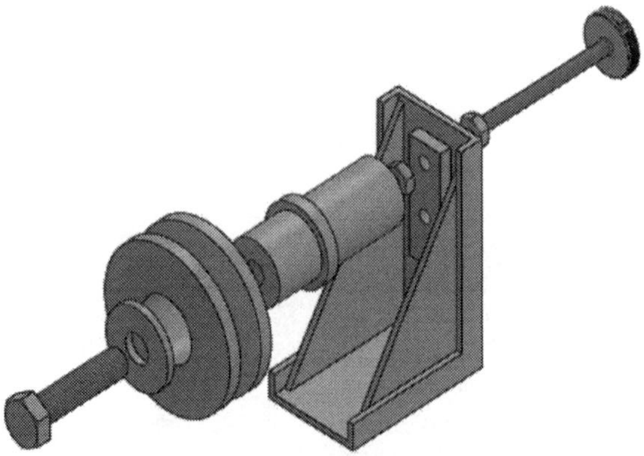

Figure 9-19 *The exploded view of the model*

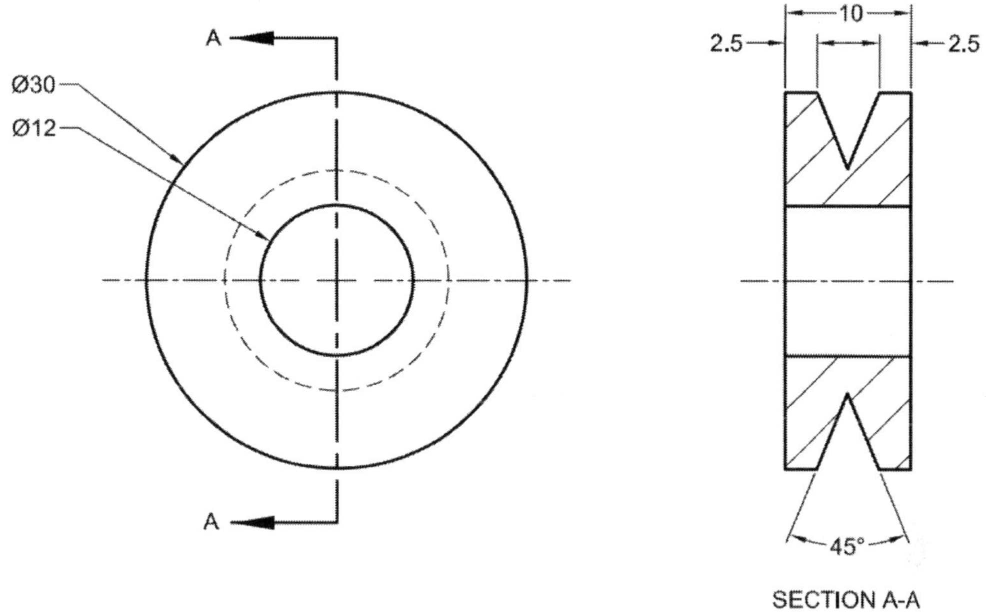

Figure 9-20 *The model of the pulley*

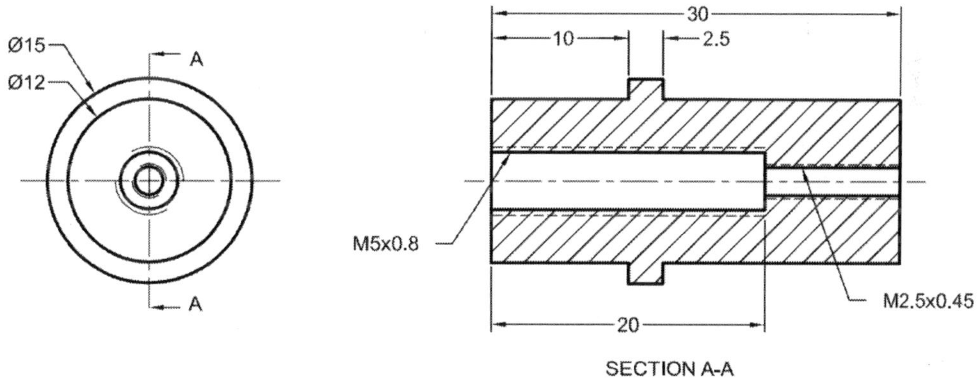

Figure 9-21 *The model of the bushing*

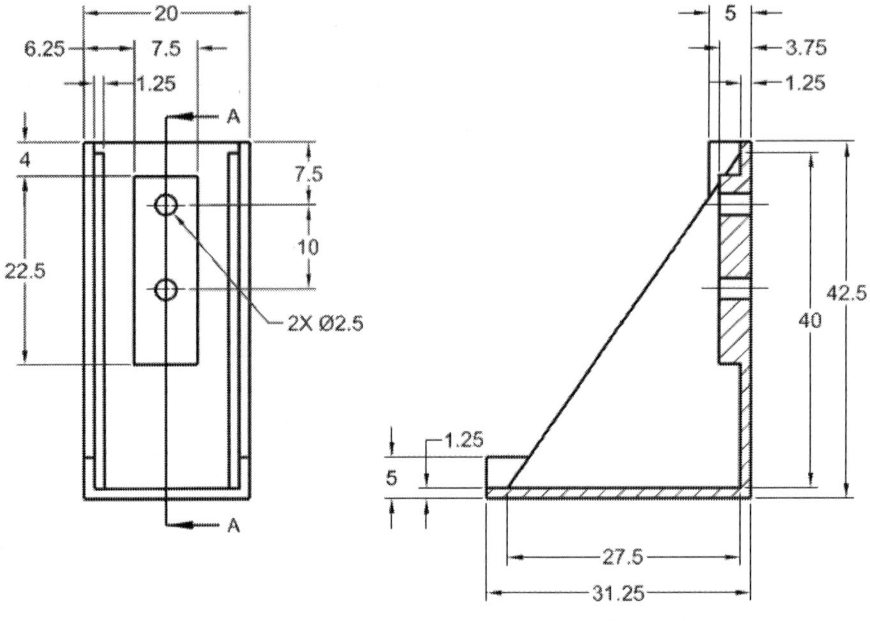

***Figure 9-22** The Left side and sectional view of the bracket*

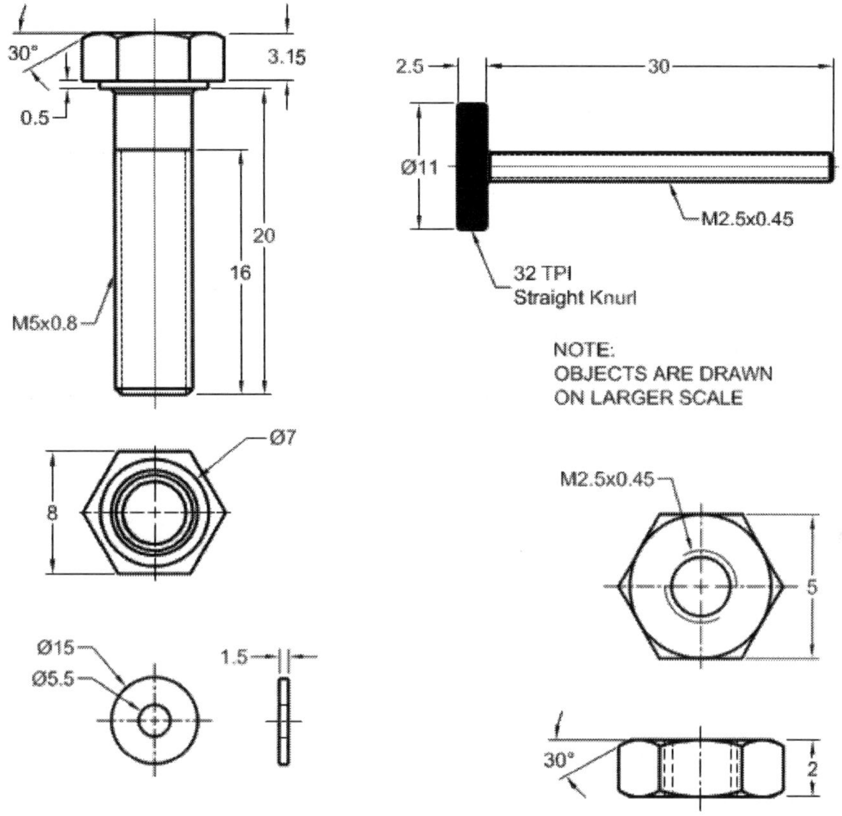

Figure 9-23 *The dimensions of the washer, cap, screw, turn screw and nuT*

Part List			
ITEM	QTY	NAME	MATERIAL
1	1	Bracket	Steel
2	2	Nut	Steel
3	1	Bushing	Steel
4	1	Turn Screw	Steel
5	1	Pulley	Steel
6	1	Washer	Steel
7	1	Cap Screw	Steel

Figure 9-24 *Parts list and material details of the model*

Answers to Self-Evaluation Test

1. a, **2**. Validation, **3**. T, **4**. T, **5**. T, **6**. T, 6. T, **7**. T, **8**. T, **9**. T, **10**. T, **11**. T

Index

F

G

H

I

J

K

L

M

W

Y

Other Publications by CADCIM Technologies

The following is the list of some of the publications by CADCIM Technologies. Please visit *www.cadcim.com* for the complete listing.

AutoCAD Textbooks

- AutoCAD 2023: A Problem-Solving Approach, Basic and Intermediate, 29th Edition
- Advanced AutoCAD 2023: A Problem-Solving Approach (3D and Advanced), 26th Edition
- AutoCAD 2022: A Problem-Solving Approach, Basic and Intermediate, 28th Edition
- AutoCAD 2021: A Problem-Solving Approach, Basic and Intermediate, 27th Edition

Autodesk Inventor Textbooks

- Autodesk Inventor Professional 2023 for Designers, 23rd Edition
- Autodesk Inventor Professional 2022 for Designers, 22nd Edition
- Autodesk Inventor Professional 2021 for Designers, 21st Edition

AutoCAD MEP Textbooks

- AutoCAD MEP 2023 for Designers, 7th Edition
- AutoCAD MEP 2022 for Designers, 6th Edition

Solid Edge Textbooks

- Solid Edge 2023 for Designers, 20th Edition
- Solid Edge 2022 for Designers, 19th Edition
- Solid Edge 2021 for Designers, 18th Edition

NX Textbooks

- Siemens NX 2021 for Designers,14th Edition
- Siemens NX 2020 for Designers,13th Edition

SOLIDWORKS Textbooks

- SOLIDWORKS 2023 for Designers, 21st Edition
- SOLIDWORKS 2022 for Designers, 20th Edition
- SOLIDWORKS 2021 for Designers, 19th Edition
- SOLIDWORKS 2020 for Designers, 18th Edition
- Learning SOLIDWORKS 2022: A Project Based Approach

CATIA Textbooks

- CATIA V5-6R2022 for Designers, 20th Edition
- CATIA V5-6R2021 for Designers, 19th Edition

Creo Parametric Textbooks

- Creo Parametric 9.0 for Designers, 9th Edition
- Creo Parametric 8.0 for Designers, 8th Edition

ANSYS Textbooks
- ANSYS Workbench 2022 R1: A Tutorial Approach
- ANSYS Workbench 2021 R1: A Tutorial Approach

Autodesk Alias Textbooks
- Learning Autodesk Alias Design 2021, 6th Edition
- Learning Autodesk Alias Design 2015, 5th Edition

AutoCAD LT Textbooks
- AutoCAD LT 2023 for Designers, 15th Edition
- AutoCAD LT 2022 for Designers, 14th Edition
- AutoCAD LT 2020 for Designers, 13th Edition

AutoCAD Civil 3D Textbooks
- Exploring AutoCAD Civil 3D 2023, 12th Edition
- Exploring AutoCAD Civil 3D 2022, 11th Edition
- Exploring AutoCAD Civil 3D 2020, 10th Edition

Autodesk Revit Structure Textbooks
- Exploring Autodesk Revit 2023 for Structure, 13th Edition
- Exploring Autodesk Revit 2022 for Structure, 12th Edition
- Exploring Autodesk Revit 2021 for Structure, 11th Edition

AutoCAD Electrical Textbooks
- AutoCAD Electrical 2023 for Electrical Control Designers, 14th Edition
- AutoCAD Electrical 2023: A Tutorial Approach, 4th Edition
- AutoCAD Electrical 2022: A Tutorial Approach, 3th Edition

Coming Soon from CADCIM Technologies
- Flow Simulation Using SOLIDWORKS 2023
- Finite Element Analysis Using ANSYS Workbench 2023

Online Training Program Offered by CADCIM Technologies
CADCIM Technologies provides effective and affordable virtual online training on animation, architecture, and GIS softwares, computer programming languages, and Computer Aided Design, Manufacturing, and Engineering (CAD/CAM/CAE) software packages. The training will be delivered 'live' via Internet at any time, any place, and at any pace to individuals, students of colleges, universities, and CAD/CAM/CAE training centers. For more information, please visit the following link: ***https://www.cadcim.com.***

Made in the USA
Monee, IL
05 November 2026

67997468R00173